WITHDRAWN

THE ELUSIVE SCIENCE

Seymour H. Mauskopf is professor of history at Duke University. Michael McVaugh is professor of history at the University of North Carolina.

THE ELUSIVE SCIENCE

Origins of Experimental Psychical Research

Seymour H. Mauskopf
and Michael R. McVaugh

Afterword by J.B. and L.E. Rhine

The Johns Hopkins University Press
Baltimore and London

Copyright © 1980 by The Johns Hopkins University Press
All rights reserved
Printed in the United States of America

The Johns Hopkins University Press, Baltimore, Maryland 21218
The Johns Hopkins Press Ltd., London

The Library of Congress Cataloging in Publication data will be found on the last printed page of this book.

BF
1028
M38

To two patient wives

So far from aiming at any paradoxical reversion of established scientific conclusions, we conceive ourselves to be working (however imperfectly) in the main track of discovery, and assailing a problem which, though strange and hard, does yet stand next in order among the new adventures on which Science must needs set forth, if her methods and her temper are to guide and control the widening curiosity, the expanding capacities of men....

Our only paradox, then, is the assertion that we are not paradoxical....

F.W.H. Myers (1886)

The great weakness of the case for thought-transference is that the accounts of it are so rare. Why don't the apparent cases come in faster, now that so many of us are on the watch?... for whilst additional proofs are waited for, questions get prematurely closed and forgotten; and in this case that seems to me a consummation which one ought to try as long as possible to postpone.

William James (1889)

Contents

Preface — xi

1
The Problems of Psychical Research in the 1920s — 1

2
The Growth of Experimentation
in the Psychical-Research Societies — 25

3
Early Psychical Research
in American Universities — 44

4
A Career in
Psychical Research—J. B. Rhine — 71

5
Extra-Sensory Perception
and Contemporary Psychical Research — 102

6
Parapsychology in Its Public Aspect — 131

7
The Articulation of Parapsychology — 169

8
The Psychical-Research Societies
in the Mid-1930s — 204

9
Parapsychology and
Professional Psychology, 1934–38 — 240

10
Towards Professional Acknowledgment — 273

Epilogue	298
Afterword by J. B. and L. E. Rhine	307
Appendix: Glossary of Parapsychological Techniques	311
Notes	313
Index	359

Preface

This book began in a casual question from us to Drs. J. B. and L. E. Rhine some nine years ago: Had they retained any correspondence or records bearing on their forty-year career in parapsychology, and would they be willing to allow us (the historians of science in the Duke area) to examine and perhaps organize them? Noncommittal though it was, this inquiry was not without its own background. Although neither of us had had any involvement in parapsychology before coming to teach in North Carolina, we had not been in the state long before the local presence of parapsychology and of the Rhines began to impinge on our consciousness—through student interest in the subject, intensifying in the late 1960s; through the occasional public lectures that J. B. Rhine still gave; and through conversations, where the mention of parapsychology usually drew slighting comments from our colleagues. Parapsychology and the Rhines were brought to our attention almost more insistently when we were away from the immediate area. We began to realize, at first amusedly and then more thoughtfully, that the subject and the couple were among North Carolina's best-known features.

It was our training as historians of science, however, that was most important in arousing and disciplining our interest in this subject, sensitizing us to the possibilities it held for significant research. By the 1960s the profession of history of science had developed in sophistication in two important respects that determined our approach. First, historians of science had largely abandoned the "presentist" viewpoint of scientific development, still so common among scientists, in which the scientific activity of the past is judged by how well it accords with contemporary theory and practice. Second, these historians were coming to focus their research on the social, institutional, and cultural life of science and scientists rather than merely on "science" as an abstract set of theories and techniques. These trends had already led to markedly new historical interpretations of earlier periods of scientific development. One group of historians, for example, had taken a renewed interest in those sciences of the Renaissance (particularly alchemy) that grew marginal in the seventeenth century and had argued convincingly that these must be taken into account in developing a complete picture of scientific activity during the Scientific Revolution. And in order to take these enterprises into account, they had pointed out, it was necessary to begin by trying to comprehend them—and the interest in them of intelligent and even brilliant minds—in terms of their own milieu in the sixteenth and seventeenth centuries. While not forgetting that they were already becoming superseded in this period, it was essential to study them without letting a presentist judgment intrude.

Such examples made us recognize the potential importance of the history of parapsychology. At the moment, parapsychology is marginal to the scientific mainstream in much the way alchemy already was in the seventeenth century; like alchemy, moreover, this field nevertheless has attracted the serious attention of a number of outstanding scientists. *Unlike* alchemy, however, it is by no means clear that parapsychology is fated to remain a marginal science; this will be an issue for the future to decide. And the history of science shows that whichever way that decision goes, it will still be important to have a thorough historical understanding of the field in the various stages of its evolution. It was with these historiographical thoughts in mind that we approached the Rhines with our question about their papers.

Little did we realize how naïve our question was! The Rhines had already turned over to the Duke University Library an immense quantity of documentary material, including both laboratory notes and records and a truly unparalleled file of correspondence, and they very graciously allowed us complete access to it. As we soon discovered, there exist other remarkably rich archives for the history of organized psychical research (as parapsychology was first called and often still is), especially in New York and London. It was not long before we had decided ourselves to undertake the project we had originally meant to make possible for future historians. Our preliminary reading immediately forced us to abandon any thought of preparing a complete history of psychical research. We chose instead to focus upon the tradition of experimental study of such psychic abilities as telepathy (in which a mind gathers information from another mind without sensory mediation) and clairvoyance (in which a mind gathers information about things or events without sensory mediation). While much else in psychical research can broadly be considered "scientific" and even "experimental" (for example, the rigorous observation of mediums), the experimental investigation of these two abilities has come to have most in common with the traditions of Western science since Galileo, in particular in its concern to develop replicable techniques and quantitative methods. We saw an opportunity to prepare a detailed study of a period of particular significance for this field: the period (1915–40) when it made its first sustained bid for acceptance by, and incorporation into, mainstream science. We were further attracted by this prospect because many of the protagonists in this movement were still alive and could discuss their work with us directly.

The succeeding years have been fascinating: they have convinced us that we were right in believing in the utility of such a study, and, almost incidentally, they have matured our understanding of how the historian of science should work. It has become our contention that during the quarter-century after 1915 psychical research was transformed from what had been a rather disorganized amateur activity, mixing spiritualism with attempts at experimentation, into a more coherently structured professional and research enterprise and began to gain, not general acceptance, but a degree of toleration from psychologists and other scientists. While work in the subject was international, it appears to us that the activity of J. B. Rhine at Duke was critical in bringing about this transformation. His first published experimental work of 1934 on "extra-sensory perception" (ESP) made an unprecedented claim for the psychical-research community, the claim to have developed a simple experimental technique for demonstrating telepathy and clairvoyance on something like a replicable basis. Rhine's particular experimental approach soon became the basis for much work in parapsychology, both in the United States and abroad. It also won considerable publicity in the mid-1930s as a technique that had at last demonstrated psychical phenomena scientifically, and this led many mainstream scientists for the first time to examine carefully the evidence for telepathy and clairvoyance.

We begin our book with an overview of psychical research in the 1920s, especially the controversies over subject matter and organization that then perplexed the field. In the second and third chapters, we examine the specifically experimental studies of telepathy and clairvoyance carried on during that decade by psychical researchers and academic psychologists. Chapters four and five study Rhine's earliest work against this background, both his career as it led up to his 1934 monograph, *Extra-Sensory Perception,* and the effect that that book had upon psychical researchers elsewhere. The sixth and seventh chapters consider, respectively, the organizational growth and the intellectual development of American parapsychology in the remainder of the 1930s, while the eighth chapter attempts to explain why the field was unable to win academic establishment elsewhere. The final chapters detail and analyze the reception of parapsychology by American psychologists, the sharp criticism of 1937–38 and the measure of recognition given it by 1939–40. Since the prospects of 1940 for acceptance and scientific incorporation of parapsychology have not really been fulfilled in the subsequent decades, we examine in an epilogue the development of the field during the 1940s, trying to draw some conclusions as to the historical circumstances that promoted its consolidation in the 1930s but thereafter limited its further advance.

A few words are necessary here to clarify the assumptions of this study. First, in accordance with our previous remarks about the historiographical treatment of earlier marginal sciences, we try to comprehend parapsychology in its own terms—that is to say, we take its subject matter and conclusions seriously. We do not adopt the position that psychical researchers' interest and commitment to the field is merely an irrational lapse, a delusion, or irresponsibility, as so many of their critics seem automatically to have assumed. On the contrary, we accept their seriousness of purpose, and we attempt to consider their activities in the same spirit, no less than we would if the field had already established itself in the scientific mainstream. This involves not merely presenting the internal content of parapsychology, its methods and theories, carefully and sympathetically; it also involves paying close attention to those external factors that have come to occupy so much of the attention of historians of science: organizational and institutional structure, educational and career opportunities, financial support, general relationship with the scientific and the nonscientific public, and so forth. Indeed, it means examining parapsychology as we would *any* specialty beginning to develop normally out of the natural-scientific background, in the same way that other emergent specialities (such as physical chemistry and x-ray crystallography) have been studied. At the same time, we have taken care to remain conscious of the peculiar differences, as well as the similarities, between parapsychology and other sciences: the disquiet with the implications of science for human values that has led so many researchers into the field; the attempt to maintain conclusions at odds with the accepted scientific world view; the difficulties in devising a rigorous methodology of testing and experimentation and of avoiding suspicions of experimental error or fraud; and the perennial problem of devising a truly replicable experimental demonstration of psychical phenomena.

Nearly a century ago, in the first years of organized psychical research, F. W. H. Myers attempted to minimize such differences. Despite the worries of scientists, he denied that there was any paradox involved in the investigation of psychical phenomena—or rather, he added, the only paradox was the absence of paradox. But the serious and painstaking investigations of psychical researchers since then have made it clear that there is in fact something paradoxical about the field, in that it forcefully defends the primacy of the experimental method but has so far been unable fully to develop a research program in the manner of the other experimental sciences. It is this unusual like-and-not-like character of parapsychology that has led us to call it an "elusive science" in our title, so as to make plain from the beginning our understanding of its uniqueness vis-à-vis the natural sciences. Our attempt to maintain in balance both of these somewhat contradictory viewpoints constitutes what claim we can make to objectivity in our treatment. No doubt this position will please neither the vigorous defender nor the sharp critic of parapsychology, but it seems to us the most satisfactory and the most fruitful one for historians of science to adopt.

It is certainly a legitimate question to ask us whether by now we *do* in fact believe in ESP. It is undoubtedly the question that has most often been put to us in the years we have been at work on our project. The answer is that we are prepared to believe that many of the significant results reported in the parapsychological literature are real and not simply artifacts of fraud or poor experimental technique (the two criticisms most often used to explain away ESP success). Certainly there have been instances of cheating by subjects or misrepresentations by experimenters, but they have not been the rule, and in any case they are paralleled by analogous cases of misrepresentation in other experimental sciences. Such cases have not warranted a blanket distrust of experimental science, and we see no reason either why one should conclude, as some critics would assert, that every report of successful parapsychological experimentation must be called into question. What these instances of significant results may mean is, however, something on which we feel able to make no judgment. Whether they are, for example, freaks of random distributions or the product of some as yet indescribable "para-normal faculty" is, as far as we can see, not yet decidable.

As we have already indicated, what has made possible a finely structured study of this crucial period in psychical research is the existence of several important manuscript collections, at psychical-research societies and at academic institutions. Most valuable for our purposes have been the archives of Dr. J. B. Rhine, now in the Manuscripts Division of Duke University Library, to which Dr. Rhine was kind enough to give us unrestricted access. Almost as valuable have been the papers of the American Society for Psychical Research (New York City) and the Society for Psychical Research (London). To them and to other institutions whose papers we have used we wish to express our thanks: the Harry Price Library of the University of London; the Archives for the History of American Psychology, of the University of Akron; and the archives of Clark, Duke, Harvard, and Stanford universities.

One other factor encouraging us to undertake this project was the thought that we might be able to discuss with many of the participants in this history their involvement in psychical research. We have been fortunate enough to carry on many such discussions, which have made possible a fuller interpretation of documentary evidence. We are more grateful than we can say to the following people, who were willing to speak or write candidly about their experiences: Katherine Banham, Hans Bender, Theodore Besterman, William Beven, James C. Crumbaugh, Laura Dale, K. M. Goldney, Joseph Greenwood, Donald R. Griffin, Rosaline Heywood, Hudson Hoagland, Gertrude Johnson, Jocelyn Kennedy, John Kennedy, G. W. Lambert, A. J. Linzmayer, Harold McCurdy, J. G. Miller, E. B. Newman, Fraser and Betty Humphrey Nicol, Dorothy Pope, Gertrude Schmeidler, B. F. Skinner, Ernest H. Taves, R. H. Thouless, Donald West, Dael Wolfle, and George and Sara Zirkle. We particularly appreciate the help of J. B. and L. E. Rhine, J. G. Pratt, and Gardner and Lois Murphy, who spoke with us fully and patiently on repeated occasions; their cooperation has been invaluable. Tape recordings of interviews with many of these individuals were carried out with the assistance of the Oral History Program of the University of North Carolina (Chapel Hill) and have been deposited in the Southern Historical Collection there. To our deep regret, not all those who helped us in our undertaking have lived to see its completion.

We are especially pleased that the Rhines were willing to add a brief afterword to our book. At least one other recent work on the history of contemporary science, Robert Olby's *The Path to the Double Helix,* has included comments by one of its protagonists (Francis Crick), and it seemed to us that it might be similarly appropriate to incorporate the Rhines' reaction to our study. Their remarks are of great interest, not least for making plain the difference in purpose that is likely always to separate the historian of science from the investigators whose work he studies.

It is in keeping with the like-and-not-like character of parapsychology that our work should have been supported in part by two very different sources. Much of our research in Boston was made possible by travel grants from the Richard Hodgson Memorial Fund of Harvard University. We feel proud to share this distinction with many of the most important figures in the history of parapsychology, a number of whom are discussed in the pages below. Grants from the National Science Foundation made possible nine months of leave time for each of us; no less valuable than the freedom this brought was the encouragement it provided at an early stage in our research.

Finally, let us thank all our parapsychologist and psychologist friends and our colleagues in the historical profession—too many to name—who have heard us discuss our ideas on innumerable occasions over these last nine years.

THE ELUSIVE SCIENCE

CHAPTER ONE
The Problems of Psychical Research in the 1920s

I

In 1922 there appeared a book with the ambitious if obscure title *Traité de métapsychique,* written by the physiologist Charles Richet.[1] At 71, Richet was one of France's most eminent biologists, one of her half-dozen living winners of the Nobel prize (his had been awarded in 1913 for his researches on anaphylaxis); he was also France's leading spokesman for psychical research, and the *Traité* was a survey of that field. Richet had been perhaps the first French scientist to argue seriously the merits of studying clairvoyance (or *lucidité,* as he called it), in the 1880s, and he had collaborated with Pierre Janet in the study of telepathy under hypnosis.[2] In the 1890s he had lent his support to the foundation of the *Annales des sciences psychiques,* which until World War I was the major French publication devoted to the discussion of psychical phenomena. In 1905, in his presidential address to the British Society for Psychical Research (SPR), he had coined the term *métapsychique* as a French equivalent to "psychical research."[3] Given his scientific standing and his long involvement with psychical research, Richet was well placed to attempt a synthesis of the mass of information that he and other researchers had been collecting for more than forty years.

Oliver Lodge, another distinguished scientist with a strong commitment to psychical research, remarked that Richet viewed his *Traité* as something of a textbook for a new science.[4] Though Lodge may have slightly overstated Richet's ambitions, the *Traité* was at least meant as an extended prospectus that might attract future researchers to the field and, more generally, make possible a wider audience for psychical research within the scientific community by convincing scientists that this radically unorthodox subject could be fruitfully studied by their own methods. The *Traité* was "to give to metapsychics a place among the old sciences, while imposing upon it the rigor, the authority, and the logic which give the old sciences their form."[5]

Richet stressed the increasingly scientific character of psychical research in a brief historical sketch of its development. He presented it as having undergone four rather Comtian evolutionary stages: a Mythical stage, from antiquity to the end of the eighteenth century; a Magnetic stage, inaugurated by the investigations of Anton Mesmer and the popularization of animal magnetism; a Spiritist stage, ushered in in 1847 by the mysterious rappings evoked by Kate and Margaret Fox in New York State, which soon gave rise to the séance, the medium, and the quasi religion spiritualism; and, finally, a present-day Scientific stage, which had dawned with the work and publications of the chemist William Crookes (1869-72), Richet's own work on hypnosis (1875), and the founding of the Society for Psychical Research (1882).[6] It was of course with the results of the investigations of "scientific" metaphysics that Richet was particularly concerned in the *Traité*.

But could psychical research—metapsychics—really be characterized as a science in the early 1920s? And could the *Traité* function as Richet hoped it would? A textbook for a science embodies the consensus of a community of investigators to a new generation of students. It explains fundamental concepts, illustrates the general laws that have been established, and describes the application of the routine research techniques to those problems that the field has identified as its main concerns. Our question must thus be whether psychical researchers had yet arrived at such a consensus about their subject matter, its methodology, and its theoretical interpretation. To put it in the language of recent philosophy of science, was it yet possible to offer a "paradigm" (Kuhn) or "research program" (Lakatos and Laudan) upon which psychical researchers could express general agreement? We will find that in the 1920s no such ground for agreement existed.

Early in the *Traité,* Richet concisely defined metapsychics as "a science dealing with mechanical or psychological phenomena due to forces that seem to be intelligent, or to unknown powers latent in human intelligence."[7] This division of psychical research into mechanical and psychological phenomena is reflected in a corresponding division of the *Traité:* into *objective* metapsychics, involving the production of physical phenomena, and *subjective* metapsychics, dealing strictly with the revelation to human minds of information by other means than through the senses. To subjective metapsychics pertained "cryptesthesia" (another of Richet's neologisms, comprising telepathy and clairvoyance),[8] the use of the divining rod, veridical hallucinations, and premonitions; to objective metapsychics belonged movements of physical objects (telekinesis), the production of "ectoplasm," or materialization, and hauntings. All of these phenomena had been repeatedly reported since the 1870s and before—Richet himself claimed to have witnessed manifestations of cryptesthesia, telekinesis, ectoplasm, and premonitions—though many took place only in the presence of a specially gifted individual (a "medium") in the course of a sitting or séance. Psychical researchers had identified a succession of particularly impressive mediums since the mid-nineteenth century. Some, like Eusapia Palladino, had been notable for their production of physical phenomena, while others, like Mrs. Leonora Piper of Boston, were remarkable for the information they could furnish to their listeners through the mediation of a "spirit control," who appeared when the medium passed into a trance state. Many other phenomena, such as hauntings, veridical hallucinations, and, to a degree, monitions, could befall almost anyone and were usually known only through the occasional personal anecdote. Still other phenomena, like automatic writing, telepathy, and clairvoyance, seemed to be manifested both by mediums in trance and by ordinary subjects under a variety of conditions, sometimes spontaneously and sometimes on demand.

In his *Traité*, Richet set out what seemed to him the best available evidence for each type of phenomenon, in order to demonstrate its reality and justify its serious and systematic study. He was convinced that psychical researchers had brought to light and were continuing to investigate unquestionable facts, "facts so numerous, so precise, and so evidential that I do not see how any unbiased man of science can cast doubt upon all of them if he consents to look into them."[9] No doubt psychical researchers generally would have agreed. But there was no comparable agreement upon which phenomena were best established and best suited to study; nor was there agreement upon what sort of explanation for the phenomena was most persuasive. The latter disagreement was the source of a particularly sharp division within psychical research in the years immediately following World War I, turning on the question, Are the reported psychical phenomena sufficient to prove the reality of post-mortem survival and the continuing existence of discarnate spirits?

The assumption that the mind and personality survive the death of body and brain had been the basis for the spiritualistic movement of the mid-nineteenth century, and the SPR was formed in the 1880s in the hope that this assumption might be empirically demonstrated. The leaders of the society conducted careful censuses and collected accounts of apparently veridical hallucinations of the dead and dying during the 1880s and 1890s, but their reports, though suggestive, did not conclusively resolve the question of survival.[10] Early in the new century a different type of evidence in support of survival began to accumulate. "Cross-correspondences," unlikely yet closely related themes and references cropping up in the automatic writings of widely separated and independently working automatists, strongly implied to those open to the spirit hypothesis the activity of a single organizing intellect supplying the automatists with material by some supernormal means.[11] The survival hypothesis gained in desirability during the war, as many psychical researchers lost close members of their families. Oliver Lodge, for example, had been active in psychical research since the 1880s, but his belief in post-mortem survival was confirmed when his son, Raymond (killed in September 1915), made apparent contact with him through the medium Mrs. Osborne Leonard.[12] When Richet's *Traité* appeared in 1922, survivalist beliefs of some sort, circumspect or enthusiastic, were widespread in England, on the Continent, and in the United States.

Richet himself did not share this belief, however. In his *Traité*, he insisted vigorously that it was unnecessary to assume the survival after bodily death of individual personalities as discarnate spirits in order to explain cryptesthesia in mediums, even though such personalities might appear to speak through the medium (as did "George Pelham" and "Phinuit" through Mrs. Piper) using characteristic mannerisms of speech and revealing information that could not possibly have been known to the medium. As a physiologist, Richet considered it "as impossible to admit the persistence of the function (mind) without the organ (brain) as the renal secretion without the kidney";[13] it was far more reasonable to suppose that by some unknown means mediums in a state of trance are able to acquire knowledge of the life and personality of a dead man—from objects or individuals—and then flesh this knowledge out in the typically clumsy and often erroneous utterances and actions of mediumistic controls.[14] Cryptesthesia may be presumed to account for all mental phenomena; even physical phenomena, such as telekinesis and materializations, "may be attributed to energies of human origin."[15] What this cryptesthesia might be remained unanswered, and Richet conceded the fact in a number of passages that echo the positivist heritage of French science: "I do not believe in the spirit hypothesis, I believe in the X hypothesis, which will probably be far superior, and will overcome us with admiration. Unfortunately, I do not know what it is, and I cannot formulate it.... But, in waiting for its proofs, I will repeat with the great Newton: Hypothesis non fingo."[16] He agreed that his supposition of vast but unknown mental energies was almost incredible; nevertheless, it was preferable to an hypothesis that to a modern physiologist seemed almost to embody a logical contradiction.

Though Richet was not alone in his rejection of the spiritist hypothesis,[17] he was certainly in the minority, and his alternative was admittedly unsatisfying. His critics among the psychical researchers included scientific figures like Lodge, who took a certain pleasure in demonstrating the philosophical weaknesses and inconsistencies of Richet's professed positivism and who criticized the unwillingness of science to look for explanations. Lodge complained that "cryptesthesia" was nothing but an impressive word hiding "nebulous ignorance."[18] Subsequently he pointed out that Richet's apparent positivism was a sham, that in fact he depended on a host of assumptions deriving from orthodox physiology. In particular, he took Richet to task for maintaining the necessity of psychophysical parallelism—his conviction that the action of the mind had to have concomitant physiological processes in the brain—which, in Lodge's sarcastic phrase, "raised the brain to the position of a fetish."[19] Lodge and other spiritist critics attacked Richet's casual assumption of vast powers in the human mind in preference to the spirit hypothesis. To take Richet's position seriously, they argued, would grant virtually divine omniscience to the mind; if the subconscious mind of a successful medium were supposed capable of selecting from the infinitely numerous pieces of information telepathically available to it just those facts relevant to the present question, no limits could be set to what it might learn.[20] Disarmingly, Richet agreed that many of these difficulties with his position were real, but he remained firm, unwilling to go beyond "the dominant, incontestable fact: knowledge of reality by means other than the normal means."[21] The disagreement between spiritists and nonspiritists was emotional and based on prior assumption, and it could not be resolved rationally. The underlying division in the 1920s over the hypotheses and explanations proper to psychical research was thus making nearly impossible the practice of a unified science that Richet had hoped his text would inaugurate.

A second issue still largely unresolved in the early 1920s, hardly less fundamental to psychical research, concerned the choice of phenomena for study. Richet's distinction between subjective and objective metapsychics—between mental and physical psychical phenomena—had long been well-established, and at first there had been no feeling that one class of phenomena was more trustworthy than the other. In the 1890s, for example, the members of the SPR had investigated Eusapia Palladino's physical mediumship with the same enthusiasm that they were devoting to the mental mediumship of Mrs. Piper.[22] Eventually, however, the SPR leadership began to concentrate almost exclusively upon mental phenomena such as telepathy and the cross-correspondences—in great part surely because the physical phenomena were by far the easier to produce fraudulently, as had been recognized time and again. However, research on physical phenomena continued on the Continent. Albert von Schrenck-Notzing in Germany and Gustave Geley in France both studied a series of mediums (most notably perhaps, "Eva C." [Marthe Béraud]), and by 1920 they were enunciating a complex theory of the vitalist origins of the ectoplasm that these mediums appeared to produce.[23] No doubt Richet's acceptance of such physical phenomena in the *Traité* disappointed many of his English readers.

Those readers would also have found themselves unable to accept Richet's conviction that *lucidité*, or clairvoyance, was a more basic phenomenon than apparent cases of telepathy. In the *Traité,* Richet had defined the general term "cryptesthesia" quite neutrally and subsumed telepathy and clairvoyance under it.[24] The term "telepathy" had been introduced by the English investigator F.W.H. Myers to mean "the supersensory transference of thoughts and feelings from one mind to another," while "clairvoyance" had come to be used for the acquisition of knowledge of information about material objects and events by other than sensory means.[25] Richet's fundamentally mechanistic orientation in physiology made it difficult for him to credit the reality of the former, since in his view it meant that the mind receiving the information would have to interpret in verbal terms the cerebral cellular vibrations of the source and, moreover, separate the information out from the myriad other vibratory "messages" encoded in the same locale. He thought it far more likely that the mind was perceiving the objective reality that had originally evoked the thought, though he agreed that the two hypotheses might be logically indistinguishable.[26]

In this respect, however, Richet belonged to a minority. Particularly among the English, it was telepathy, not clairvoyance, that was taken most seriously. The early SPR investigations of post-mortem survival and of veridical "phantasms of the living'' had seemed to yield clear evidence of pure mind-to-mind (or spirit-to-mind) communication. On the other hand, what little evidence for pure clairvoyance had been published seemed unconvincing, and the survival hypothesis in conjunction with evidence for telepathy made it difficult to see why clairvoyance was a necessary hypothesis at all; "for of whatever nature the thing known may be, it is almost impossible to ensure that knowledge of it is not possessed by some other mind, and, if such knowledge does exist in any other mind, whether incarnate or discarnate, telepathy may be the means through which the medium becomes aware of it.''[27] Moreover, the unconcealed materialist assumptions of Richet's scientific training that had led him to accept clairvoyance struck many psychical researchers (even other scientific figures, such as Oliver Lodge) as unwarranted and unsatisfying.[28]

One final issue was almost as divisive of psychical research as the survival question, though it was openly addressed less often: In what sense, and to what extent, should psychical research attempt to make itself "scientific"? The SPR had originally been founded on the conviction that scientific scrutiny ought to be as productive when applied to psychical phenomena as when applied to the purely physical realm and had attempted to model its research upon that of contemporary science.[29] The community of psychical researchers clearly prided itself upon the fact that men like Crookes, Lodge, and Richet—from mainstream science—could actually accept the reality of the phenomena in question. Over and over again, in presidential addresses and other public statements, we find the leaders of the field appealing to the history of science in order to show that scientific study will eventually justify those who have conscientiously looked for the facts without letting prejudice blind them.[30]

On the other hand, the adoption of a "scientific" attitude clearly involved a number of dangers, all of which were causing considerable disquiet to many in the 1920s. As we have just seen, and as several psychical researchers remarked, to use modern science to explain metapsychics necessitates distorting the facts: the analogies, the working hypotheses that seem natural and are most directly testable by contemporary science may well be invalid or misleading.[31] Richet's case illustrated to a great many spiritists the extent to which science could actually blind even the sympathetic student to the truth by constraining him to accept nothing that would violate a cosmos of vibrations and materialism.

The usefulness of science lay rather in the universality of its method; scientific theories and laws might be questioned or abandoned in favor of new truths, but it was always the same proven method that was responsible for the resulting progress. As most psychical researchers understood this methodology, it implied meticulous observations or fact-gathering with extraneous circumstances rigorously supervised and controlled and unwavering commitment to the facts thus established. This essentially "natural-historical" approach to science, one through which virtually all scientific investigation passes, had been what Richet had in mind even in discussing "experimental cryptesthesia." For the natural-historical method was better suited than strict experiment to much of the material of psychical research, since the subject was concerned with the classification of spontaneous psychical incidents and with mediumistic studies, in which the phenomena were unpredictable and the governing variables unknown.

By the early 1920s there were already psychical researchers who chafed at the need for rigor even in this natural-historical approach, feeling that so much had already been established by the SPR's critical methods that standards could be relaxed to make possible a wider examination of the evidence. Such researchers also claimed that the cautious and professedly skeptical attitude of the "scientific" investigator tended to inhibit the performance of their subjects.[32] Not unnaturally, then, the prospect of a truly *experimental* approach to psychical research—testing a narrowly defined problem, controlling all variables, and gathering a specified class of data (quantitative, if possible)—was regarded rather doubtfully.

To be sure, the leaders of psychical research had encouraged experimentation in the 1880s and 1890s, years that had seen a number of restricted experimental studies of telepathy that explored the problems of experimental design and evaluation. Such studies had not proven remarkably fruitful, however, and had largely been supplanted by exploratory investigations of automatists and mediums in the early twentieth century—especially since a tightly controlled experiment seemed likely to prevent the phenomena it was designed to explore. René Sudre, something of a disciple of Richet's, vividly expressed the concerns of many thoughtful psychical researchers:

> In the highly praiseworthy desire to strip from this research its mystical or worldly character, we have copied physics or physiology laboratories, we have set up a great empty room, covered with ceramic on floor and walls, and have heaped up there steel and copper instruments in bizarre shapes: photographic chambers, electrical apparatus, and so on. When a subject enters such a room, he has the impression of coming into a surgical clinic or into a torture chamber, and that is enough to sterilize him. We cannot misunderstand the psychology of metapsychic subjects any more disastrously. To obtain the maximum of what they can give, we must on the contrary impress them favorably, and create for them an atmosphere of warmth and confidence. A study, a room furnished with taste where one regularly works—these are the best laboratories for metapsychics.[33]

This capital necessity, he concluded, could be reconciled with the needs of science by fitting out a comfortable room with all the necessary equipment hidden from the subject's view or set up in an adjoining room. Nevertheless, preconceptions like this made the typical laboratory research of the experimental psychologist difficult to undertake or to imagine. The world of psychical research in general had a much less circumscribed understanding of "conducting an experiment": it meant simply to put to the test, with as much rigor and observational control as the subject would permit, anyone who purported to be able to produce supernormal physical or mental phenomena. In almost every case, these "experiments" were designed only to demonstrate the existence of psychical phenomena, not to probe into the conditions that permitted their display; to collect proof rather than to test research questions.

There are occasional exceptions to this generalization. As we will see, in the early 1920s the French engineer René Warcollier was deeply immersed in a series of long-distance tests of telepathic transmission of images and phrases.[34] More widely known, perhaps, was the model experimental study reported by H.I.F.W. Brugmans to the First International Congress of Psychical Research in 1921. Brugmans and two colleagues in the psychology laboratory at Groningen had studied a twenty-three-year-old physics student, Van Dam, seating him in a curtained booth and placing themselves in a soundproof room above, from which they could look down on him unseen. Van Dam had been told to indicate one of forty-eight spaces on a sort of checkerboard, which (unknown to him) had been randomly selected beforehand by the experimenters; he had 60 successes out of 187 trials rather than the 4 that chance would allow. Brugmans further reported that a study of Van Dam's failures revealed that his guesses were much closer to the correct space than chance would have allowed; that success *increased,* if anything, with increased distance between subject and experimenters; that different experimenters had equal results in trying telepathically to guide the subject's hand; and that giving the subject alcohol increased the rate of success.[35] Brugmans's study received general praise, and not only scientists like Richet but more suspicious individuals like René Sudre could still acclaim it warmly.[36] But few references to "experimentation" in the early 1920s reveal this interest in sustained concentration upon small, well-defined problems.

Complicating the attitude of psychical researchers towards science as a model was their suspicion of professional scientists and their unwillingness to accept the latter's judgment as a standard. J. G. Piddington, one of the leaders of the SPR in the 1920s and a specialist in clarifying the classical and literary allusions in the cross-correspondence texts, expressed this distrust openly in his presidential address of 1924:

> Training in physical science and the pursuit of it, as it seems to me, do not necessarily make a man a better critic of the evidence collected by this Society than do some other forms of education and employment.... Men of science are not, so far as I know, specially expert in assessing the value of human testimony; nor are they specially expert—perhaps rather the contrary—in literary matters.... I do not, then, believe that official science is our proper court of appeal; and I do not know of any formal body which could assume that function with authority.... It is, I believe, only among what for want of an exacter definition I have called "educated people" that a competent tribunal will be ultimately found.[37]

Some researchers, themselves scientists, had been more hopeful. William McDougall had argued four years before that psychologists at least were well aware of the deficiencies of a rigid materialism and would be sympathetic to the implications of psychical phenomena; what held them back was their fear to risk

The Problems of Psychical Research in the 1920s

The experimental situation for H.I.F.W. Brugmans's Groningen experiment. The experimenters, in a room above, select one of forty-eight target squares by lot and attempt to communicate this selection to the subject below, who is to point to a square on the board in front of him; the experimenters observe through a window in the ceiling.

the reputation of their own relatively new and unproven subject, as well as the fear that "the least display of interest or acquiescence on their part may promote a great outburst of superstition on the part of the public, a relapse into belief in witchcraft, necromancy, and the black arts generally, with all the moral evils which must accompany the prevalence of such beliefs."[38] McDougall, convinced that scientists' objective judgments could only be helpful to the field, looked forward to enlisting them in psychical research; but his belief was not widely shared.[39]

It is thus very difficult to recognize in the psychical research of the early 1920s the sort of programmatic consensus that any science presupposes. The field was divided, not over trivial points of interpretation or technique, but over the major constituent elements of a coherent discipline: subject matter, assumptions, methods, and standards of evaluation. Richet had written confidently of the "scientific stage" into which psychical research had passed fifty years before, but his confidence is belied by the disagreements that occupied the field. Despite his expressed intent, his *Traité* had little chance of becoming—and did not become—a spokesman for a unified subject; nor is it likely that any attempt at a synthetic study could have done so. There are reasons why the 1920s might be expected to have been a decade favorable to the serious scientific examination of the field: the combination of the disorienting social and cultural consequences of World War I with the revolutionary discoveries of contemporary physics, undercutting a comfortably closed and materialistic world view, could well have led some scientists to take up the study of psychical phenomena.[40] Yet this happened only rarely. In the end, scientists proved unwilling to grant that the material psychical researchers had so long accumulated was of adequate evidential value to establish the validity of the phenomena.

This is well illustrated by the reception given Richet's *Traité* by the French scientific community upon its presentation to the Académie des Sciences. Such a work by a Nobel laureate could scarcely be ignored, but reviewers did not fail to comment on what they perceived as its lack of hard evidence. The psychologist Henri Piéron wrote in amazement that "after having read this *Traité*, where innumerable anecdotes are set out, one cannot escape—or at least I could not escape—a feeling of real stupefaction. How can we call a collection in which *not one actually demonstrated fact* appears a scientific treatise?" He kindly attributed its looseness to Richet's own *génie,* good at hypotheses but definitely not at experimental rigor.[41] The stir caused by the *Traité* was enough, however, that in the year of its publication two psychologists and a physiologist from the Sorbonne—Piéron, Georges Dumas, and Louis Lapicque—organized an experiment at the physiology laboratory of the Faculté des Sciences to study the objective phenomena allegedly produced by the physical medium Eva C. Fifteen sittings were held, without results except to leave the suspicion that Eva might have produced ectoplasm by regurgitating previously swallowed material. Negative results prove nothing, as Richet pointed out sharply, but they left the feeling that the positive results previously reported would not be easy to substantiate.[42]

Of all the discussions of Richet's work within the academic community, the most thoughtful and most interesting is surely Pierre Janet's long review in the *Revue philosophique.*[43] Janet was perhaps the most eminent psychologist in France, but, more than that, he was peculiarly well qualified to judge a work on metapsychics because so much of his own psychological research had been carried on in abnormal psychology, including some studies of instances of hypnosis during the 1880s that seemed to have been induced telepathically.[44] In his review Janet shows himself quite sympathetic to Richet's general aim of rendering a scientific account of the evidence for metapsychics. Much of the review is given over to a detailed presentation of Richet's arguments, all set out with perfect fairness and often indeed an approval for which Janet does not feel it necessary to apologize; on the contrary, he classes himself, neither with the narrow-minded partisans of official science nor with the spiritists, but in a third group, in which he says he is by no means alone.

> These are the men who already know most of these facts, who have already studied them with great interest, who from time to time have thought that they too had observed analogous phenomena. Unfortunately the descriptions that they had read had not seemed to them entirely convincing, their own observations had seemed to them incomplete and insufficient in some respect, and whatever their own inclination to admit some of these facts, they had maintained some doubts as to the actual reality of metapsychic phenomena. More men of this kind than Richet imagines have thrown themselves upon this book, and have read it to the end with a great sympathy, hoping to find there the conviction that they lacked.[45]

Unfortunately, Janet continues sadly, these potential converts are, after reading the *Traité,* "bored, unhappy, and, without really knowing why, a little more skeptical than before." Janet interpreted this disquiet to arise principally from issues of methodology and evidence. The evidence with which the book is overflowing consists of a vast number of suggestive cases widely variable in evidential value, almost none (by Richet's own repeated admissions) wholly satisfactory. Richet would never have proceeded like this in presenting a biological fact; there he would have set out a single best experiment rather than a mass of feeble cases intended to reinforce one another.[46] Besides, Richet's own strictures simply did not go far enough. Faced, for example, with a photograph that reveals a piece of "ectoplasm" as actually a creased, two-dimensional surface—evidence that might arouse the gravest suspicion in the scientist-reader—he acknowledges the appearance of fraud but insists that there are reasons why ectoplasm might look this way. Such an attitude is bound to leave the sympathetic reader not merely uncomfortable or embarrassed but mistrustful as well, for it suggests that his guide has become a believer, abandoning science along with his critical spirit.[47]

Richet's *Traité* was indeed largely anecdotal in style, recounting much "natural-historical" evidence, and moreover was not very critically argued. Lodge had indicted the work for these shortcomings in his review for the SPR, but he had also pointed out that the *Traité* was meant to be a general text, that it was designed to provide a broad introduction to the field but not to provide specific details.[48] This was also Richet's defense in his immediate reply to Janet in the *Revue philosophique:* his book was meant to be didactic, and had he meant it as a monograph, it would have looked quite different. In any case, Richet argued, would the "open-minded" *really* have been convinced by a tiny number of absolutely irreproachable experiments? Would they not complain that so small a number of cases, however perfect, might well be due to chance or hidden error and refuse to commit themselves to a belief in cryptesthesia on so limited a basis? The psychical researcher, he went on, must adopt a different approach. The evidence he can study will appear spread over a continuum of quality, ranging from indistinct, merely suggestive indications of supernormal faculties to the best evidence that a Mrs. Piper, say, can produce. The student who has had experience of this highest degree of achievement will be able to recognize in the lower ranges of the continuum partial successes that more or less clearly resemble perfection. Moreover, contrary to what Janet supposed, this is exactly comparable to what happens in orthodox biological experimentation: the biologist interprets imperfect or incomplete experimental results as strengthening the conclusions of a unique, irreproachable experiment, so long as they tend in the same direction and fit his normal expectations.[49]

In speaking of "normal expectations," we put our finger on the crucial matter separating the scientist from the psychical researcher. The laboratory scientist with little or no experience of psychical research will inevitably approach the field with an understandable skepticism, and hence he will attack the continuum of evidence from the opposite end. He will begin with cases that psychical researchers themselves have exposed as instances of fraud; will then dismiss those cases that, while not proven to be fakes, look fraudulent and could easily have been faked; and will finally be led to condemn by extension those cases that appear to be watertight. Richet's conviction that a few perfect experiments would have carried no weight was probably quite sound. It is far more likely that faced with an apparently watertight experiment, Janet's scientists would have denied the testimony of the witnesses to the event—Janet himself had already criticized Richet for depending on the testimony of other authorities, even the firsthand testimony of such observers as Crookes and Myers.

This is a problem nearly impossible to resolve, then or now. Does not every scientific argument depend upon the reader's acceptance of the validity of the author's testimony and of his cited sources? Yet in so unorthodox a field as psychical research, not even direct testimony is enough to carry conviction of the achievements of a Mrs. Piper. For scientists, belief depends upon first-person experience, and no one else's assertions can be taken as proof. But in this case, then, why would the mainstream scientist take up psychical research? He cannot detect the phenomena until he is willing to grant their existence and is thus able to relax his professional criteria for judgment; but he almost certainly will not grant their existence without first having experienced them.

We have treated the exchange between Janet and Richet at such length because it is so revealing of the difficulties psychical research had to face in order to be accepted by mainstream science. Once again Richet's assessment of the field as being in its "scientific stage" seems to have been overoptimistic, for surely in one sense a subject cannot be a true science until it is acknowledged as such by other scientific disciplines. And if Janet remained unmoved, how could less sympathetic scientists whose orientation and research were further removed from the field be expected to welcome it as a fellow to their own? Clearly, the "natural-historical" approach to psychical investigation was, in this respect at least, inadequate. What was needed was something closer to the experimentation practiced in the other sciences, in which psychical phenomena could be shown to be to some extent replicable, lawful, and perhaps measurable. Given the limitations inherent in the phenomena as they had so far been studied, and given the ambivalence that many psychical researchers still felt towards the experimental ideal, this was not possible in the early 1920s.

II

The very suggestion that Richet's *Traité* might have functioned as a textbook for psychical research presumes the existence of a community of psychical researchers. We have already examined some of the intellectual issues debated by that community in the early 1920s and have come to understand its internal divisions. But so far we have said nothing about its organization and membership, and it is important to understand these structural features of psychical research as well.

In dating the origins of "scientific" psychical research to the 1870s, Richet was associating the field with a decade that had seen a particularly rapid growth of scientific specialization and the crystallization of several independent disciplines. The discipline that would come to have the closest ties to psychical research, psychology, made its own academic debut at this time with the establishment of Wilhelm Wundt's laboratory and institute at Leipzig in 1876 and of American departments in the next decade. The fields of abnormal psychology and psychopathology were also beginning their development as scientific and professional specialties to be distinguished from the disreputable activities of mesmerists and the clinical concerns of alienists. Indeed, Richet's own work on hypnosis played a role in leading the psychiatrists of the Paris and Nancy schools to at last accept hypnosis and hysteria as genuinely *psychological* phenomena.[50]

The first psychical-research organizations were established at almost this same time—the Society for Psychical Research in London in 1882 and the American Society for Psychical Research in 1884. However, the field's amateur character made it structurally quite distinct from the other new scientific disciplines, in that it lacked an academic base and therefore the prospect of professional careers. In the 1920s, forty years later, it remained an essentially amateur enterprise, carried out by devoted individuals who had been forced to make their livelihoods and careers in other, more acceptable fields. Charles Richet and Oliver Lodge are good examples, except that both were professional scientists. For the societies included all ranks of the middle and upper classes: academics, doctors and lawyers, the employed, and the independently wealthy. Most of the members, however, had little active involvement in psychical research besides paying their dues and communicating an occasional psychic incident for publication. The societies were really led by a small number of committed people whose articles dominated the literature of the field and who thus became—and remain for us—its spokesmen.

The oldest of these societies, the SPR, was still preeminent in the 1920s. Founded principally to ensure the systematic investigation of spiritualist, mesmeric, and psychic phenomena, it had included among its charter members many of the outstanding figures of late Victorian society and had possessed a wide foreign membership as well. It had been the hope of many of its founders that it might some day prove possible to demonstrate empirically the reality of postmortem survival, and a general survivalist tone remained characteristic of the society's leaders, although it was muted by their strong conviction that psychical research required the most cautious and considered judgments. The SPR published short communications and business in a monthly *Journal* and longer studies in its *Proceedings;* furthermore, the long monographs of its founders—F.W.H. Myers and Edmund Gurney, for example—come close to being official publications of the society. It had originally assigned the supervision of research to a variety of investigative committees, but these had been abandoned, and in 1922 the official research of the society was supervised by its newly appointed Research Officer, E. J. Dingwall, whose principal concern was the investigation of purported mediums.

The leadership of the SPR in the early 1920s centered upon Eleanor Mildred Sidgwick, whose longevity, acuity, and social position had made her the symbol of the national tradition of organized psychical research. A member of the distinguished Balfour family (her brother Arthur had been prime minister from 1902 to 1904) and the widow of Henry Sidgwick, the Cambridge philosopher who had been one of the chief agents in the founding of the society, she herself had enjoyed an active academic career. She had done original scientific work with her brother-in-law, the physicist Lord Rayleigh; and she had been treasurer of her Cambridge college, Newnham, and subsequently its second principal. She was also remarkable for her incisive and thoughtful writings on psychical research, in which she had been involved since the 1880s. She had been editor of the society's *Proceedings* and *Journal* for a decade at the turn of the century; she sat on its governing council from 1901 until her death, serving as its president in 1908/9; and from 1907 until she retired in 1932, the society's jubilee year, she held the office of Honorary Secretary.

Closely associated with Mrs. Sidgwick in directing the affairs of the SPR was a small group of individuals, some of whom lived with her at the Balfour home at Fisher's Hill, Woking (her brother Gerald Balfour, J. G. Piddington, and Alice Johnson), and including among others two younger figures, Helen and W. H. Salter. Something of a "closed corporation," these individuals sat on the council, shared the administrative positions of the society (including the editorship of its publications), and thus in effect determined the policy of the SPR throughout the 1920s. The society continued to include literary and scientific celebrities in its membership, as it always had done: most active in the early 1920s were the psychologist William McDougall, who left Oxford for Harvard in 1920; the philosopher F.C.S. Schiller, who had returned to Oxford from Cornell in 1897;

Oliver Lodge; and Arthur Conan Doyle.[51] Nevertheless, such figures were markedly less influential in the society's affairs than the Fisher's Hill leadership. Despite their fame, both Lodge and Doyle were regarded with some suspicion by conservative members of the society because of their firm advocacy of the survivalist hypothesis.

The French counterpart to the SPR, the Institut Métapsychique International, was founded only in 1919 by a wealthy businessman, Jean Meyer. Charles Richet, of course, had long been a spokesman for psychical research in France, and Richet continued to lend his prestige and moral support to the movement in the 1920s, but he was not actively involved in the IMI. The institute had come into being as a result of Meyer's acquaintance during the war with two physicians, Rocco Santoliquido and his associate, Gustave Geley. The three soon found themselves in agreement that systematic research into psychical or metapsychic phenomena would be valuable; at the war's end, therefore, Meyer established the institute for just such systematic research, placing Geley at the head.

In many respects the IMI was analogous to the SPR. Both attempted to report careful investigations and to set standards for the field; both held regular colloquia for their members upon current topics of research; and both were entirely dependent upon the donations of their members for their activities. The French counterpart of the SPR *Proceedings* and *Journal* was the institute's bimonthly *Revue métapsychique,* which provided the French with a forum for publication and served as a listening post upon international psychical research. Like the English society, the IMI was directed by a very small number of individuals; indeed, in its first years the institute spoke almost exclusively in the voice of Geley, who was editor of the *Revue métapsychique* as well as director of the institute.

There was no organization comparable to these societies in postwar Germany. Psychical research in that country was dominated by the personality of Albert Freiherr von Schrenck-Notzing, who had begun his own studies in the field in the 1880s. His younger colleague in psychical research, Rudolf Tischner, summed up Schrenck-Notzing's orientation in the judgment that "he was essentially an observer and preferred to pursue demonstrational investigations; carrying out a series of quantitative experiments, or putting questions to nature, was remote from his interests." A wealthy marriage not only made possible Schrenck-Notzing's own investigations of mediums, it allowed him to subsidize (and control) the periodical *Psychische Studien* (renamed *Zeitschrift für Parapsychologie* in 1926) until his death in 1929. Tischner and the Tübingen philosopher Traugott Oesterreich tried to bring German psychical research out from under Schrenck-Notzing's influence at the end of the 1920s, hoping to find backers for a "Deutschen Gesellschaft für Parapsychologie" along the lines of the SPR, but the economic crisis of the early 1930s aborted these plans.[52]

The organization of psychical research in America was unsettled. The American society (the ASPR) had never enjoyed the same broad support that its

English counterpart had. Organized late in 1884 at the prompting of one of the SPR founders, the physicist William Barrett, the American society's original membership was led by a number of critical and tough-minded academic scientists, men who were much less interested in proving survival after death than in investigating the claims of spiritualists (with whom they almost immediately classed the SPR itself). The one major American scientist with warm sympathy towards psychical research was the psychologist William James, and he soon turned to the English society as more truly congenial to his interests. The ASPR leaders lost interest in psychical research when they found they could establish no positive facts, and with funds in short supply the ASPR was disbanded—or rather absorbed by the SPR—in 1889. Ironically, James and an investigator for the SPR, Richard Hodgson, had just begun the study of perhaps the most impressive mental medium in history: an American, Mrs. Leonora Piper. Despite enthusiastic reports from England about this subject for study, however, the ASPR was not revived for nearly twenty years. It reappeared in 1907 as "Section B" of an "American Institute for Scientific Research" proposed by James Hervey Hyslop, a former professor of philosophy at Columbia (1889–1903), who had studied psychology with Wilhelm Wundt at Leipzig.[53] For a decade Hyslop ran the ASPR virtually single-handedly, but in 1917 he appointed as his assistant Walter Franklin Prince, an Episcopal minister with training in abnormal psychology and a deep interest in psychical research, who for more than fifteen years would represent American concerns for a "scientific" approach to the subject.

Prince's involvement with psychical research went back to 1910, when as rector of All Saints', Pittsburgh, he had encountered a bizarre display of multiple personalities in a young woman, Doris Fisher. Prince and his wife took Doris into their home for support, treatment, and study, eventually adopting her as their daughter. One of the personalities manifested by Doris displayed what seemed to Prince undoubted telepathic abilities; another claimed to be literally a "guardian spirit" of the young woman and indeed exhibited a maturity quite out of keeping with the rest of her behavior; and Prince interpreted this latter personality as an independent if incorporeal entity that had somehow become associated with Doris's psyche. His subsequent psychical investigations reflected the broad range of spiritistic interests opened up for him by the Doris Fisher case—mediumistic inquiries, studies of automatic writing supposedly from long-deceased authors—in which his concern for careful and rigorous inquiry earned him the approval of the leaders of the SPR (in 1930 he became the society's first American president since William James).[54]

When Hyslop died in June 1920, Prince became assistant (acting) director of research and editor of the *Proceedings* and *Journal* of the ASPR—publications that Hyslop had revived. Miles Menander Dawson, a New York lawyer, became secretary, and Gertrude Ogden Tubby, long an associate of Hyslop's, became assistant secretary. The effective operation of the society in fact devolved upon

Prince and Miss Tubby. It was nominally governed by a board of trustees—comprising eight business and professional men, whose president, W. C. Peyton, had held the post for several years—but this group had never exercised anything more than a loose general supervision over the society (still officially the "Institute"). In May 1921 the trustees elected William McDougall to fill Hyslop's place at the head of the society.

It is during McDougall's presidency that we find our first indications of something that might have been predicted: that the failure of psychical researchers to reach a consensus on the intellectual substance of their field could have unfortunate consequences for the institutions that were attempting to coordinate research. This happened first in America because both consensus and institutions were particularly weak. In 1920 those in the United States who were attentive to psychical research included a small number of invariably skeptical or hostile academic psychologists; a much larger group of investigators convinced of the truth of the survival hypothesis, among whom Hyslop had been most prominent until his death; and a few (like Walter Franklin Prince) occupying the prudent middle ground shared by the SPR leaders. McDougall in effect tried to bring the two extremes together and in the process forced them still further apart. The same potential for polarization and institutional fragmentation was present in the other societies, too, but nowhere else was it as fully realized.

When McDougall was elected ASPR president, he had just arrived in the United States from Oxford to fill Hugo Münsterberg's chair in psychology at Harvard. He had been chosen president of the SPR in the previous year (1920/21) and seems to have wanted now to associate the American society more closely with the best of the English tradition. In W. F. Prince and a few other members of the American society he found a nucleus that shared his sympathy for the Sidgwick conception of psychical research: low-key, rigorous, conservative, and respectable. The name of the American "Institute" was now officially changed back to the original American Society for Psychical Research, and exchanges between the English and American groups, rare since the deaths of Hodgson (1905) and James (1910), began to recur. McDougall and Prince were beginning to establish connections between American psychical researchers and their counterparts abroad.

The early 1920s were propitious years for doing this, since it was at this time that an attempt was being made to consolidate the entire field of psychical research by instituting regular international congresses, with national committees to maintain interim communications and handle arrangements for impending congresses. The first of these congresses, held in Copenhagen in 1921, was organized in the hope of creating "a central international office, which [was] to bring into contact experimenters on [the] same lines in different parts of the world."[55] At McDougall's urging, Prince was sent as the U.S. representative, his expenses met by private subscription and a small grant from the ASPR.[56] In

The First International Congress for Psychical Research (Copenhagen, 1921). In the first row, Helen Salter is third from the left; Walter Franklin Prince, sixth; Albert v. Schrenck-Notzing, eighth; Gustave Geley, tenth and last. In the second row, H.I.F.W. Brugmans is second from the left; Hereward Carrington, fourth; W. H. Salter, tenth (next to last). In the back row, René Sudre and Carl Vett are respectively third and second from the right-hand end.

Copenhagen, Prince was able to meet and establish close ties not only with English psychical researchers but with researchers on the Continent as well. Some months later René Sudre wrote to ask Prince whether a regular exchange of publications between the French and American societies might now be established, and Prince replied in a manner that shows how seriously he took the idea of international cooperation.[57] Prince had demonstrated his interest in creating a truly international movement at Copenhagen by his attempt to get the congress to establish a standard terminology for psychical research, a project he continued to promote at future meetings—at Warsaw in 1923, through the American delegate Gardner Murphy, and again personally in Paris in 1927.[58]

The community of spiritualists in the ASPR, Hyslop's heirs, may have felt somewhat restive because of this reorientation of the society towards the English model. They seem also to have been disturbed by McDougall's plans for the future work of the society, which included a more active program of research and the establishment of ties with orthodox science and the mainstream of American academic life. Certainly the trustees were somewhat taken aback, and McDougall was evidently a little bitter at their initial lack of enthusiasm for his proposal to begin a fund-raising drive to subsidize research.[59] It fell to Prince to direct this drive, since McDougall had to spend most of his time in Cambridge rather than in New York, the headquarters of the society, and had excused himself from the campaign. There is nevertheless some evidence that the campaign met with a slow but positive response.

McDougall tried to associate the ASPR with the academic world at the same time by setting up an Advisory Scientific Council; he hoped that the participation of scientists in this council would encourage the public to accept the society's activities as worthwhile.[60] The first members of the council included a half-dozen academic psychologists: J. E. Coover (Stanford), Morton Prince (Harvard), Charles L. Dana (Cornell Medical College), Frederick Peterson (Columbia), L. T. Troland (Harvard), and Joseph Jastrow (Wisconsin).[61] Prince, Dana, and Jastrow all had belonged to the original ASPR in the 1880s; Coover and Troland, who had both carried out some unsuccessful tests for telepathy, exemplified the limited place psychical research had acquired in the modern psychology laboratory.[62] At its second meeting, in November 1921, the council decided to hold monthly meetings at which experimental work could be reported and discussed; it also undertook the collection of anecdotal material, purportedly supernormal impressions received in dreams or in the waking state.[63] Psychologists were joined on the council by enthusiasts and committed spiritualists—Elwood Worcester of Boston, for example, and W. C. Peyton—but in spite of this wide spectrum of attitudes the council functioned well; even the unremittingly skeptical Joseph Jastrow was soon brought round to admitting the value of the council's work.[64]

The discontent of ASPR members with McDougall's policies was revealed dramatically in the spring of 1923. At a meeting of the board of trustees on 6 April a new president—the Reverend Frederick Edwards—was chosen to replace William McDougall, apparently to the utter surprise of McDougall and his friends, who tried perplexedly to learn where Edwards stood.[65] It was not long before Edwards made it clear that his interests were quite different from McDougall's. Virtually his first two acts as president were to assume the editorship of the ASPR publications from Prince (making Miss Tubby associate editor) and to allow the Advisory Scientific Council to lapse. At the same time, he announced plans for the development of state and local branches and councils of the society to function under the loose leadership of the New York office.[66]

The coup was ostensibly intended to help smooth out the executive operations of the society. Since Hyslop's death the day-to-day responsibilities for the institution had been shared by Prince and Miss Tubby, two incompatible individuals. It must be admitted that Prince was not an easy man to get along with. Upright and incorruptible though he was, he was also acerbic and sometimes rude, capable of taking and giving offence with equal ease. He was especially sensitive to supposed encroachments on his prerogatives and utterly contemptuous of anything that suggested to him gullibility or incompetence. Miss Tubby, too, was insistent upon her rights, as well as proud of her investigative work. She carried out mediumistic studies on her own, at considerable expense, and had consistently interfered with Prince's publication policy.[67] McDougall, in Cambridge, may not have known how deep the trouble ran, for he did not move to control it; the suggestion (made after the coup) that McDougall had been replaced in order better to control the New York office thus may have some merit. Prince himself, however, believed that Miss Tubby had engineered the whole affair.[68] He claimed, too, that neither McDougall nor Margaret Deland, both trustees, had ever received notice of the crucial April meeting; if true, this strengthens the possibility of an organized coup.[69]

There were other reasons why some members of the ASPR might have felt dissatisfied with McDougall. Certainly those who had been drawn to psychical research by Hyslop would have been unhappy with McDougall's attitude, for the two disagreed in two important respects. Hyslop had been far more convinced of the soundness of the survival hypothesis than was McDougall, for one thing. While he did not reject the possibility of telepathy, he tended to consider the word a label for those few inexplicable facts that the spiritist hypothesis could not yet encompass but might well soon manage to include. Nor did he believe in the value of an experimental-scientific approach to psychical research, as McDougall certainly did. Hyslop stressed that if telepathy existed, it was an isolated phenomenon to be found only in "the mediumistic type," and that experiments designed to test for it among the mass of "normal people" were unlikely to succeed. In any case, he believed, the physiology of perceptive response to external stimuli, the findings of laboratory science, were absolutely irrelevant to the study of psychic communication.[70] Views like Hyslop's, which ran counter to the interests of Prince and McDougall, were being expressed by many ASPR members in the early 1920s and underlay, for example, the critical review given Richet's *Traité* in the ASPR *Journal* in 1922.[71]

In the spring of 1923 McDougall addressed such issues openly, insisting on the need to impose scientific rigor on a psychical research that had hitherto been somewhat antiscientific or at least antiexperimental. In January the ASPR *Journal* had reprinted a lecture by McDougall entitled "The Need for Psychical Research." Aiming against the overcredulous spiritualist, McDougall singled out Sir Arthur Conan Doyle, who had just completed a successful American lecture tour. How could Sir Arthur, who represented the best of spiritualism but who had abandoned interest in research, be won back to support and cooperation? Not by lowering standards of investigation; "we must continue to run the risk of estranging them by the rigidity of our scientific principles."[72] McDougall was answered in April by George E. Wright, organizing secretary of the London Spiritualist Alliance, whose reply to McDougall's insistence upon maintaining scientific standards is particularly revealing: "Spiritualism appeals to all classes of the community. Hence the majority of its adherents must needs be persons who lack the capacity to understand and to appreciate the necessarily difficult methods of analysis and argument in Psychical Research.... I have no doubt that Dr. McDougall's way is the right way to bring conviction to men of science. But it has very slight effect on the general public."[73]

In that very month the ASPR coup took place, and Frederick Edwards (now editor of the *Journal* as well as president of the society) began to make statements denying any inconsistency between the aims of spiritualists and those of the ASPR, while by implication distancing the society from the academic world and from psychology in particular.[74] Since Arthur Conan Doyle had returned to the United States on a new lecture tour to publicize spiritualism, Edwards encouraged him to write a reply to McDougall's criticisms, which appeared in the June *Journal*. Edwards had heard Doyle speak twice in April and wrote warmly of Doyle's appeal for the general public. In particular, Edwards praised Doyle's public defense of the reality of physical phenomena, most notably ectoplasm. It was Doyle, Edwards wrote happily, who was gradually convincing the American mind that ectoplasm existed and was causing the number of disbelieving " 'cotton-wool' professors" to dwindle. "This country," he concluded, "owes a great debt to Sir Arthur."[75]

These exchanges well illustrate the new regime's anti-intellectual, antiacademic, "populist" conception of psychical research and make clear how far removed it was from the ideal to which McDougall and Prince were devoted. It should therefore be no surprise that the American psychical-research community was beginning to split apart in mid-1923. Walter Franklin Prince wrote privately of the new orientation of the society—"All this slop in a scientific journal!"[76]—but he was not in a position to launch any sort of public protest against its policies. Instead, he maintained a vigorous dissatisfaction in correspondence with other disaffected members; and there seem to have been many of them, for the ASPR lost 108 members in 1923.[77] The center of disaffection was Boston, where McDougall, Elwood Worcester, Margaret Deland, Gardner Murphy, and a few wealthy patrons were to be found. McDougall and Margaret Deland both resigned from the board in June,[78] and at the end of the summer the Boston group framed a remonstrance to the ASPR, objecting to the new policy of identifying with such "unscientific," spiritualist organizations as the British College of Psychic Science and insisting on reforms.[79] The veiled threat in the background was the alternative of a second, Boston-based society competing for public support; Prince had privately indicated his willingness to work for such a Boston society as long as he could be assured of financial security.[80] Some of the new ASPR trustees—such as Waldemar Kaempffert, formerly editor of the *Popular Scientific Monthly*—seem to have wanted to carry out the reforms, but in the end little was done.

What triggered the final split was the "Margery affair" and a new figure, J. Malcolm Bird. Bird belongs to that class of psychical researchers who combine contradictory extremes: possessed of an excellent mind, with good intentions, a critical disposition, and a sincere interest in psychical research, yet capable of blind enthusiasm leading to excess and often giving the impression of seeking notoriety. Bird had had training in mathematics and then in the late teens had joined the *Scientific American,* supervising those fields not covered by anyone else on the staff—relativity, for example. By 1922 the magazine had decided that psychical research deserved regular coverage, and Bird (along with W. F. Prince and an independent investigator, Hereward Carrington) was contributing articles on the subject. In December 1922 the magazine announced a "psychic phenomena" contest: twenty-five hundred dollars was to be given for the first satisfactory psychic photograph, and the same amount for a manifestation of physical psychic phenomena under test conditions.[81] In 1923, riding the crest of public interest, the journal sent Bird abroad to gather firsthand impressions of European psychical research. The trip was made at the invitation of Sir Arthur Conan Doyle, with whom Bird returned when Doyle began his American tour in early April. Bird's experiences at such institutions as the British College of Psychic Science impressed him, on the whole, though he was cautious in expressing his judgments.[82]

Late in 1923 there appeared an exciting claimant for the *Scientific American* prize: Mina Crandon (the wife of a Boston surgeon), known under the name Malcolm Bird gave her, Margery.[83] Margery allegedly went into communication with her dead brother Walter when in trance, and "he" then produced both vocal and physical manifestations of quite an extraordinary range. The *Scientific American* committee judging the contest—William McDougall, Hereward Carrington, the physicist Daniel Frost Comstock, the magician Harry Houdini, and Bird—agreed to take up her claims and began sittings with her in April 1924. Almost immediately violent disagreements erupted within the committee. McDougall claimed to have been seriously skeptical of Margery's "gifts" in November 1923, when he had had a private sitting, and his negative attitude hardened with continued exposure to Margery. Bird, on the other hand, rapidly developed into Margery's principal defender.

It is not difficult to see why the new leaders of the ASPR now began to regard Bird with favor. He was a friend of Conan Doyle's, a sympathizer with the spiritist hypothesis, an increasingly vocal supporter of physical phenomena, a man possessed of undoubted scientific qualifications (if they mattered), and a strident opponent of the "gas house gang"—McDougall and the Boston group.[84] Frederick Edwards was by no means personally convinced of the validity of Margery's claims,[85] but he approved of Bird, and finally, early in 1925, Malcolm Bird quit his connection with *Scientific American* and was hired as the ASPR Research Officer for physical phenomena. It was a new position; the trustees simply divided the old post of Research Officer into two halves, leaving Walter Franklin Prince as Research Officer for mental phenomena. In a sense the move was reasonable, for Prince had indeed been looking for an assistant, but Bird would not have been Prince's choice, and he could stand the strains no longer. By the end of February, Bird was on the research staff of the ASPR, and Walter

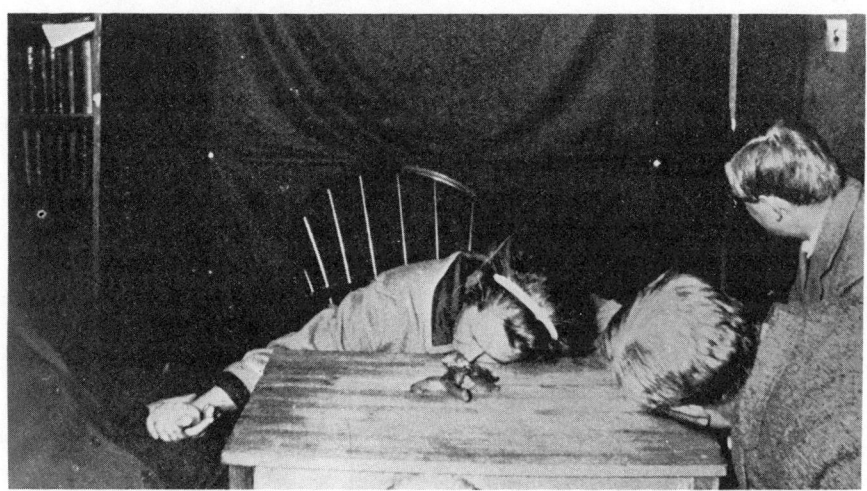

A séance with Margery of 1925. William McDougall (nearest the camera) and E. J. Dingwall are observing the ectoplasm seemingly proceeding from her mouth. (Courtesy of American Society for Psychical Research.)

Franklin Prince had a contract as Investigating Officer of the new Boston Society for Psychic Research. Edwards exulted shortly after Prince's departure for Boston, "It has been a long struggle but we have won."[86]

Perhaps; but the victory was Pyrrhic, for it destroyed any chance of the amateur societies' being the locus in which serious psychical research might be carried out in America. The ASPR became entangled in the increasingly messy business of the Margery mediumship, committing itself to her and eventually ostracizing even Bird when in 1930 he admitted that he had long believed her to be a fraud.[87] *Scientific American* never did award her—or anyone—its prize. The Boston SPR, on the other hand, was simply not wealthy or big enough to do systematic research. Walter Franklin Prince tried to reassure his friends that the small membership of the Boston society was really perfectly in keeping with its antipopulist, anti-ASPR origins.[88] But Prince was its only effective researcher, and he was forever searching for respectable outside material that he could publish to supplement his own writings. If McDougall had been as active in the Boston SPR as he had been in the American society, things might have been different; but by 1925 McDougall had in effect given up on societies as the vehicle for his concerns and was looking hopefully to the American university as a more promising context for scientific psychical research. It was a decision with enormously important consequences for the field.

Nowhere else, as we have said, did the internal disagreements over the assumptions and methods of psychical research have so disastrous an effect upon the amateur societies; nevertheless, the problem was not a uniquely American one. No institutional framework, national or international, was yet entirely capable of satisfactorily harmonizing the various concerns of its members—Richet would have had irreconcilable critics of his *Traité* as well as defenders in every society. A firmer agreement on matters of method and of content would be required before psychical research could—if it chose—think of itself as a science.

CHAPTER TWO
The Growth of Experimentation in the Psychical-Research Societies

While psychical research was not divided as sharply in Europe as it was in America, it certainly came under increasing strain as the 1920s wore on. Everyone was sadly conscious that its continuity with the great creative days of the late nineteenth century was coming to an end; one by one the surviving leaders of the field who had shaped and guided it for so long were disappearing. Germany's most prominent psychical researcher, Albert Freiherr von Schrenck-Notzing, died in 1929. In France, the three founders of the Institut Métapsychique passed away within a few years of one another—Geley in a plane crash in 1924, Santoliquido in 1930, and Meyer in 1931. In England, William Barrett, a founder of the SPR and perhaps its original inspirer, died in 1925; A. J. Balfour, Mrs. Sidgwick's brother and another of the society's founders, died in 1930, as did Arthur Conan Doyle. A very few of the prominent figures from the past lived on—Richet died in 1935 at eighty-five, Mrs. Sidgwick in 1936 at ninety, Lodge in 1940 at eighty-nine; nevertheless, the active leadership of European psychical research was passing rather suddenly to a new generation.

This transfer of leadership to a new group was made more complicated by an intensification of the old disagreements over the aims and methods of psychical research. In addition, the late 1920s also witnessed the new seriousness of a small number of investigators who undertook a variety of narrowly experimental studies of purely mental phenomena. By the early 1930s the English psychical-research community in particular was far more seriously and openly divided than it had ever been.

Less than ten years before, in 1924, in his presidential address to the SPR, J. G. Piddington had hinted at the existence of the same polarity in England as was then breaking apart the American SPR.

> Sir Oliver [Lodge] once divided the members of this Society into a Right and a Left wing. The Right comprised those of cautious and conservative views who uphold the strictest methods of investigation; the Left comprised those who believe more than the Right think is good for them, . . . and who would not view with alarm some relaxation in the rigorous methods of investigation hitherto pursued. . . . For my immediate purpose, then, Sir Oliver's dichotomy will not serve; and to bring out what it is that separates the particular section of members of whom I am speaking from those opposed to them, I shall call the latter the High-and-Dry School and the former . . . the

Not-High-and-Dry School. The dividing line here has nothing to do with how much or how little of the phenomena under investigation one accepts as supernormal, and concerns only opinions as to methods of investigation and standards of evidence. The Not-High-and-Dries, if I do not misrepresent them, take the line that so much has been established beyond cavil that we can now safely relax to some extent the stringent cautions and the very high standard of evidence on which the Society has hitherto insisted. . . . The High-and-Dries take the line that we cannot now, or for a long time to come, and probably ever, afford to lower our "evidential" standard, or modify our methods—our admittedly irritating and meticulously wary methods—of investigation.[1]

These distinctions sometimes made for strange alliances. Piddington would certainly have included himself and the Fisher's Hill group in both the Right and the High-and-Dry school and placed Arthur Conan Doyle and many spiritualists on the other side of both divisions. Yet E. J. Dingwall, the young man who had become the society's research officer in 1922, while especially insistent upon rigorous methods of investigation, was far more interested in physical mediumship than was the conservative leadership—making him a man of the Left, perhaps, but a very High-and-Dry. Piddington's sensitivity to these issues foreshadowed the attacks that would come to be made upon the traditionalist center of the society by those in disagreement with one or another of their policies.

The first formal rift in the structure of organized psychical research in England came in 1925, when a "National Laboratory for Psychical Research" was established, equipped with a considerable amount of laboratory apparatus and possessed of a very impressive library. Despite its name, this was essentially the operation of a single wealthy man, Harry Price, who was especially interested in testing physical mediumship.[2] Price was somewhat similar to J. Malcolm Bird, in America, in combining what appeared to be a genuine commitment to scientific and experimental method with an enthusiasm for physical phenomena that paralleled the extremes of spiritualist credulity. Also, like Bird, Price was very much a publicist, even a showman, in the cause of psychical research. He offended the sensibilities of many; moreover, his National Laboratory seemed to repudiate the SPR stance on physical phenomena in favor of an inadequately critical attitude. As Dingwall (himself still by no means unsympathetic to such phenomena) wrote bitterly to Price: "The thing is as clear as daylight. There are roughly two ways of inquiring into mediumistic phenomena—the L[ondon] S[piritualist] A[lliance] cum B[ritish] C[ollege] [of] P[sychic] S[cience] cum S[tead] Bureau way and the S.P.R. way. You prefer the one and I prefer the other."[3] It was not long before a few English researchers began to hint that Price might not be above "helping along" the phenomena by *staging* apparently positive results.[4] Nevertheless, there remained many interested in psychical research for whom Harry Price's National Laboratory appeared to be a valuable alternative to the SPR, carrying on studies (often successful) of phenomena like hauntings or physical mediumship, which the older institution had decided to ignore.

In 1927 the council of the SPR decided not to reappoint E. J. Dingwall as research officer, apparently feeling that his investigatory interests were beginning to diverge from their own.[5] Dingwall soon became an outspoken critic of the SPR as an institution that had lost its effectiveness by abandoning "the canons of scientific procedure" and was becoming susceptible to spiritist control.[6] The Sidgwick leadership in effect replaced Dingwall with Theodore Besterman, who had joined the society in 1925, having just turned twenty, and the next year had volunteered to organize the society's collection of books. Made librarian in 1927, he began to have a voice in SPR affairs, and in 1929 he was selected to replace Helen Salter as editor of its *Journal*. By this time his association with the Fisher's Hill group was a very close one, and he expressed his satisfaction with the traditional policies of the society in an address to the Fourth (Athens) International Congress in 1930, in which he insisted that psychical research was steadily winning some acceptance by the academic world and by "the average educated mind" and stressed the continuing need to gather experimental data. He pointed to Brugmans's experiment and to the cross-correspondence material as well-nigh conclusive evidence of supernormality, contributing to a "record of progress ... with which we have no great reason to be dissatisfied."[7]

Besterman's increasingly important role in the society caused additional problems. Among other things, he had taken on some investigatory responsibilities—supervising some small experimental studies and examining the claims of occasional mediums abroad—and in 1933 he was given the title of Investigations Officer; there were those who thought it a mistake that Dingwall, who had had a scientific background, should have been supplanted by Besterman, who did not. More than this, as editor of the *Journal* Besterman had immediately antagonized a great many people by what seemed to them his unwarranted and offensive incredulity about most psychical phenomena. It was not merely the cocksure disbelief but the derisive tone in which it was expressed that offended these people; an extreme sample of Besterman's style (though from a few years later) is his review of Alexander Cannon's *Powers That Be* (1934), here quoted in its entirety from the SPR *Journal:*

> Dr Cannon is, or was until recently, associated with one of the L.C.C. lunatic asylums. He is also, so he tells us on his title page, Kushoo Yogi of Northern Thibet and Master of the Fifth of the Great White Lodge of the Himalayas. A work entitled *Powers that Be* and written under such auspices must attract, even command, attention.[8]

When we consider, too, that in 1930 Besterman was only twenty-five and that many of his targets had been wholly committed to psychical research for more years than he had lived, we can understand the almost incoherent rage with which they so often pronounced his name.[9] Yet Besterman was in effect the protégé of the SPR leadership,[10] and the disaffected could do little but resign—or join Harry Price.

The first open challenge to the Sidgwick group came in early 1930 when a caustic review by Theodore Besterman of a very uncritical book on some recent spiritualist physical phenomena led Arthur Conan Doyle to resign his membership in the society in order, as he put it, to protest the unscientific work of a group that for a generation had attempted to hinder every serious student in the field.[11] He elaborated on this theme in a circular letter sent to all members of the society, singling out Mrs. Sidgwick and Dingwall for criticism along with Besterman and suggesting that the SPR be removed from their control.[12] The current president (L. J. Jones) and honorary secretaries (Mrs. Sidgwick and W. H. Salter) as well as Besterman defended the society on the grounds that its work in the last ten years with Mrs. Leonard, with cross-correspondences, and with telepathy had been new and important pieces of research, tending to support, for many, the likelihood of post-mortem survival. While admittedly its investigations of physical phenomena had unfortunately been infrequent, this was owing to the reluctance of physical mediums to accept the high standards insisted upon by the SPR.[13] The furor over Doyle's resignation was followed in winter 1930/31 by an attempt by Harry Price to consolidate his National Laboratory with the SPR; he publicized his offer to the membership generally at the same time that he presented it to the council.[14] When the council declined his proposal of an amalgamation, Price and some other members tried to raise support for a contrary movement by the entire membership at the SPR's general meeting in February 1931, but this attempted coup came to nothing.[15]

The report of the council for 1931 stated that Doyle's "attack on the Society failed completely, only six Members and one Associate, none of whom had ever taken an active part in the Society's work, resigning on the ground of agreement with this circular." But this was less than completely candid. During 1930, 77 other resignations had been submitted, and since 1920 the total membership had fallen from 1,305 to 954.[16] The council laid the blame on the economic depression, but it is quite likely that for many the depression merely confirmed decisions that growing discontent had already suggested. The discontent burst out again later in 1931 when H. Dennis Bradley sent to each member of the society an announcement of his resignation, entitled *An Indictment of the Present Administration of the Society for Psychical Research.*[17] The *Indictment* gave a very full statement of the reasons for the dissatisfaction that he and numerous others had come to feel: Theodore Besterman's rudeness and incapacity; the inertia of the SPR leadership; its hostility to physical or overtly spiritualistic phenomena; its unjustified rejection of Harry Price's offer. Accounts of Bradley's accusations soon found their way into the press and stirred up still more bitter exchanges. English psychical research was threatening to follow the American example by breaking up into discordant communities.

In just these years of mounting tension increased attention was being given by a few French and English psychical researchers to well-defined experimental studies—many of them, perhaps, seeking some common ground in calm, objective research. Generally speaking, these individuals were relatively young; primarily interested in mental phenomena such as telepathy and clairvoyance and in the possibility of looking for these powers, not in the especially gifted mediums, but in the general run of mankind; not concerned to uphold or decry the survivalist hypothesis; and possessed of an orientation (often derived from scientific training) that suggested to them the merits of a precisely delimited research program rather than a diffuse, "natural-historical" survey. For a brief period it seemed almost as if their investigations could provide the foundation for a new consensus among psychical researchers.

The experimental testing of mental psychic abilities had, of course, a history that went back to the origins of organized psychical research; Charles Richet had been one of the pioneers. But it had become comparatively infrequent in the first two decades of the twentieth century, as different approaches came to appear more profitable. One of the few individuals to pursue such work consistently was a French chemical engineer, René Warcollier, who apparently began his experimental work about 1904, when he would have been thirty. Warcollier in a real sense bridges the gap between the experimental work of the late nineteenth century and its revival in the mid-1920s, for his work spanned the entire period. During the period 1905–15 Warcollier had published short papers in the *Annales des sciences psychiques,* and he had virtually completed the manuscript of a book when World War I broke out. This book was finally published in 1921 as *La Télépathie; recherches expérimentales.*[18] Richet himself provided the preface, which began, "Voici un livre de science, et de vraie science."

> It is not easy to read his book; it is not a work of popularization, but a profound study that requires perserverant attention. . . . But the great advantage of books of science, like this one, is that the reader, having all the necessary evidence before his eyes, can make up his own mind.[19]

Richet no doubt saw *La Télépathie* as being an exemplar of the kind of scientific monograph that the developing science of *métapsychique* ought to be—and soon would be—producing.

Warcollier began his book by considering spontaneous telepathy, surveying the studies that had already been made (by the SPR, for example) and presenting a general interpretation of the telepathic process based upon this material, an interpretation that was certainly shared by many other students of telepathy and can ultimately be traced back to F.W.H. Myers. In brief, it supposes that telepathy involves the emission of some sort of force from the agent's subconscious, and only figuratively the transmission of its actual contents: the images that the percipient sees derive from the stock of his own subconscious, stimulated or called up by the agent's activity. The process is most often an unconscious one; the volition of the agent clearly is not necessary, and indeed transmission seems to occur best when both agent and percipient are in a relaxed state, semiconscious, in trance, or asleep.[20]

Warcollier went on to discuss what experimental telepathy can show. He summarized the successful experiments on transmission of playing cards or numbers carried on with unselected subjects by the SPR and the ASPR, contrasting with these the failures of "laboratories for telepathy set up at great expense" in American universities. It is possible, he admitted, that their negative results may indicate that persons with telepathic ability are relatively uncommon; but this is not an entirely satisfactory explanation, since in other experiments there has been good success, and it is more likely that their method has been wrong. Sometimes it is clear that their procedure is so highly mechanized as to prevent the experimental subject from assuming a properly relaxed frame of mind; sometimes it fails to recognize that telepathy may occur only when an agent and a percipient have somehow been sensitized or attuned to each other, just as wireless transmitters and receivers have to be built for one another. Thus such negative reports are not conclusive, and it remains quite possible that telepathy may be common to all men under the proper conditions.[21]

Warcollier's own telepathic experiments were of a different nature, modeled consciously upon the chemist's procedure of considering a single entity in a wide range of experimental situations.[22] He described some experimental work with school children and friends involving attempted transmission of objective material, but he insisted upon the importance of escaping the monotony of card-guessing by imbedding the objective material in a game.[23] In general, however, the heart of his research lay in work that involved himself as percipient with friends as agents. From 1906 on, he tried to develop within himself an awareness of the image-content of his subliminal mind and to see whether that image-content would match the material—first images and then actual drawings—being sent by an agent. The results, he admitted, were difficult to evaluate precisely, as others who had tried the same sort of experiment had recognized. All he could do was try to compare qualitatively "the resemblance between each particular design and its reproduction with the resemblance between another design from another series, chosen at random, and that same reproduction."[24] He presented a great many illustrative cases; usually there would be at best partial success, only fragments of the original design coming through, but occasionally a completely successful transmission would take place.

It was actually these imperfect transmissions that Warcollier found most interesting, since they could illuminate the process of psychic transmission.[25] He argued that they indicated that what was communicated in telepathy was, not an idea, but sensation or form, either fragmentary or entire; the sensory elements that strike the agent's eye can be passed to his subconscious and thence in whole or in part to the percipient's subconscious without the agent understanding what he is seeing. In both the agent's and the percipient's mind, however, some sorts of deformations occur that psychologists have long recognized to affect normal perception, because of the association of ideas. Oliver Lodge describes a case where the agent tried to "send" a particular silver teapot, thinking all the while how much it looked like a duck; the percipient saw a silvery duck. The percipient presumably picked up the agent's associations, but the associations present in his own mind subliminally may also have distorted or transmuted the sensory elements as they proceeded up from his subconscious to his conscious mind.[26]

Warcollier commented bemusedly on how far his research had led him from the physical side of telepathy to the psychological side, to recognize how much contemporary psychology might be able to contribute to metapsychics. He remained very much the physical scientist, however, in his attempts to find some sort of model that by analogy might help illuminate telepathy. In many respects his position is identical with that which Richet would be enunciating in the *Traité*. Thoughts per se cannot be transmitted; we can only imagine the emission of vibrations from one brain to another. The analogy to wireless is a natural one to the physicist or the engineer. Like wireless, telepathy requires some sort of harmony, or tuning, between stations; just as wireless is affected by climatic conditions, telepathy can be influenced by observers or strangers. But the analogy cannot be pressed too far, since telepathy uses no code, does not seem to operate at a determinable speed, and apparently does not obey the inverse square law (though Warcollier was willing to suggest that this last might simply be due to the extreme sensitivity of the human brain). In any case, keeping the lessons of modern physics constantly in mind allows us to understand the subtlety of matter and to understand how the vibrations of external bodies can be reproduced in our minds and can influence by induction other nervous systems in other brains.[27] For Warcollier, as for Richet, it was the assumption of psychophysical parallelism that made sense out of telepathy. But Warcollier went much further than Richet was willing to go in trying to provide a physical model that might help to relate psychical phenomena. No model from one scientific domain ever fits perfectly in another; this model—or this *combination* of models—was certainly demanding of serious consideration as being the most comprehensive and sophisticated attempt to explain telepathic phenomena since the theories of F.W.H. Myers.

Despite Richet's approval, *La Télépathie* did not inaugurate a tradition of experimental telepathy in France. René Sudre, reviewing the work for the *Revue métapsychique*, criticized what he took to be an overanalytical approach in Warcollier's decision to isolate telepathy for concentrated experimental study.[28] Gustave Geley, then director of the IMI, was himself inclined to stress the importance of physical phenomena rather than of mental. It was only after the death of Geley in 1924 and his replacement by Eugène Osty that Warcollier began to find a cordial reception at the IMI, and he began to give conferences there while carrying on experimental work under its auspices;[29] during the later 1920s, Warcollier published yearly accounts of his continuing research in the *Revue métapsychique*. Much of this work was built upon a new experimental tactic, an attempt to demonstrate long-distance telepathic transmission between large groups. Some of the individuals who assisted him were his own friends; some he had been put in touch with by Osty; a few had been introduced to him by Carl Vett, the moving force behind the International Congresses of Psychical Research. The most important member of this last group was a young American psychologist, Gardner Murphy, who had become deeply interested in psychical research and who was the only man in the United States carrying on work analogous to Warcollier's. Murphy and Warcollier met at the Warsaw congress in 1923, and from then until 1925, when Murphy became ill and discontinued his experimental activities, Warcollier's Paris group made attempts to communicate with a group of "receivers" that Murphy had formed in New York.[30] The results impressed Warcollier, at least, as going beyond what chance would have provided, and for his part Murphy remained a firm admirer of Warcollier's approach to the study of mental phenomena.[31] Even so, and despite the continued support of Osty and Richet,[32] Warcollier found no one interested in undertaking experiments comparable to his own.

In England, Warcollier's work was praised by its reviewer for the SPR *Journal,* Helen Salter, who began: "This book well deserves the tribute paid to it by Prof. Richet... of being a scientific contribution to the subject." She obviously found inadequate his materialist need to interpret telepathy as a psychophysical process rather than as a purely psychic process, but she singled out for positive comment his experimental inquiry into the psychological factors affecting transmission.[33] She spent some time in describing one of his card-guessing games, designed to mitigate the tediousness of the simpler forms of telepathic experiment, perhaps in the hopes of stimulating further experimentation in England. Six months later a mathematician wrote the *Journal* that the mathematics of the card-guessing game were not so straightforward as they seemed to be,[34] and nothing more was heard of it. It is tempting, however, to suspect that Warcollier's treatment had something to do with the series of informal experiments that was begun in the SPR secretary's room in 1923, experiments whose object was "not so much to increase the evidence for telepathy as to try to obtain the kind of evidence that may throw light on the conditions under which telepathy is most likely to operate [, b]earing in mind the generally inhibitory effect of anxiety, or of too much conscious attention, on the emergence of subconscious ideas...."[35]

These experiments became the first stage of a new research effort at the SPR. One of the leaders in this effort was Ina Jephson, an intensely interesting figure who, apparently lacking scientific or mathematical training, was still able to attack the experimental issue profitably and probingly and in a way very different from that which had become characteristic of SPR tradition. With Isabel Newton, secretary of the SPR, Miss Jephson began the experiments trying to identify the psychological conditions favorable to thought-transference, and carried them on until June 1925. Two years later she reported on this first work (which used pictures, objects, ideas, diagrams, numbers, and playing cards variously as subjects for mind-to-mind communication), concluding that on the whole it had provided no support for thought-transference. In the course of these experiments, however, Miss Jephson had begun to see that card-guessing success need not automatically be interpreted as telepathy:

> As time went on I became more and more puzzled as to why we should always assume that the impression came from the agent's mind and not from the picture or card itself. Several small incidents seemed to point to this last conclusion. On one or two occasions I felt, when I was the receiver or percipient, that the impression of the chosen object came to me before the agent had looked at it, or else, as on one occasion when picture postcards were the objects chosen, that my impression tallied with a neighbouring card unseen by the agent, instead of the one at which the agent had been gazing. It was borne in on me with more and more force that it was quite an unfounded assumption that the agent was necessary at all. I think we are much too apt to put down to telepathy, occurrences which are due to clairvoyance alone.... This problem of telepathy versus clairvoyance has always seemed to me the central problem of psychic research, and these experiments were devised very largely in the hope of emphasising how unnecessary is this hypothesis of "telepathy" and of throwing some light on the problem as a whole.[36]

She began in March 1924 to do clairvoyant experiments with playing cards, first with herself as subject and then with others.

In deliberately choosing to investigate clairvoyance, Miss Jephson was going against collective opinion in England, where (as we have already seen) telepathy had always been considered the fundamental mental phenomenon. "Agent" and "percipient" had been consciously employed by the earliest English researchers to describe what appeared to be taking place in thought-transference, and it was difficult thereafter to question an increasingly ingrained terminology, as would have been necessary if clairvoyance were to be taken seriously. For to view clairvoyance as a real phenomenon that might be involved in apparent instances of telepathy is of course to destroy the characterization of "agent" and "percipient" as active and passive figures in a transmission of thought.

Outside of England there was more interest in clairvoyance and less resistance to the notion that "percipient" might be a misconceived term. Charles Richet, for instance, who believed firmly in clairvoyance, had no doubt that "percipience" was the truly active faculty. The papers published by René Warcollier in the *Revue métapsychique* show that he also was being drawn round to this conviction. In *La Télépathie* he had already pointed out that telepathy apparently depended, not on the volition of the agent, but on that of the percipient. By 1924 Warcollier was declaring that "le percipient parait même plus important que l'agent";[37] and in a paper of 1928 Warcollier had come round entirely to this point of view, arguing, on the nature of "telepathic accord," that the percipient was probably the crucial figure for the researcher to study and claiming that in any case the notion of spatial direction—*from* and *to*—is probably mistaken, that what is involved is probably better seen as a meeting in the psychic realm.[38]

During the 1920s there were signs that while the society had not abandoned its conviction that telepathy was the type-phenomenon, it, too, was beginning to concede that Myers's distinction between agent and percipient might be incorrect and that something other than "transmission" from the former to the latter could be involved. A 1915 paper by Mrs. Sidgwick discussing the trance phenomena of the medium Mrs. Piper gives an early hint of this;[39] in a later paper (published in 1923) in which she classified the backlog of cases of telepathy that had been reported since the publication of *Phantasms of the Living,* Mrs. Sidgwick called attention to semiexperimental cases in which one party, whether agent or percipient, is trying to evoke an impression, while the other party to the experience is unconscious that anything is going on. In cases where the percipient is taking the active role and is successful in receiving impressions from another party, the traditional idea of a transmission is considerably weakened.[40] Mrs. Sidgwick concluded by recognizing that the initial assumption of a transmission had been suggested partly by experimental and spontaneous case reports but partly too by an analogy with the communication of radiant energy, and she proposed that the time had come to try new analogies: she offered one not quite so materialistic as the last, that of minds being in contact with one another—a physical analogy, to be sure, but an elastic one and one that did not involve distinguishing the supposed functions of agent and percipient.[41] Contact of a mind with a physical entity, however, as in clairvoyance, was a possibility that she still did not raise.

In a sense, therefore, the program Miss Jephson undertook in 1924 of carrying out tests for successful card-guessing by clairvoyance involved a reconsideration of basic theoretical assumptions, as well as a commitment to a definite experimental task. The first results of this new series were at the very least suggestive: she discovered what she took to be a tendency common to all subjects to succeed with the early guesses but to fall below chance towards the end of a series of guesses, a tendency that she interpreted as suggesting "some element other than chance at work . . . a faculty which was liable to fatigue or confusion, or to some deterrent element."[42] It was then necessary to establish a scoring system that would allow the experimenter to determine just how successful a particular guess was—to measure, for example, how much more surprising it is to guess, for an unseen queen of diamonds, the queen of hearts rather than the 5 of clubs; the statistician R. A. Fisher, who was just publishing his epochal *Statistical Methods for Research Workers,* worked out a table of scores that could be used to weight the relative success of different approaches to complete success.[43] Miss Jephson's experiments continued to bear out the decline effect, and in 1926 the society published an appeal to interested parties to take part in her work by reporting their own clairvoyant experiments, conducted along the lines she had laid down.[44]

It was the sum total of this final series on which Miss Jephson reported in

1928. Two hundred forty people had supplied the basic data: each made five series of five guesses, drawing a card from a deck of fifty-two before each guess and returning it to the deck afterwards. Her subjects were selected to a limited degree—they included those with some previous success or with a professed interest in the experimental research—but they were by no means professional mediums; yet their average score on each of the five guesses was well above mean chance expectation. Miss Jephson admitted in her paper that the genuineness of these recorded efforts might be called into question, that some of her subjects could have doctored their results in any of a number of ways, since their work had not been witnessed; but against this possibility she pointed out that runs witnessed by her and by others gave essentially similar results and that in particular the "fatigue curve" remained in evidence in all cases. In further confirmation of her results, she called attention to the report by G. H. Estabrooks of his work at Harvard, published by the Boston SPR the previous year, which gave similar results both as to degree of success and as to presence of a decline curve; she identified a similar regularity as mentioned in an 1884 paper by Charles Richet.[45]

Miss Jephson probably entertained the hope that a key to future psychical research might be available in this simple, replicable experimental situation, a form of investigation that had long been neglected. Her success led her to speculate, not about general problems, but about ways in which the model might be further extended and the phenomena elaborated experimentally. On the one hand, she perceived almost as clearly as Warcollier the advantages of associating clairvoyance with accepted psychology and cast her interpretation of the "fatigue curve" into a form that made the parallel with the science clear: she called attention to its resemblance to the curve familiar to industrial psychology, where output diminishes during a task only to improve with the prospect of rest.[46] On the other, she pointed out areas in which her results indicated a need for further experimentation. An agent had been present in Estabrooks's experiments but not in hers—did his somewhat greater degree of experimental success indicate that two sources of information (the card and the agent's mind) were better than one? And would not this imply that "clairvoyance" should include the perception of thoughts as well of objects and that a separate "telepathy" was really redundant? Such speculations, she agreed, could be resolved only by experiment, and she encouraged its pursuit.[47]

In its general sophistication of interpretation, its use of mathematics, its promise of replicability, and its apparent success, Miss Jephson's work was outstanding and bound to attract attention from those who were scientifically oriented within psychical research. An enthusiastic response to the paper from Charles Richet has survived, expressing delight in its support for his belief that clairvoyance subsumes telepathy.[48] From the SPR membership Miss Jephson received more thoughtful criticism, particularly from the few researchers who were more interested in developing laboratory experimentation than in continuing anecdotal or mediumistic studies or who found her attempted mathematization stimulating. Their reactions combined serious questions on points of interpretation with an eager acknowledgment that her investigations might well open up new areas of research. H. F. Saltmarsh expressed some reservations about her judgment "that if clairvoyance be extended to 'include the perception of mental images as well as material objects the hypothesis of telepathy becomes redundant,' " as well as about the idea that "the factor which causes the deterioration of the later guesses is fatigue"; but he insisted upon "the great interest which I feel in your work & my conviction of its value" and suggested a number of experiments that could be carried out to identify the psychological elements involved in clairvoyant perception and to distinguish between distraction and fatigue as the cause behind her decline curve.[49] W. Whately Smith, who had been interested for more than a decade in mathematical and physical techniques for studying mediums, criticized her intensive discussion of the best individual performances and her concomitant neglect of the overall aspect of the six thousand trials, and he expressed more sympathy than she had for "pure psychical action" as an alternative to clairvoyance; but he too had a number of experiments to suggest, including a means of determining the emotional response of the subject to each of the cards in a deck of fifty-two by using a bridge galvanometer so that an individual's innate card-guessing preference might be taken into consideration.[50] There is very little doubt that Miss Jephson's paper aroused hope and interest in much of the SPR in that it promised to provide a firm foundation for work upon which the skeptical and the merely cautious alike could collaborate.

Published almost simultaneously with Miss Jephson's report was another SPR account of an experimental inquiry into mental phenomena that also seemed to hold some promise. This work had been carried out by V. J. Woolley, who had used the BBC in February 1927 to communicate with subjects in a "telepathic" experiment.[51] The idea (probably suggested by American radio experiments of 1924) had been proposed by Harry Price to the BBC in 1925, but nothing had come of it then; only when Woolley took up the idea at E. J. Dingwall's suggestion did something happen—perhaps because Woolley was honorary research officer and a member of the SPR council.[52] Agents at the SPR offices were given five objects to transmit at successive five-minute intervals: a playing card (2 of clubs), a Japanese print, a bunch of flowers, another playing card (9 of hearts), and Woolley himself in a grotesque mask. At the same time, Oliver Lodge instructed wireless listeners to record whatever impressions they received. In this way Woolley hoped to discover whether telepathy could be seen to operate in the absence of any personal contact between agent and percipient and to canvass a very large mass of people for successful percipients with whom further work could be carried on.

Nearly twenty-five thousand sets of impressions were sent back to the SPR, but the results were inconclusive and brought home more heavily than ever the need to devise some sort of technique by which actual success could be established precisely and then compared to degree of success expected by chance. The two tests with playing cards satisfied the first criterion but not the second, for, as Woolley recognized as soon as the results began to come in, people simply do not choose playing cards at random. Hence there was no possible way of fixing the degree of success vis-à-vis chance except roughly, by comparing the number of successes in one test with the number of times the same card was guessed in the test where it was not the target. This comparative approach permitted some judgments to be made, and it also permitted the experimenter to evaluate success in guessing nonobjective materials in the light of a control, a familiar problem for the SPR from its studies of mediums. Thus, using a similar procedure, 1,020 people were shown to have identified the third subject correctly as flowers, 523 the fifth wrongly as flowers; and while 236 identified the fifth subject correctly as "someone masquerading," no one so misidentified the third subject. These results, suggested Woolley when he published them in 1928, did seem to indicate a supernormal faculty on the part of a few (swamped by the failures of the majority), and he arranged to study some of the more successful listeners in greater detail.[53]

Taken with Warcollier's work, these two studies were strongly suggestive of psychical abilities in normal, unselected subjects, and they were undoubtedly highly encouraging to those who, like Smith and Saltmarsh, hoped to see psychical research given a more experimental-scientific tone. More generally, they must have pleased all those who were beginning to wonder whether psychical phenomena could ever be evoked under satisfactorily rigorous conditions. For the SPR to have sponsored a convincing experimental demonstration of telepathy or clairvoyance would have furnished an excellent reply to those critics who, like Dingwall, were beginning to argue that the society was neither progressive nor scientific. Hence the society's leadership began late in the 1920s to show some interest in the experimental activities of a few investigators, even though the majority of its members still found mediumistic studies or the natural-historical approach more congenial to their generally survivalist interests.[54] It did not seem inconceivable that a recognized "science" of psychical research might at last be established upon investigations such as these into mental phenomena, if they continued to achieve positive results.

It thus proved the more disappointing when repeated attempts to replicate the Jephson work, carried out by a painfully careful experimenter, all proved unsuccessful. The principal figure in these uniformly negative investigations was a forty-year-old mathematician on the staff of Queen Mary College (University of London), S. G. Soal. Soal exemplifies the new hypercritical, almost obsessively skeptical side of psychical research perhaps even better than E. J. Dingwall, and it was this attitude, combined with his mathematical qualifications, that made him such a devastating critic of Miss Jephson. Skeptical as he was of most claims, however, he accepted in principle the possibility of supernormal mental abilities and was insistent that they should be looked for with the methods of experimental science. During the next ten years, in fact, Soal was to become perhaps the most prominent English practitioner of the experimental approach to psychical research, and his consistent (and loudly announced) failure to demonstrate psychical abilities in his subjects undoubtedly helped discourage most other English investigators from attempting experimental projects of their own.

Soal had first become interested in psychical research by reading Oliver Lodge's *Raymond* during the war, and when shortly afterwards his brother Frank was killed near the Somme, he resolved to try to contact his dead brother once he was out of the army.[55] Eventually, a series of sittings in 1921-22 with the medium Mrs. Blanche Cooper convinced him that mediums could occasionally communicate veridical material concerning the dead, though Soal then saw no way to decide between the spiritist and the telepathic explanation.[56] Shortly thereafter he began to look into the possibility of devising an absolutely fraud-proof experiment to reveal telepathy and gave particular attention to long-distance experimentation, in which fraud or unconscious direct communication between agent and percipient was inherently less likely.

At this time—the mid-1920s—the publication of René Warcollier's experiments on telepathy was beginning to have a stimulating effect upon English psychical research, and it is of interest as illustrating Soal's tough-mindedness to summarize his quite critical attitude towards this work, as expressed in letters of 1930. He could speak with relative warmth about Warcollier's energy, ingenuity, and enthusiasm and his having "brought considerable powers of introspection and psychological insight to bear on the interpretation of his results." The experiments themselves, however, were too loosely reported to make his demonstrations of telepathy itself entirely convincing. If some sort of control were set up, or if conditions were clearly enough reported to allow the reader to judge what Warcollier claimed as a success (and how often such successes occurred), then this work might be more compelling; but apparently Warcollier was often content to claim as successes instances that were far from clear-cut.

> Even when the percipient has not guessed the object chosen by the agent, it is often possible for the experimenter to claim a partial success on the grounds that there is some associative connection between the percipient's impression and the actual object.... If, as M. Warcollier seems to believe, telepathy often works by very obscure associations apparent only to the percipient himself, then it is difficult to see how science is going to deal with it objectively at all.[57]

Soal was thus from the beginning clearly anxious to establish a procedure that would allow both objective determination of success and (as far as possible) the elimination of personal interaction. He helped V. J. Woolley score the results of the first BBC test in February 1927, and that summer he began a second series of long-distance experiments with those percipients who seemed to have had some success. From October 1927 to June 1928 Soal arranged for weekly sessions in which two or more agents meeting together in London focused on three objects at specified five-minute intervals; the objects usually were not materials susceptible to statistical treatment, but rather "picturesque toys, working models, animals, flowers," chosen, when possible, to arouse amusement or excitement in the agents. On any given evening, the percipients ranged in number from nineteen to sixty-two; after February 1928 they included eight members of Warcollier's group in France, but in general they were distributed throughout Great Britain. Except for a half-dozen instances, there were no ties of acquaintance between agents and percipients. These experiments provided some material "suggestive of supernormality" but nothing really compelling, certainly nothing in the nature of regular success for any particular percipient.[58]

In mid-1928, therefore, Soal began a vastly more ambitious program of experimentation, one that would employ a larger number of percipients and would center on material that could be treated statistically. A radio appeal located 470 participants; between October 1928 and March 1929 they took part in weekly sessions essentially like those of Soal's earlier series but involving in addition to material objects some other items as targets (geometrical figures, three-digit numbers, letters, playing cards) suited to mathematical treatment. In addition, Soal devised variations on his basic technique that were intended to test a number of Warcollier's hypotheses: that increasing the number of agents should increase the number of successes in communication, for example, or that inducing the percipients to fix their mind on a common object would produce "mental contagion" and lead to subsequent common guesses among them.[59]

In the midst of this second telepathic series conducted by Soal, Miss Jephson's report on clairvoyance was read to the society (June 1928) and published in its *Proceedings*. Soal's own training and experience immediately brought certain questions to his mind. Since no previous experiments had ever hinted at so widespread a faculty of clairvoyance in man, was it not more likely that some systematic flaw had affected the unsupervised conditions of the Jephson experiment? Soal suspected that such results were quite likely due to conscious or unthinking distortion of the experimental procedure by participants—for example, waiting until a success in guessing took place before beginning to record a series of guesses.[60] R. A. Fisher had suggested just such a possibility upon first seeing Miss Jephson's data.[61] In any case, if the conclusions were true, they were important enough to warrant repeating the experiment under more nearly controlled conditions. Hence as soon as Soal's second experimental series on telepathy had come to an end, and before he had had a chance to calculate its results, he, Miss Jephson, and Theodore Besterman were given the sanction of the council "to repeat the experiment under stricter conditions in order to confirm, if possible, Miss Jephson's results and perhaps to discover one or two percipients with outstanding clairvoyant powers."[62] The subjects of the most recent Soal experiment made up the bulk of the percipients used, but the experiment was also carried on by Walter Franklin Prince, in Boston, with 95 percipients and by Gardner Murphy, at Columbia University, with another 157.

Each percipient was sent a set of playing cards every week for five weeks, each card being enclosed in its own opaque, sealed envelope. The cards were guessed without opening the envelopes, and cards and recorded guesses were then sent back to the experimenters. The results were scored by Fisher's method during the summer of 1929—Soal complained that "this work alone occupied me for 6 to 8 hours a day during several weeks of the long vacation"[63]—and published in the *Proceedings* early in 1931; the results of over nine thousand guesses provided "no confirmation of the hypothesis that the faculty of clairvoyance is a normal faculty possessed in a slight degree by the majority of civilised persons. . . . " The card-guessing scores of only ten of the original percipients exceeded two standard deviations from the expected chance value, and when these individuals were retested, none again scored significantly above chance.[64]

The 1931 paper was a joint effort of the "committee" of three: Besterman described the methodology, Soal set out the statistical results—and Miss Jephson provided a "theoretical analysis," which must have been a bitter task. She was forced to admit that there was no sign of clairvoyance in the new experiment; worse yet, what she described ironically as "a fatal industry" had led her to examine the data for signs of her cherished fatigue curve, and again she was forced to admit that it could not be borne out, that chance alone was in play in determining variations from first guess to last guess. She conceded the likelihood of some systematic selection factor having produced the extra-chance results of her initial experiment—and yet having herself scored legitimately above chance, and having observed others do so while manifesting the expected decline curve, she stubbornly refused to consider the question closed. She offered one way out, one further topic for experimentation: Might it not be that the new playing cards of the second, joint experiment precluded the clairvoyant process? Clairvoyants traditionally assumed "that an object can carry with it impressions relevant to it"—perhaps a thought is "permanently dynamic" (in this of course she was being forced to abandon her initial, firmly materialist orientation) and becomes associated with material objects; then the used cards employed in the first experiment might have had impressions associated with them, and her successful subjects might have gained their information from these impressions. She suggested that one might see whether used playing cards yielded better results in such a test; this was her one despairing way out of the chance results so carefully and unequivocally tabulated.[65]

With the clairvoyant experiment concluded by late 1929, Soal returned to the scoring of his own second and more ambitious experimental series designed to test for a telepathic faculty; his results followed those of the joint experiment into print and appeared in the SPR *Proceedings* for 1931/32.[66] Once again they were entirely negative. There was no sign in the data of a general supernormal perception of any of the classes of objects that had been employed as targets, and the Warcollier hypotheses as to factors tending to strengthen telepathic communication were not confirmed either. Worse yet, not only had the experiment failed to demonstrate a general faculty of supernormal perception, it had been unable to turn up one single indisputably gifted percipient. Soal drew the obvious conclusion: that the Mrs. Pipers of the world were exceedingly rare. The increasing tendency to assert the commonness of psychical ability was totally unjustified; experiments purporting to demonstrate this were based on loose and flexible criteria of success lacking any definite standards of comparison and in general were prompted by the experimenter's own "will-to-believe," "the most injurious influence in psychical research today." With this sweeping criticism and his claim that under normal conditions the ordinary person never experiences telepathy, backed up by the most careful and extensive telepathic and clairvoyant experiments ever attempted up to that point, Soal was calling into question the very validity of a nascent experimental science of metapsychics.[67]

English psychical researchers responded to Soal's papers with stunned silence. It was difficult in the extreme to rebut his conclusions. He had carried on his work in a manner entirely faithful to the objective, rigorous standards on which the SPR historically had prided itself, and under its auspices. Moreover, he could not be dismissed as merely closed-minded or unsympathetic, for he had long shown a very broad interest in the subject (Gardner Murphy has left a record of the summer of 1929, when he, Soal, and Warcollier met for a few days in an orchard in France to discuss the mathematical and experimental techniques proper to the study of telepathy).[68] The SPR was left in some confusion as to how to respond. The only escape from these results seemed to be to argue "that the results by no means rule out the occurrence of telepathy and clairvoyance, all that they do is to emphasize the rarity of such events under experimental conditions," as the Oxford psychologist William Brown put it.[69]

Soal's completely negative experimental results came as a considerable shock to the psychical researchers in France as well. Warcollier's continuing series of publications had helped to make widespread the belief that telepathy was a universal human attribute and had strengthened the expectation that it could be studied experimentally. Osty, in 1931, had expressed his complete conversion to this point of view in a complacent paper explaining how to recognize and to develop the faculty of "connaissance supra-normale." When Soal's work appeared a year later, it was the immediate stimulus for a much longer review article by Osty of the general subject "télépathie spontanée et transmission de pensée expérimentale." Here Osty confronted the problem of why, fifty years after the initial clearly positive work of the SPR, Soal's careful experiments should have had no success, and he tried to ease the doubts of his readers.[70] He recapitulated the experimental work of half a century—the failures of Soal and of American academics, as well as the successes of the SPR and Warcollier. Soal, he pointed out, had criticized Warcollier's results as lacking any basis for calculating the relation of successes to failures and had argued that if a statistical evaluation were possible, Warcollier's successes would not surpass chance expectation. On the other hand, Warcollier had explained to Osty that his success was due to the fact that the members of his group had been very good friends, while Soal's percipients had no such ties to him or to his agents; Soal had ignored the importance of establishing conditions favorable to thought-transference.[71] Was there any merit to Warcollier's argument?

To try to decide this question, Osty discussed his own experiences with brilliantly gifted individuals like the Polish engineer Stefan Ossowiecki. These percipients had repeatedly shown that they could perceive the thoughts of others, he noted, but it was clear that there was indeed some sort of psychological variable that determined their success; Ossowiecki, for example, had to put himself into a special psychophysiological state of trance and then place himself in rapport with the subject.[72] In addition to the Ossowieckis, Osty insisted, there were far more people who could not pick up mental representations selected experimentally but who could delve in others' thought unconsciously, especially when they had the additional advantage of not concentrating on dull, objective material, as Soal's subjects were. For telepathy to be successful, the information to be transmitted had to have deep personal and emotional significance. Hence Soal's failures could be explained in any of a number of ways, all of which came back to this: his careful experimental method was not suited to the phenomenon he hoped to observe.[73]

Neither English nor French objections had any force for Soal, who grew steadily more skeptical and recalcitrant. In early 1934 he presented a review of "Experimental Telepathy and Clairvoyance in England, 1881–1933"—not to the SPR, but to Harry Price's National Laboratory.[74] Here he caustically attacked the outstanding telepathic experiments of the psychical-research tradition, pointing out how easily they could have been produced by fraud or by hyperaesthesia and insisting that careful controls and absolute precision in the reporting of conditions would be required before any positive conclusions as to the presence of telepathy could be drawn. He commented bitterly upon the effect of his own negative results upon the SPR, complaining that the little group at the head of the society had become cool to him since he had spent their money to arrive at a conclusion hostile to the hypothesis of telepathy, and had never asked him to read another paper.[75] Nowhere did Soal bother to respond to the arguments of the French that his sort of objective, depersonalized experimentation was sterile and bound to fail.

Soal then turned to scrutinize the Jephson experiments as the most ambitious of those supposed to demonstrate clairvoyance and called attention to the utter failure of the attempt at their replication. The need for controls here, he said, is also obvious in the case of book tests with mediums, which are not so well controlled nor so strictly judged as they should be and are appraised with wishful thinking, not scientific caution. The policy of the SPR was set, not by its members of real scientific distinction, but by people with basically literary interests who hoped to find evidence tending to establish the likelihood of postmortem survival. Collaboration with a psychological laboratory was necessary in

The Growth of Experimentation in the Psychical-Research Societies 43

order to make sure that men with a scientific background and skeptical rather than credulous by inclination scrutinized the experimental work—on both these points Soal criticized the SPR for replacing E. J. Dingwall by Theodore Besterman as research officer;[76] then perhaps English experimental psychical research would produce convincing results. For Soal still did not deny the possibility of telepathy: he concluded by describing the Groningen experiments of Brugmans et al., conceding that it was not easy to find fault with their research.[77] But in spite of his professed willingness to find something of merit in Brugmans's work, Soal had done his best to destroy the recent surge of interest in experimental telepathy.

It is impossible to say what would have happened if new research had not almost immediately intervened, but there are signs that—at least in France—Soal's statistically based conclusions were forcing metapsychists to abandon the laboratory approach, to turn back to the study of the brilliant individual in preference to that of humanity. Moreover, some of them seem to have felt it necessary to start all over again by proving the very existence of telepathy, which had so long been taken for granted. Soal's mathematics had made such an impression that two different papers in the *Revue métapsychique* tried to turn the weapon against him and to show that the only proper use for mathematical treatment was in establishing one absolutely airtight case of telepathy or clairvoyance. One of these, published in 1934 by Charles Richet, then at the end of his life, involved a recent reading by the clairvoyant "Pascal Forthuny" (Georges Cochet). Forthuny had described a number of details concerning an absent invalid known to the sitters but not to him, and by multiplying together the probability of each of these details having been guessed by chance, Richet was able to calculate the compound probability of the description's success as one in ten million. Astronomical figures like this, he argued, were equivalent to certitude.[78] But in using statistics to identify the astronomically unlikely, the French were abandoning the assumption that Warcollier and Jephson had shared and for a time had made to seem quite plausible—that supernormal powers of this sort were widely distributed in the population and could be investigated systematically. Had psychical researchers abandoned any hope of discovering, studying, and replicating mental phenomena in a broad spectrum of the population, they would have found it difficult ever to press their claims to serious scientific consideration for their field. But in 1934 more ambitious and insistent claims than ever before were advanced for both the widespread distribution of mental psychic abilities and for experimental success in studying them. These claims came, not from one of the psychical-research societies (although they were published there), but rather from a source and indeed a tradition of investigation that we have not so far considered: academic psychology in the United States.

CHAPTER THREE
Early Psychical Research in American Universities

I

In 1898 Harlow Gale at the University of Minnesota made a survey of the place held by psychical research in American universities, inquiring of eleven psychologists at as many schools. The replies he received revealed that psychical research was enjoying positive treatment from psychologists at only two universities (from William James at Harvard and from Gale himself at Minnesota) and was being rather cursorily investigated by philosophy professors at two others (by J. H. Hyslop at Columbia and by an unnamed associate professor at Pennsylvania). At Chicago it received "some attention in connection with the regular psychological work, but it is given no important status." At Brown, Yale, and an unnamed "larger" university it found only "critical" or negative discussion. At Smith the subject had been abandoned by Norman Gardiner. At Clark University the president, G. Stanley Hall, admitted an interest in psychical research but gave no evidence that he said anything substantive about it in the classroom. Finally, at Cornell not only was nothing mentioned on the subject but the psychologist (presumably E. B. Titchener) "will not even have the S.P.R. *Proceedings* in the library lest they should inflame the imaginations and corrupt the minds of the students." Despite the recognition given it by James, Hyslop, and Gale, psychical research certainly had not yet managed to win admittance to the then rapidly developing professional field of academic psychology.[1]

Almost twenty years later, in 1917, Hyslop himself conducted a similar survey of twelve American universities that seemed to show that, if anything, the position of psychical research had deteriorated. The earlier supporters of the field had disappeared from American universities: James was dead, Gale had left academic life, and Hyslop had moved from Columbia to the ASPR. At best, the subject was touched on occasionally in university courses on abnormal psychology or hypnosis.[2]

Yet in fact the potential for serious scientific inquiry had by this time become far stronger than ever at three particular schools: Clark, Stanford, and Harvard. In the previous decade all three schools had received special bequests for the furtherance of psychical research; and at Stanford and Harvard research fellows had already been appointed in the field. In 1911 Thomas Welton Stanford (the younger brother of the founder of the California school, who had amassed a fortune in Australia and was intensely interested in spiritualism) had offered fifty thousand dollars to his brother's university (which he had never seen) for the establishment of a division of psychical research in the psychology department. With a certain misgiving on the part of both the departmental chairman, Frank Angell, and the president of the university, David Starr Jordan, the bequest was accepted, and a recent Stanford Ph.D. in psychology, John E. Coover, was named Fellow in Psychical Research in 1912.[3] The Harvard situation was a little different. There psychical research had been provided for by funds donated in 1911–12 in memory of Richard Hodgson (d. 1905), the Australian-English psychical researcher who had done so much to revitalize the field in America in the late nineteenth century. In 1916 Mrs. John Wallace Riddle augmented the original bequest on condition that research be implemented immediately, and the first beneficiary was Leonard Thompson Troland, a twenty-eight-year-old specialist in physiological optics who had been a Ph.D. student of Hugo Münsterberg's.[4] Both psychologists, as it happens, published a first report on their investigations in psychical research in the very year of Hyslop's survey, 1917.

The professional setting in which Coover and Troland went to work on psychical research was on the whole not one favorable to positive conclusions. American psychology had grown rapidly as an academic field in the late nineteenth century. A symbolic starting point is sometimes found in William James's foundation of a laboratory for experimental psychology at Harvard in 1876, but a more important impetus was perhaps the large-scale experimental program being implemented at virtually the same time by Wilhelm Wundt in Leipzig. By the early years of the next century psychology had become firmly established in American universities, with flourishing graduate programs at Harvard, Johns Hopkins, Cornell, and Chicago making particularly strong contributions to its continuing vigorous growth. The Wundtian approach to psychology, although perhaps the most important element in the developing academic field, was never entirely dominant, and by 1915 American psychology exhibited a rich and complex theoretical and methodological picture. Nevertheless, the main emphasis in the United States fell on experimental psychology, as distinct from philosophical, abnormal, or clinical psychology. This inevitably had an important bearing on the reaction of psychologists to psychical research as an enterprise to be carried on within the universities. While psychical research could (as we have seen) enter the classroom through courses in abnormal psychology, it was only able to become a research subject insofar as it accepted the strategy and outlook of the experimental psychologist.

Joseph Jastrow, of the University of Wisconsin, made just this point quite forcibly in 1900, when psychical research was beginning to lay claim to academic respectability:

> If the problems of Psychical Research, or that portion of the problems in which the investigation seems profitable, are ever to be illuminated and exhibited in an in-

telligible form, it will only come about when they are investigated by the same methods and in the same spirit as are other psychological problems, when they are studied in connection with and as a part of other general problems of normal and abnormal Psychology. Whether this is done under the auspices of a society or in the psychological laboratories of universities is, of course, a detail of no importance. It is important, however, what the trend, and the spirit, and the method, and the purpose of the investigation may be; as it is equally important, what may be the training, and the capabilities, and the resources, and the originality, and the scholarship of the investigators.[5]

Frank Angell reiterated this argument in introducing J. E. Coover's account of his experimental research at Stanford, insisting specifically that the would-be psychical researcher would need to devote his undivided attention to the subject and would require "special extensive training in the psychology of motor automatisms and of subliminal impressions, in the ideational and affective processes underlying belief and conviction, in illusions of perception and the value of evidence."[6] However, these explicit prescriptions as to the professional qualifications necessary to the psychical researcher did not mean that American psychologists were favorably disposed toward devoting their own laboratories to psychical research. The insistence of a Jastrow or an Angell upon psychological training for the psychical researcher barely veiled a generally negative attitude towards the field.

This mixture of professional standards and hostility typical of American psychology is epitomized by no one better than by G. Stanley Hall, a man unusual for his catholicity of psychological interests and indeed one who had helped to found the original ASPR: "Give us one little fact, ever so little, that we can freely test and reproduce one a year in our laboratory. We will cross seas to see it, will acknowledge our mistaken skepticism, and confess telepathy, and turn the research of one laboratory at least in a new direction."[7] Hall's irony and sarcasm illuminate a basic circularity in the attitude of psychologists that was nearly impossible to break. For while they insisted upon using the rigor of experimental psychology to "test" psychical phenomena, they were sufficiently skeptical a priori of the existence of any phenomena as to be quite sure that such tests could yield only negative results. We can see this same confusion of attitudes in more detail in a number of fairly elaborate arguments that American psychologists framed against psychical research in the 1880s and 1890s—their very elaborateness indicates how seriously the field was then being taken.[8]

G. Stanley Hall himself wrote one of the first critiques of the subject in 1887, a review of the first six volumes of the SPR *Proceedings* and of Gurney, Myers, and Podmore's *Phantasms of the Living*.[9] The first part of the review was given over to criticisms of the experimental techniques and methods of reporting used in these initial studies. Hall's criticisms were scathing: the experimenters had omitted crucial details of their test situations, had provided incomplete protocols, and had in general exhibited a remarkable naïveté concerning the possibility that hyperaesthesia and fraud might explain their subjects' successful guesses. Moreover, an experimental psychologist would have considered it important to record the details of the mistaken guesses: were there any patterns in them that made them correspond more or less closely to one or another mode of normal perception? For example, in card-guessing situations, guessing the color correctly more than twice as often as the suit would suggest something like visual transfer of color; confusing the digits *nine* and *five* would suggest auditory transfer, since to the ear (but not the eye) these figures are alike; and confusing 2 and 3 or 6 and 8 would suggest something like muscle-reading, since the initial stroke in the formation of the numbers in each pair is similar. Hall's point in all this was to emphasize what he considered the unprofessional and casual techniques employed by the English experimenters. In particular, the extent and degree of human sensitivity and hyperaesthesia reported in the psychological literature seemed to Hall to invalidate the entire experimental basis of telepathy.

Hall's position thus was that the investigation of psychical phenomena lacked the rigor that experimental psychology could give it. But at the same time the only way Hall could envision this rigor being brought to bear was in a destructive role. For he was convinced that telepathy did not exist; it was simply an artifact created by the credulity of the investigators and the chicanery of the subjects. The reasons for his conviction seem to stem in part from his scorn for the spiritist associations of psychical research. That a telepathic hypothesis did not necessarily entail spiritism (and might even be an alternative to it) would have seemed to him a quibble—and not unjustly, for the psychical-research tradition had been started by men many of whom were looking for empirical proof of personal survival. A more specific reason for his disbelief in telepathy—and it was on telepathy that other psychologists, like Hall, tended to focus—was that it seemed to undercut the psychophysiological basis of mental activity. Unlike later opponents, who usually rather vaguely objected that telepathy lacked a physical mechanism, Hall set out his critique in detail. Telepathy, he argued, was in conflict with the law of "isolated conductivity," whereby sensory signals passing along one nerve fiber could not jump across to another nerve fiber no matter how close they were. Hall asked rhetorically: "Is it likely that a neural state should jump from one brain to another, through a great interval, when intense stimuli on one nerve cannot affect another in the closest contact with it?"[10] Given not merely the difficulty of a possible neurological mechanism for telepathy but indeed its inconceivability (at least as far as the neurology and psychology of the late nineteenth century were concerned), even the use of statistics to try to assess the degree of improbability in successful card-guessing and the like did not impress Hall.

Hall's sentiments found parallels in other detailed criticisms of psychical research during this period, criticisms aimed at different features of the subject. The Harvard experimental psychologist Hugo Münsterberg developed an elaborate refutation of psychical phenomena based upon his acceptance of psychophysical parallelism, the philosophical position then dominant among experimental psychologists as a means of coping with the mind-body problem. This view maintains that the mental or "psychic" activities of a sentient mind are paralleled exactly by the physical or physiological activities of the corresponding brain. The two sequences of activity never interact one with the other; they only coincide. Münsterberg saw psychophysical parallelism not simply as a convenient philosophical stance for the psychologist, but as a necessary one, since the ultimate purpose of science was to construct a mechanical world view freed from subjective considerations. In terms of this position the failing of psychical research was not that it purported to investigate phenomena whose scientific explanations were as yet unknown but that in the form in which psychic phenomena were set forth—without a neurophysiological counterpart—these explanations were unknowable. Therefore the phenomena had to be rejected by the scientist from the start.[11]

To the methodological and philosophical objections of Hall and Münsterberg, Joseph Jastrow added a professional one: by its prominence in the public mind and by its popular association with psychology, psychical research contributed to a false public image of the field of experimental psychology. Psychologists could not afford to be misinterpreted, to be regarded as collectors of ghost stories or investigators of mediums; a sharp line had to be drawn between the two fields.[12] Considering all these grounds for suspicion, it is little wonder that the 1898 and 1917 surveys showed little support for psychical research among American psychologists.

To be sure, there were some exceptions to this general hostility, the most notable being William James. James's attitude towards psychical research was just the opposite of Hall's and Jastrow's. Unlike them, he was benevolently open-minded to the possibility of supernormal psychic phenomena and even to the survivalist hypothesis. His subtle, complex, holistic psychology was much more conducive to tolerance on these matters than was the more sharply structured, analytical psychophysical parallelism of the dominant introspectionist tradition. James participated actively and enthusiastically in the work of the British and American societies for psychical research and was a close personal friend of F.W.H. Myers and Richard Hodgson; and he devoted much time and energy to studying and publicizing the psychic abilities of Mrs. Piper. He did not make an issue of the necessity for psychical research to be housed in a laboratory of experimental psychology in order to be "properly" carried on; he conducted no formal laboratory studies and seemed quite content to base his own conclusions on his personal investigation of mediums like Mrs. Piper or upon the reports of his English colleagues.[13]

The only other important American psychologist to express any sympathy for psychical research was James Rowland Angell. In his *Chapters from Modern Psychology* (1912), Angell admitted the existence of "a very respectable body of evidence" tending to show that telepathy was an occasional reality.[14] He came further into the open in a private letter acknowledging receipt of the 1917 monograph of the Stanford Fellow in Psychical Research, J. E. Coover: "I have often myself been tempted to undertake some of the lines which you have followed, but without special financial resources, it has seemed to me quite hopeless. Added with this, there is, of course, a certain scientific risk involved in attacking these fields so long taboo to the 'scribes and Pharisees' of our psychological community."[15] Yet Angell's wistful comment makes vividly apparent the strength of the forces working to exclude the investigation of psychic phenomena from orthodox psychology in America, and the difficulties involved in adopting it as one's subject of research. A William James could espouse the field with comparative impunity, but quite clearly a J. R. Angell (who was then dean of the University of Chicago) did not believe *he* could.

In this setting, therefore, the bequests made between 1907 and 1912 to Clark, Stanford, and Harvard to subsidize academic psychical research forced a difficult decision upon administrators and psychologists at these universities. Should a subject so obviously undeserving really be dignified by scientific study, despite the waste of time and funds it would involve? The bequest of five thousand dollars to Clark—from J. A. Battles, in memory of his father-in-law, Joseph Smith—had required the income to be expended on "lectures on the subjects of spiritism, or occult or mystic psychic phenomena, telepathy, somnambulism and kindred subjects, or if in the judgment of the faculty the income may be used in experimental tests to scientifically explain unusual physical or psychical phenomenal apparitions."[16] G. Stanley Hall was then president of Clark University, and he was somewhat nonplused by the gift. He arranged for a first series of lectures to be given at Worcester in January and February 1914—by people as diverse as the psychical researcher Hereward Carrington and Harvard's abnormal psychologist Morton Prince—and gave a small grant to a colleague wishing to investigate an ostensible case of telepathy.[17] However, he may have become wary of involving Clark further in a subject capable of discrediting his own field of psychology so easily. When in 1916 Charles Cross at M.I.T. inquired of Hall about psychic research at Clark, Hall replied coldly, ". . . we would much prefer to have you use the phrase 'Psychological Research' rather than 'Psychic Research' in connection with the Smith-Battles Fund here, as we do not indulge in what is generally known as psychic research here."[18] For the next decade and more, Clark made no expenditures at all from the Smith-Battles income.

At the other two universities the situation was somewhat different. There the monies were applied to systematic psychical research, but research that conformed closely to the attitudes that were dominant in contemporary psychology. It was carried on as experimental psychology, in the laboratory and not in the séance parlor, and test materials both manageable and measurable were employed. Moreover, not unexpectedly, at both schools it was carried on by men already somewhat skeptical of the phenomena they were supposed to study. If it would be unfair to say that Coover and Troland carried out their tests with the hope of disproving telepathy, it would be a distortion to say that they expected to find any sign of it.

It was at Stanford that to all appearances the task of making psychical research the subject of truly scientific investigation was accepted most wholeheartedly. John E. Coover, made fellow in psychical research and associate professor of psychology in 1912, at age forty, retained his title until his retirement in 1937; and all during that quarter-century, he was identified in the university catalog as teaching courses on the literature and achievements of the field. Yet Coover's own serious experimentation was apparently limited to the first five years of his fellowship, during which he carried out a program of research published as an enormous monograph of over six hundred pages in 1917. Thereafter he seems to have carried on no further investigations, almost surely because he—like Hall and Jastrow—could no longer take the subject seriously.

Just how deep Coover's skepticism went at the outset is difficult to say. He claimed that he had outgrown a naïve childhood belief in psychic phenomena, and yet in announcing the monograph to his patron, Thomas Welton Stanford, Coover outlined a project of buttressing spiritualist belief with the philosophical idealism he professed.[19] This, of course, may have been written to flatter Stanford's interests, for less than a year later Coover wrote to the president of Stanford, "I should have been better satisfied . . . with an opportunity to put part of my time upon research the material of which is not so meagre and elusive, not so offensive in the nostrils of my fellow psychologists, and more directly applicable to problems in psychology, education, or psychotherapy."[20] Perhaps the clearest view of his eventual position is contained in an article published in 1920 in the *Homiletic Review*. In that article, Coover adopted a stance on the mind-body problem verging on psychical monism, which he viewed as flowing out of the philosophical tradition to which he assigned Wundt, Külpe, Titchener, and others in experimental psychology. From this standpoint, he insisted, there need be no philosophical incompatibility between the thrust of science and one's acceptance of spiritualism or supernormal psychical phenomena (as Hall or Münsterberg would have objected); the difficulty with psychical research was that its hypotheses simply had not yet been established according to the canons of scientific proof.[21] This article, to be sure, was published after his own initial investigations had been concluded; whether it may be taken at face value and, if so, whether it also represents his attitude in 1912 is not easy to decide.

In any case, whether or not Coover pursued his research between 1912 and 1917 with a genuinely open mind, he certainly carried out an exceedingly elaborate and meticulous program of experimentation, in which he brought to bear the full arsenal of experimental psychology upon the problem of telepathy; not merely the massive monograph but the laboratory notebooks for this research testify to his care and thoughtfulness.[22] Apparently the first project to be taken up was a study of "the feeling of being stared at," which was followed by more elaborate statistical tests involving lotto-block- and playing-card-guessing on the part of student subjects as well as of professional and private "psychics." The experiments with playing cards were by far the most ambitious of these. Coover pointed out that playing cards possessed obvious advantages for the experimental psychologist—they were readily available, convenient to manipulate, and susceptible of relatively easy statistical evaluation.[23] Thus we can study the peculiar features of this first experimental foray into academic psychical research very efficiently by concentrating upon Coover's work with card-guessing.

It is throughout entirely clear that the techniques of experimental psychology remained at the front of Coover's thought, as the procedure he laid down for "experimenter" (agent) and "reagent" (percipient) testifies. The setting was to be formal, cool, and orderly: his back to the experimenter, the reagent closed his eyes and "assumed a calm, receptive, quietly expectant state of mind."[24] In order to test for different modalities of thought-transference, the experimenter, whose task it was to shuffle, cut, and expose the bottom card, acted upon the target card in ways that varied according to the outcome of a die throw. If the die came up with an odd number—1, 3, or 5—the experimenter would glance at the card and try to convey it to the reagent via a mode of thought-transference determined by the particular number: 1 meant visually (trying to convey the image of the card); 3, kinaesthetically ("feeling" the movement of the vocal organ without imagining the sound); and 5, visually, kinaesthetically, and auditorially (with the experimenter imagining himself and others to be shouting the nature of the card). Besides this, "the experimenters held in consciousness a determined attitude of will that the content should reach the reagent." If the die came up with an even number, a "control" situation supervened whereby the experimenter, although turning up the card, did not look at it and remained unaware of its nature. The reagent, of course, had no idea whether he was guessing a card unknown to the experimenter or one whose nature the experimenter was trying to convey telepathically.

The reagent recorded his guesses of color, number, and suit separately, in order to help assess which of these factors might be most easily transferred. Then he performed an elaborate introspection on his mood as well as on the nature, vividness, duration, persistence, spatial attribute, and certainty of the communication purportedly received. A general description of the subject's mental and physical state was recorded for each experimental session, and the reagent was encouraged to write in additional comments to supplement his answers to the specific queries put to him. This was the basic experimental format, on which Coover rang a few changes: in the first thousand experiments he systematically increased the distance of the experimenter from the reagent, and in another twenty-five hundred experiments he varied the time interval during which the guess was to be made.[25]

Coover quite obviously wanted his study to be a model of the application of experimental psychology to psychical research, in which all possible factors could be statistically correlated. It was also meant to be a model of introspective

Wundtian-Titchenerian psychology. While Coover made no reference to neurological or psychodynamical hypotheses that might be used to explain telepathy, his reliance upon introspection still makes it plain that he was not without his theoretical assumptions: specifically, that mental processes are conscious processes (as opposed to unconscious) and that there exists a psychic structure correlative with physical (neurological) activities that can be revealed by introspection. In this respect Coover was very much in the mainstream of what had been and what to a degree still was orthodox psychology in 1917.

Coover's methods were those of the experimental psychologist in another respect: instead of analyzing the results of individual reagents, he pooled all his data for purposes of statistical evaluation. This again was in line with the increasingly quantitative approach of psychology, in which the idiosyncrasies of individual subjects in themselves were generally of no interest to the psychologist; what was of interest was their statistically describable distribution about a collective norm. This particular technique also had some relevance for the contemporary psychical-research tradition, as Coover well knew. As we have seen, one question widely debated by psychical researchers had been whether psychical ability is rare or generally distributed among mankind; the main proponents of the former position had been James and Lodge, and of the latter, Charles Richet. Coover's approach was thus one that tested Richet's hypothesis, in that if telepathy were indeed a general property, it would be expected to appear as a statistically significant distortion of the aggregate data.[26] He did not go on to examine the Jamesian hypothesis by further testing *individual* subjects who scored high in his first experiments.

Coover's techniques reflected both the interest of the psychologists and the prevalent view of psychical researchers in one final way, in that he concentrated exclusively upon telepathy (which he took implicitly to be a process initiated by the experimenter) and showed a peculiar blindness towards the possibility of clairvoyance. It will be recalled that Coover had used a "control" in his card-guessing runs: when the experimenter's die came up even, he remained ignorant of the card for that particular case, while the reagent performed his usual guess and introspection. Coover seems to have felt that this control could help distinguish telepathy from mere "lucidity."[27] As he went on, however, he treated clairvoyance, or lucidity, not as something to be experimentally differentiated from telepathy, but as a steady background noise that need not be taken into account. This view that clairvoyance is a constant noise is of course a testable hypothesis: one could easily compare the reagent's successes at guessing cards *not* imaged by the experimenter with mean chance expectation in order to see whether some psychic faculty were operating. But Coover did not do this; rather, he treated the control data arbitrarily as normal chance results and judged the results of his correlation of the "Card Imaged" data and introspection factors in terms of their closeness to or discrepancy with the "Card Not Imaged" data, supposed to be the product of chance.[28]

Coover's conclusions at the end of this card-guessing experiment were fairly restrained; he summed up the data as showing only that no trace of thought-transference had been found in his subjects and that therefore if it existed, the telepathic power must be relatively rare.[29] But his inability to accept *any* sort of telepathy is quite apparent in a lengthy appendix to the monograph with the revealing title "Grounds for Scientific Caution in the Acceptance of the 'Proof' of Thought-Transference," in which he reviewed the work of the British SPR so as to show the methodological weaknesses in its supposedly most telling experiments, emphasizing that *"technical experience in experimental psychology is requisite both for the control of the experiments and for the interpretation of the results."* Nevertheless, he ended by insisting upon the open-mindedness of psychologists. They would not dogmatically insist "upon the laws of the 'isolated conductivity' of the neurons"; if telepathy would present itself with unimpeachable evidence, it would be admitted.[30] An attempt to sound more unbiased than psychologists of the previous generation is clear from the reference to "isolated conductivity," which recalls G. Stanley Hall's arguments of the 1880s; but for all its air of objectivity and dispassion, Coover's *Experiments in Psychical Research* reinforced the traditional negativism of academic psychology towards psychical research.

It would be delightful to know what Thomas Welton Stanford made of this monograph, the only child of his endowment. Coover himself somewhat defensively expressed the belief that Stanford would be satisfied, since "like any other intelligent and sound-minded business-man," he would appreciate solid scientific work even if it did not yield the results he hoped for.[31] Stanford died in August 1918, however, without passing any judgment that has survived on Coover's study. It is certainly doubtful that this massive report brightened the last months of his life.

Coover sent copies of the monograph to a number of well-known psychologists, psychiatrists, and psychical researchers, and the reactions of many of these have been preserved; we have already had occasion to quote J. R. Angell's response. The man who gave the most appreciative reply, however, was not Angell, who was guardedly sympathetic to experimental psychical research, but E. B. Titchener of Cornell, who was implacably hostile to it. Titchener's reply was fuller than any other; in fact Titchener was the only respondent who gave evidence that he had actually read the monograph closely. If Coover had hoped for approval of his conclusions from the dean of experimental psychologists, his hopes were certainly realized. Praising Coover for having made an "admirable beginning" and for having "given us ammunition for a long time to come," Titchener went on to outline his personal animus against psychical research:

> I believe with Faraday that science is a strictly limited and circumscribed affair, implying a 'narrowness' of attitude, while yet it has given gifts to our civilisation that the researchers have not paralleled in one single item: and I should like to have the scientific man's definition of science, with all its limitations, forced upon the

spiritists,—documented historically and justified (if necessary) by results. I should like to bring home to them, not only that no scientifically-minded psychologist *does* believe in their 'telepathy,' but also that no scientifically-minded psychologist *can;* if he did, he would not be scientific; and if they get results, then the results cease (so far as science can assimilate them) to be telepathic. There is very much more in this than a mere quibble about words; there is the whole consistency of science.[32]

Titchener suggested that Coover (or others) might publish in other debunking studies (as he took Coover's project to be) something in the way of a scientific manifesto, "not offensively aggressive, but put perhaps in the form of an historical introduction, which might appeal at any rate to the better educated of the believers." He ended his letter with renewed congratulations and a wish for more such "effective" studies—but while Coover did raise suggestions for further investigation and continued to lecture on, and write an occasional piece about, psychical research, he did not publish any further experimental results in the field. The Psychical Research Monograph No. 1 never had a sequel.[33]

The Harvard endowment for psychical research, the Richard Hodgson Memorial Fund, differed from the Stanford bequest in one extremely significant respect: it was designed from the outset, not to subsidize a permanent member of the faculty, but to be applied towards research problems of outstanding potential and interest. The Hodgson Fund was thus not something that could be easily drawn upon, and this special status assured its future connection with experimental studies to a far greater degree than proved true for the Stanford money. A first suitable recipient for the fund was not found until 1916/17, when L. T. Troland was granted research support totaling $843.94.[34]

Troland had shown some signs of interest in psychical research while still a doctoral candidate at Harvard, but he had not really been sympathetic. He had published an early article, "The Freudian Psychology and Psychical Research," intended to provide a "naturalistic" (as opposed to "spiritistic") explanation of most psychic phenomena—telepathy, hallucinations, monitions, and mediumship—in terms of Freudian psychodynamics.[35] Specifically, he argued that atavistic animal abilities and repressed desires, contained in the subconscious but unperceived by the conscious mind, might explain something like telepathy: what might be operative is a sensitivity of the percipient's subconscious to subtle gestures and changes of expression of the agent, this being a possible example of "animal capacities [which] have undergone suppression in the course of racial evolution." The article was quite loosely constructed and argued at a general, hypothetical level. No specific case of psychic phenomena was treated; Troland simply claimed that most cases *could* be accounted for by some such assumption.

Apparently Troland was predisposed towards psychical research in still another way: he (like Coover) was a philosophical idealist, a "psychic monist" who believed that physical entities presupposed psychical ones. In a 1918 article on "paraphysical Monism" Troland pointed to relativity theory as overthrowing the last vestige of objective physical entities and argued that material particles and their interacting forces were but reflections of psychic atomistic entities.[36] It may well be that some sort of philosophical idealism is a requisite to any serious interest in psychical research, and certainly, as Coover noted, it is not incompatible with the entertaining of spiritistic or physical hypotheses; but as the work of both Coover and Troland shows, neither does this philosophical position entail acceptance of such hypotheses.

The specific research project that Troland took up was an outgrowth of the psychological theory that he had learned as a graduate student. Coover's had been, too; but whereas Coover's assumptions were those of the introspectionist tradition, Troland's principal tool—natural enough for a psychophysiologist—was the reflex-arc concept. The psychologist, as Troland had come to understand, studies the flow of energy along a physically continuous path, from object to receptor to brain to muscle to environment. This is of course the model that at just this time was becoming so productive in the hands of the new behaviorist school, but Troland was unwilling to go quite this far in his psychophysiology.[37] When claims for psychic phenomena are interpreted in this theoretical context, the phenomena appear to be instances where energy transfer takes place even though "gaps" or breaks occur in the underlying physical continuity. The first task of the experimental psychologist who wishes to study such claims, then, is to study ordinary cases of energy transfer, where no such discontinuities exist; to introduce physical "gaps"; and then to see whether energy (manifested as information) still passes through. Troland's own investigations centered upon what he called a "split reaction experiment," analyzing telepathy as a situation in which the complete reflex-arc is split in half between the agent and the percipient, leaving a gap between. In the agent, the afferent process of stimulus reception and transmission to the central nervous system takes place when the target image is perceived; in the percipient, the efferent transmission from the brain to the motor center takes place when a note is made of the "guess" of the target image; but there is a physical gap and no communication between the two. All this is worlds away from the detailed analysis and high degree of introspection employed in the work done at Stanford.

The difference between the two approaches is just as pronounced when we consider experimental design. Coover had applied an elaborate introspective procedure to a very simple situation, guessing playing cards; Troland introduced the most fully mechanized test situation that had yet appeared in psychical research—so as to promote exact reproducibility of the experimental scene and to *"eliminate the personal equation of the researcher"*[38]—and abandoned introspection! His machinery consisted of a stimulus head, made up of a stimulus field of a central light point with two identical squares placed symmetrically on either side. By means of an electromagnetic "stimulus shuffler" one or the other square was illuminated at random. The agent viewed the stimulus field; the percipient moved a switch (in case muscular action was sensitive to supernormal influence) to guess which square had been lighted up. After a specified time (forty to eighty seconds) the process was repeated. A recording device in the machine kept tallies of the square actually illuminated in each trial, the guess of the percipient, and the number of trials that had taken place. After one hundred trials, the machine shut off automatically.[39]

Even more emphatically than his contemporary at Stanford, Troland argued that the results accumulated in such an experiment had to be analyzed statistically. Alleged supernormal phenomena in particular had to be presented in mathematical form "because of their apparent conflict with recognized quantitative principles of physical science."[40] He pooled the scores from all agent-percipient pairs (without recognizing that he was thus testing for telepathy as a general human faculty rather than a rare one) and found that in 605 trials there had been 284 correct guesses, "chance" giving 302 or 303. This was 1.5 times the standard deviation *below* chance, which for Troland was no less interesting than a result above chance to the same degree, and he noted that while it was not especially significant, "the odds are about fourteen to one that the explanation is not to be found in pure chance."[41]

What are we to make of this unnecessary concession? Not, surely, that Troland took these few moderately promising results very seriously; after all, these 605 trials proved to be the beginning and end of his psychical research, small return for such an elaborate experimental situation. Indeed, Troland had probably never taken the work very seriously. His own description of the actual conditions of the experiment makes apparent a casualness that would immediately have been the target of criticism if he had claimed markedly successful results.[42] The whimsical conclusion to his report is perhaps most revealing of his attitude:

> He [Troland] feels confident, moreover, that the majority of psychical researchers will regard it as a ridiculously mechanized device for dealing with problems whose solution can only come through media of sympathy and suggestion. On the other hand, the average academic psychologist will look upon the above proposals as a waste of good technique upon a hopeless situation. However, if the task which actually lies before us is that of studying the problems in a scientific way, neither type of opinion would appear to offer us much assistance.[43]

Troland may have been entertained by the claims of psychical research and attracted by the initial methodological problems it presented to the psychophysiologist, but he was not at all prepared to see in it a fruitful or engrossing career.

Coover and Troland together represent the poles of experimental psychology in the United States in 1917: introspectionism on the one hand, near-behaviorism on the other. Their reactions to psychical research make it quite clear that the abstract objections that academic psychologists raised against the subject helped to condition the nature and results of whatever experimental work they might be induced to do in the field. Different in their approach, they nevertheless arrived together at conclusions that were very far from fulfilling the intentions in which their fellowships had been established. And yet academic psychical research in the United States was by no means aborted in their failure. The very existence of these funds at Stanford and at Harvard and the fact that they had actually been applied towards the experimental investigation of psychical phenomena by professional psychologists encouraged many psychical researchers in the 1920s to expect the increasing acceptance of their subject by the university community. At Harvard, in particular, their hopes were not entirely disappointed.

II

In a very critical review in the SPR *Proceedings* of Troland's report, the Oxford philosopher F.C.S. Schiller ended on an optimistic note: "But is it too sanguine to hope that when our present President takes over the command of the Harvard Laboratory he will provide the required 'stimulus' and Dr. Troland's researches will be continued?"[44] The reference was to William McDougall, who in the fall of 1920 became professor of psychology at Harvard University, filling the chair once held by William James, which had been vacant since the death of Hugo Münsterberg. McDougall was then in his fiftieth year, a psychologist of international reputation. Educated in the natural sciences at Manchester and Cambridge and reading widely in philosophy and psychology on his own, he had opted for a medical degree upon completing his A.B. at Cambridge in 1894. He studied medicine at St. Thomas' Hospital in London, where he came in contact with C. S. Sherrington, and pursued neurological studies back at Cambridge during the long vacations. Yet McDougall, whose lifelong interest in what he called "human nature in all its aspects" was already making itself felt,[45] had decided that neurology was not enough in itself; zoology, philosophy, and psychology—and the social sciences generally—were also necessary to a comprehensive understanding of human nature. The major influence on McDougall at this stage was William James, and there is certainly much of the Jamesian model evident in McDougall's breadth of interest and in his emphasis on the richness, complexity, and concreteness of human experience.[46]

McDougall received his B.M. in 1897, but by then he had abandoned any plans for a medical career. After taking part in an anthropological expedition to the Torres Strait and Borneo, he settled on psychology as the field in which to launch his "direct attack on the secrets of human nature."[47] After a year's study of experimental psychology with G. E. Müller at Göttingen, McDougall returned to England at the end of 1900 to accept a post at University College, London. Four years later he was appointed Wilde Reader in Mental Philosophy at Oxford.

It was during the Oxford period that McDougall's psychological interests really expanded, even though he was never fully satisfied with his situation there. Originally concentrating on the psychology of vision, he soon widened his research to include social psychology, the mind-body problem, and abnormal psychology. In each of these fields he produced an important book, occasionally even a definitive one,[48] all the while maintaining his interest in experimental psychology. In these same years McDougall was helping to organize psychology professionally in England, first by helping to form the British Psychological Society and to undertake the publication of the *British Journal of Psychology* while still at University College and then, at Oxford, by attracting about him as advanced students some of the young men who were to become the leading lights of British psychology in the next generation, including Cyril Burt and J. C. Flugel. When McDougall left Oxford for Harvard in 1920, he had become his country's most eminent psychologist by far.[49]

Ironically, the philosophical position that McDougall had been forging for himself since his early years had tended increasingly to isolate him from the body of the psychological community.[50] Neither introspectionism nor behaviorism was ever to his taste; in fact, although he was trained in medicine and neurophysiology, he was never willing to devote himself fully to experimental work.[51] While experimentation was important, the tendency to depend entirely upon its analytical strengths aroused McDougall's indignation as distorting the very human experience and behavior that psychology claimed to want to understand.

In particular, there were two features of human experience that McDougall felt to be *sui generis* and to defy all analysis into other entities: consciousness and purpose. Consciousness came to imply for McDougall the existence of a holistic mental entity, the mind, the soul. Moreover, this entity was perhaps capable of interacting with and directing processes of the brain, regardless of what the laws of physics might seem to suggest, as he argued in his famous defense of animism, *Body and Mind* (1911).[52] The purposive nature of human behavior was the other primary element revealed by experience, and purpose became the hallmark of McDougall's famous hormic psychology.[53] Goal-directedness was itself an antecedent cause of human and animal action and could not be reduced to stimulus-response pairs any more than mind could be reduced simply to neurological networks or to structures of psychic entities.

This outlook set McDougall well apart from virtually all his colleagues. That holism, teleology, and psychophysical interactionism of this kind were antimaterialistic in import—indeed, that their negations defined materialism—was something of which McDougall was perfectly well aware, and he gloried in his role as the critic of orthodoxy. In the 1920s and 1930s he made it something of a mission to combat materialism wherever he saw it. "Materialist" and "mechanist" were epithets of contempt to be affixed firmly where they belonged, no matter what the standing of the adversary.[54]

As this outlook marked him off from most psychologists, so it tended to make him sympathetic towards psychical research. In November 1901 he was elected an associate of the British SPR, and in March 1903 he was co-opted to its council, on which he served almost continuously until his election as president of the SPR in 1920. His support derived, not from mere iconoclasm, but from the real conviction that psychical research had collected solid evidence of an aspect of human nature that official science had so far ignored. In one of the last chapters of *Body and Mind* he discussed the importance of the cross-correspondences discovered in automatic writings. These materials forced the investigator to choose between one of two hypotheses—survival or telepathy. The survivalist hypothesis, which if proven would constitute an empirical verification of animism, he could not yet think fully established; but psychical research had at least established the occurrence of facts incompatible with mechanism, and hence the alternative to survival could only be telepathy, for which the evidence "is of such a nature as to compel the assent of any competent person who studies it impartially."[55] Whichever the final conclusion of psychical research, then, it was sure to be consonant with McDougall's own philosophy. More than this, as the appeal of scientific materialism increased, McDougall came to see this opposition to materialism as a moral struggle. The materialistic conception of the world could only lead to the bankruptcy of values, for without a vision of the transcendental significance of man, life became empty of purpose and meaning, as he declared in 1922: "There is no good reason to think that in the absence of such beliefs, any high moral tradition could have been evolved by any branch of the human race. Are we then justified in assuming that, if the foundations are sapped away, the superstructure of moral tradition will continue to stand unshaken and unimpaired, powerful to govern human conduct through the long ages to come? I gravely doubt it."[56] McDougall allotted to psychical research the task of stemming "the destroying tide" of materialism, since only here could "facts incompatible with Materialism" be discovered.[57]

Yet McDougall's deep interest in psychical research was never an uncritical one. In his presidential address to the SPR of 1920, when he originated the distinction between the "left" and "right" wings of the society, he insisted that he himself belonged "very decidedly to the right wing," that is, to the branch that insisted upon the need to gather unassailable evidence and to explain it as conservatively as possible.[58] A few years later, to an American audience, he presented his own assessment of what he believed had been proven by psychical research. He was convinced, he said, that all the main types of psychical phenomena—telepathy, clairvoyance, even some of the physical manifestations of mediums and survival—were now highly probable. But he remained guarded in his final judgment, willing only to commit himself definitely to the imperative need for science to investigate such phenomena seriously.[59] Throughout his career, in fact, McDougall was uncomfortably balanced between two positions. Philosophically and morally he was predisposed to accept psychical research; yet his professional training compelled him to adopt a skeptical and critical attitude towards any claim for psychical phenomena, particularly towards any spiritualistic explanation for them.

As we have already seen, McDougall was elected president of the American

SPR when he left Oxford for Harvard in 1920, and it is now possible to understand a little better his motives for wanting to make the society not merely a spiritualist organization but a scientific society with close ties to the academic world. Perhaps, too, it is possible now to share his feelings about the coup of 1923, which marked the end of his plans. From that point on, McDougall more or less abandoned the idea of the amateur society as the proper vehicle for psychical research and turned instead to the university. As the senior psychologist at Harvard, he was excellently placed for this, since he was able actively to employ the Hodgson Fund (unused since Troland's work) in support of promising experimental psychical researchers.

The first of his protégés was Gardner Murphy, a young graduate student in psychology at Columbia who had long possessed a strong interest in psychical research. Before the war, he had spent a year at Harvard working for his M.A. He had chosen to work with Troland, precisely because of the latter's activity in psychical research, and had been his assistant in his "split reaction" experiment.[60] In late 1921 Murphy came up from New York to see McDougall about the possibility of a permanent position with the British SPR. As Murphy described the scene, "He [McDougall] wheeled his chair towards me with a sort of electrical intensity, snapping into action: 'Why don't you come here?' He went on to spell out the availability of the Richard Hodgson Fund. I talked it over with my mother, thought it through, and decided I could do this without giving up the part-time position available at Columbia, and without risking my neck altogether on what might prove to be a dead end. I could identify myself with the Richard Hodgson Fellowship."[61] Murphy began to draw support from the income of the Hodgson Fund (strictly speaking, there was no "Fellowship") in the fall of 1922 and continued to do so for three years.

What set Murphy apart from Coover and Troland as an investigator was his sympathy with the aims of psychical research. This sympathy originated in his family background—he read William Barrett's *Psychical Research* from his grandfather's library when he was sixteen—but it was at school that it crystallized, in an acute spiritual crisis. Murphy (whose father was an Episcopal minister) had developed an intense religiosity during his teens, which was first seriously challenged in his junior year at Yale, in the Darwinian, "naturalistic" anthropology course of Albert G. Keller. Entering Harvard graduate school in the fall of 1916, he found Keller's outlook reinforced by E. B. Holt's course on the philosophy of nature, and late one night the next spring he decided "that I would have to give up my religious faith."[62] However, there remained psychical research, which Murphy saw as a potential hostage in the scientistic camp and through which he might eventually find his way back to religious belief. Consequently, under Troland's direction, Murphy undertook a reading program in psychical research and assisted Troland in his telepathic experiments while working for his M.A. After two years in the army, Murphy returned to graduate school in 1919—this time to Columbia, which looked more impressive to him than Harvard since Hugo Münsterberg's death and Robert Yerkes's departure—and while working in psychology, maintained his study of psychical research by following a course of reading suggested to him by Isabel Newton, secretary and librarian of the SPR. In the summer of 1921 he made a three-week trip to London and read in the archives of the SPR; he returned greatly impressed by the society's evidence for survival, and in the fall he approached McDougall at Harvard.

It is difficult today to learn exactly what work Murphy carried out in the three years he was supported by the Hodgson Fund (1922-25), because he considered his researches to be unsuccessful and never published any of them. Nevertheless, it is clear enough what Murphy hoped from the beginning to accomplish: to concentrate his training as a psychologist upon the isolated problem of experimental telepathy. "I thought," he said subsequently, "that getting telepathy under experimental control would be relatively easy: I thought the scattered, incoherent nature of the data was due to the amateurish, puerile, half-baked amateurism of the whole business."[63] He began, therefore, by searching for suitable subjects. In the December 1922 issue of the ASPR *Journal*, Murphy asked to hear from individuals who appeared to have the gifts of "telepathy; clairvoyance; clairaudience; premonitions; coincidental dreams; apparitions; automatic writing or other automatisms having supernormal features."[64] He expressed his willingness to visit subjects within a hundred-mile radius of either New York City or Cambridge. These preliminary inquiries probably resulted in the brief account of experiments with some forty-one subjects that Murphy referred to in a letter written a year and a half later. He did not say where he had conducted his testing—in Boston or New York—but the latter is perhaps more probable. In these experiments, Murphy used pictorial material for telepathic transmission. The agent and the percipient each had a group of fifty pictures; the agent selected one of them for transmission, while the percipient chose the one he had mentally "received" from his pack and noted down the reasons for his choice. Variation in the experimental context was employed to elicit the psychological conditions for success: multiple agents and percipients were used; long-distance testing was tried, together with extended periods for response; and an attempt was made to vary the emotional tone of the experimental setting.[65]

It is perfectly apparent from these materials that Murphy was organizing his studies entirely differently from both Coover and Troland. His method was not simply that of the academic experimental psychologist; indeed, he had regarded Troland's elaborate mechanism and reflex-arc theory as being much too simplistic to deal with psychical phenomena. Murphy was concentrating upon the more spontaneous material, and in particular on the psychology of the telepathic situation—that is, he was trying to understand what psychological factors contributed to the success of a telepathic transmission. The available literature seemed to show only that telepathy occurred, not what the psychological process was that caused it to take place.[66] In this respect Murphy's approach could scarcely be more different from Troland's, which treated telepathy simply as a problem in information transfer. Rather than as a simple mechanical process, Murphy interpreted it as evanescent, complex, and psychologically obscure.

Overall, these experiments yielded little concrete information. What was most startling—and disappointing—to Murphy was that there appeared to be no improvement in telepathic ability with continued practice. Two of his subjects did very well; the other regular ones did "well enough," the average score for all being one correct guess in five. The only psychological correlation Murphy was able to establish was that a state of abstraction to the point where the agent (or percipient) felt himself virtually to have left his body seemed to enhance guessing ability; otherwise the matter remained in doubt.

During his first year of Hodgson support, Murphy became increasingly familiar with the strengths and weaknesses of various experimental strategies, working out his own favored approach. In 1923 he was sent as the ASPR delegate to the Second International Congress of Psychical Research, held in Warsaw, where he was able to talk with H.I.F.W. Brugmans. On his way back, he stopped in Paris and met René Warcollier, whose approach to experimental psychical research came eventually to influence Murphy's with especial force. Perhaps encouraged by the success already reported by Usher and Burt in transmitting sense-impressions between London and Prague, the two arranged to carry out a series of long-range experiments across the Atlantic. Warcollier directed one group of participants in Paris, and Murphy directed his New York group. From 1923 until Murphy became ill in March 1925, there were thirty-five sittings, in which the Americans served twenty times as percipients and fifteen times as agents.[67]

The features of this transatlantic experiment are illustrated well in the correspondence between Murphy and Warcollier concerning a transmission of 17 May 1924. An American audience in Boston was asked to concentrate on Hamlet's phrase "Oh, that this too, too solid flesh would melt," with its accompanying emotion of distress and thoughts of suicide. Three of the Paris group obtained what Warcollier termed "curious coincidences." One (who knew only ten English quotations) wrote, "To die, to sleep, to dream, Hamlet, with an accompaniment of the memory of the scene on the platform at Elsinore." A second wrote, "Life is an apprenticeship, to fame or sorrow." Warcollier himself saw "the image of a sword in the form of a cross, with the idea of a cold, sharp blade." In the fifteen experiments in which the Americans were the agents, there were five such coincidences; in the twenty conducted in the opposite direction, there were also five coincidences.

In this example we get some insight into the advantages and disadvantages involved in the attempt to transmit qualitative or pictorial material. The most marked disadvantage is the extreme difficulty of judging the degree of closeness of identity between the agent's thought and the percipient's response and hence of evaluating a series of pictorial targets. The advantage over simpler neutral symbols—geometrical designs, say—is that if one assumes that actual communication is taking place, the pictorial target can allow the investigator to draw some conclusions about mental operations, about the association of ideas, and about the symbolic form in which emotion can be expressed; for this reason it became Murphy's first choice for experimental material in later years. In this case, as it happens, the experimenters were "doubtful whether to classify the present material as further evidence of telepathy."

A second type of long-distance telepathic experiment carried out by Murphy during his Hodgson Fund tenure involved utilizing radio broadcasts. Eugene F. McDonald, who had just founded the Zenith Radio Corporation and was already much interested in psychical research, proposed the idea to Murphy;[68] he carried out a first set of such tests in Chicago early in 1924 (whose results were never reported in detail), and another set a few weeks later, in which J. Malcolm Bird of the *Scientific American* collaborated. In this second set of experiments the radio audience was told that they were to guess a total of twelve different items.

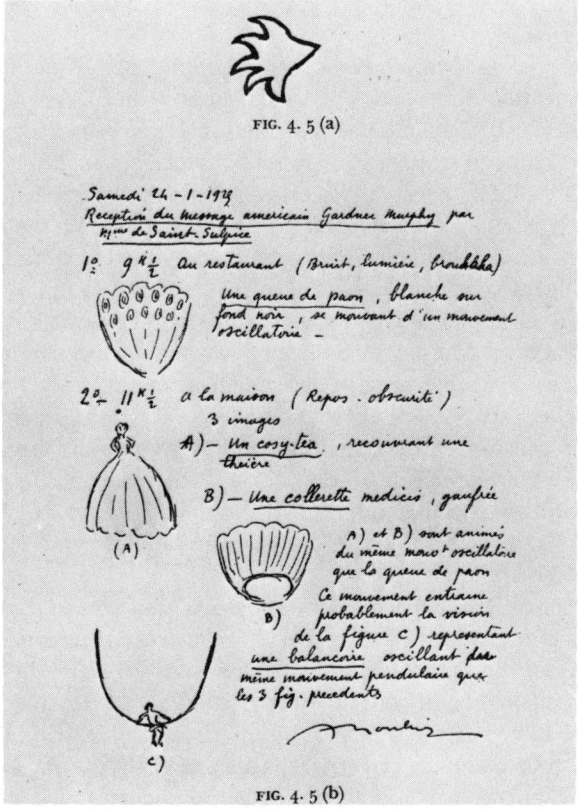

The responses of a Parisian experimenter (Fig. 4.5b) in René Warcollier's circle to the diagram (Fig. 4.5a) "communicated" by Gardner Murphy's New York group on 24 January 1925. (Courtesy of American Society for Psychical Research.)

These included a number between 1 and 1,000 (522), an American state (Indiana), four stick sketches of a person in various positions or activities, a specific object held in Bird's hand (a key), and a picture of a sporting event (a cross-country run). The point of the radio broadcast was to reach a large pool of subjects, not to test the efficiency of radio waves for psychical communication.

Some 457 listeners responded to the test. Their answers revealed interesting psychological features of tendencies in guessing—such as favoring the number 999 or naming either New York or a neighboring state, on the one hand, or a state remote from New York, such as Florida or California, on the other—but they were not encouraging as substantiating telepathy. Once again we can see in Murphy's experimental work great sensitivity to the psychology of guessing; some interesting insights into the psychology of particular responses; and no positive results. It should be added that in these radio experiments the value of a statistical evaluation of the guesses was considered, but the difficulties introduced by the recognizable patterns of guessing were judged too great to make the evaluation possible.[69]

During these years, Murphy was gaining confidence in the reality of telepathic communication from reports of spontaneous and experimental cases, even though he himself could claim no convincing success in producing the phenomenon. But, as he came to realize, in order to convince a scientific audience of its genuineness, it would be necessary to bring the psychological and physiological factors involved gradually under control, so as to achieve a replicable experiment. The rare gifted individual would have to be culled from the population at large, and stringent conditions would have to be imposed upon his laboratory testing, not all at once but little by little, keeping him interested and cooperative at all times. Moreover, Murphy recognized that a satisfactory test material had to be identified, one "capable of accurate statistical measurement, and a method which could be repeated at will under controlled conditions, capable of observation by any person properly equipped and willing to give the time." For all his study, however, Murphy was unable to decide which sort of material made for better experiments—objective targets, like the letters and numbers used by Brugmans and Coover, or subjective impressions, like the pictorial material he had used with Warcollier.[70]

Murphy's attention to experimentation did not lead him to neglect the study of promising mediums during these years. For example, in early 1923, at the outset of his Hodgson tenure, Murphy undertook to examine a medium who "talks in the personality of an American Indian, who states that he died at Winnipeg some forty years ago, and at times speaks fluently an incomprehensible dialect which he claims to be of Athabaskan stock"; Murphy approached the anthropologist Franz Boas at Columbia for help in deciding whether the dialect was genuine.[71] (Boas's conclusions have not survived.) When the *Scientific American* announced its prize for proof of psychical ability that same year, Murphy was named as an alternate to McDougall on the committee that was to judge the claimants. He was described, rather generously, as "not merely a colleague but actually a close associate of Dr. McDougall" and as "in admirable position to serve as a sort of liaison officer between the Boston and the New York members of the committee."[72] While Murphy never figured in the subsequent rabid controversy over Margery, he did have some sittings with her. The psychologist Harry Helson has described how, when a graduate student at Harvard, he acted "as assistant to Murphy who was Hodgson Fellow in psychic research" and used to visit McDougall "to talk about our last seance with Margery or about some medium I had investigated or about the phenomena at our last table-tipping session in the laboratory."[73] Murphy seems soon to have become *persona non grata* to Margery, but he had two years of occasional sittings with Mrs. Piper, then near the end of her career. These, too, were disappointing; it was his impression that her powers had greatly failed.[74]

McDougall's designation of Murphy as the second beneficiary of the Hodgson Fund had the incidental effect of giving the young man in his late twenties an unusually prominent position among American psychical researchers. As we have seen, he was the official American delegate to the Second International Congress in 1923, and he continued to serve as one of four delegates on the American committee. He played an unofficial but apparently quite important role in mustering support for the Boston group in the disaffection following Frederick Edwards's accession to the ASPR presidency. He was also named to the board of trustees of the Boston society when it finally came into existence in March 1925.

That spring, however, Murphy began a long siege of ill health: complications following influenza caused intermittent periods of invalidism and even blindness that persisted until 1934, and during that time Murphy withdrew almost entirely from psychical research. Curiously, however, the illness coincided with the efflorescence of his career in psychology. He had of course always had a foot in both camps; he had received his Ph.D. in psychology (1923) and had begun to teach in the general-studies program at Columbia while simultaneously doing research under the Hodgson Fund. Before 1925 he had published no more as a psychologist than as a psychical researcher; yet by 1929 he had written and seen into print his first two major works, his *Historical Introduction to Psychology* (which, with E. G. Boring's *History of Experimental Psychology,* published in the same year, became standard in the field) and his *Abnormal Psychology*. He had also developed new courses at Columbia and had begun to direct graduate research.[75] Conceivably this emphasis upon psychology at the expense of psychical research developed from a sense of futility over the difficulty of scientifically defining psychical abilities and experiences. In his career as "Hodgson Fellow" Murphy had witnessed some striking individual cases that seemed to point to psychical abilities, but he had not succeeded in developing a satisfactory program of experimental research. He had come to recognize, in fact, that "the complexities of the data and the uncertainty of almost every conclusion you reached" made experimental telepathy a far less simple problem than he had imagined three years before.

Although he had been associated with psychical research for many years, William McDougall had relatively little experimental experience in the field, and he may have had unrealistic expectations for Murphy's work.[76] In his recollection ten years later he was certainly a little unfair when he wrote that

> we made him [Murphy] Hodgson fellow at Harvard. His home, or his mother's, was at Concord, Mass., only 20 miles away; & we naturally expected him to make Harvard his headquarters &, as far as possible, his place of work. He first obtained our consent to spending half of each week at Columbia; & then spent practically all his time there, merely dropping in to see me for a few minutes occasionally. If he had settled down to work at Harvard, I think he might have got results, & have held the fellowship practically as long as he liked, & have acquired some useful psychology also. I am writing from memory & what I say may be inaccurate; but roughly it is correct—& the upshot for Harvard and P.R. was *nil*.[77]

But whatever disappointment McDougall may have felt in 1925 at Murphy's lack of success, it did not discourage him from taking Murphy's advice in making new disposition of the Hodgson funds. George H. Estabrooks, a graduate student in one of McDougall's courses at Harvard, had become interested in Murphy's work and had read some of the SPR publications. It was at Murphy's recommendation that McDougall arranged for Estabrooks to receive a fellowship of one thousand dollars to support his work in psychical research during 1925/26 while he was completing his doctoral dissertation. Estabrooks's work was very much in the tradition of Coover and Troland, with whose studies he claimed to be familiar. Like them, Estabrooks took as the hypothesis to be experimentally tested that telepathy was a reasonably widespread ability of otherwise "normal" people, and like them, he used an experimental situation conducive to "objective" assessment and testing material that could be evaluated statistically. Also like them, and in line with his hypothesis that telepathy was a normal ability, he pooled his data for statistical treatment without differentiating between his subjects' scores and indeed without indicating what they were. The result was the most influential series of experiments in American psychical research before 1934.[78]

Estabrooks's experiments were carried out in the Harvard psychology laboratory (then in Emerson Hall), in a room that was in fact double: an inner room shut off from an outer one by double doors. The percipient was placed in the inner room, while the experimenter and the agent(s) were in the outer room. The two rooms were connected by a signaling device comprising a clockwork apparatus in the outer room and a telegraph key in the inner one. The key was programmed to give a double click at twenty-second intervals once the mechanism was activated. The percipient was instructed first to number a piece of paper from 1 to 21 and then to note down the first playing card that he thought of when he heard the double click of the key, doing this twenty-one times. Meanwhile, the agents in the outer room shuffled and cut the deck in such a way that they were concentrating on a card at the moment when the double click signaled the percipient to make a guess. The percipients were told to remain as unselfconscious about the experiment as possible. Three series of experiments were carried out: in the first, the subjects were Estabrooks's personal friends, while in the other two, students drawn from McDougall's psychology classes were used. The results were in sharp contrast to those reported by Coover and Troland, for in each of the three series Estabrooks reported successful guessing of color and suit of a high order of significance: in his best series, the first, he found that the chance likelihood of obtaining so high a number of successful guesses was 1 in 19,200.

Estabrooks did not use the elaborate technique of introspection employed by Coover, by now quite old-fashioned, but he was far more sensitive to the possible psychological factors influencing guessing than Troland appears to have been. This led him to look for correlations between scoring and test conditions, attitudes, and so forth. Thus the fact that the first series was the best seemed to him to be best explained by his personal friendship with the subjects and by their ignorance and unselfconsciousness about the experiment. This and other correlations led Estabrooks to argue that the most favorable test conditions were those that encouraged a "quick and friendly rapport with the subjects and put them at their ease as to the experiment,"[79] while not departing from careful control.

Certainly Estabrooks looked upon his experiments as preliminary investigations meant to establish that telepathy took place, not particularly to establish what laws governed its appearance. Nevertheless, he did note a number of relationships even more specific than the general correlations just mentioned. Most striking, and most suggestive for later researchers, was the "fatigue" effect. This was his observation that the guesses made in the first half of a sequence of twenty-one yielded higher positive scores than did the guesses made in the second half. Indeed, when the guesses for each series were examined in this light, it was revealed that the second-half guesses were significantly successful only in the first series, and Estabrooks noted that a rest had been given the subjects in this series after their first ten guesses.[80] His second observation emerged from a retesting of the subjects of the first series in two widely separated rooms. The results of this test over greater distance were of no positive significance; indeed, the results of suit-guessing were decidedly below chance expectation—significantly so, had Estabrooks (like Troland) been interested to comment on below-chance results. This failure Estabrooks ascribed tentatively to the greater distance involved.[81]

It says a great deal for McDougall's influence at Harvard that Estabrooks's work was treated with serious consideration by at least one other psychologist there, Edwin G. Boring. By the time Estabrooks had written up his experiments in June 1926, McDougall had left Cambridge on a year's round-the-world trip, and Estabrooks therefore submitted a draft of his paper to Boring for criticism and apparently advice in regard to possible publication. It was a natural move,

George Estabrooks's Harvard experiment of 1925/26. On the left, the experimenter or "operator" (in this case apparently Estabrooks himself) records the card on which a "sender" is concentrating; on the right, a percipient records his guess every time a timing apparatus gives a signal. (Courtesy of American Society for Psychical Research.)

for Boring (who had come to Harvard from Clark in 1921) was the leading experimental psychologist at the university, head of its psychological laboratory, and an editor of the *American Journal of Psychology*. In view of Boring's professional standing and of his future role vis-à-vis parapsychology, his long and very careful response deserves special emphasis. Boring agreed without hesitation that Estabrooks's findings were "interesting and . . . worth publishing from the point of view of the scientist," though he raised some questions about the interpretation that Estabrooks had placed upon them and made some suggestions for changes. Furthermore, he offered to accept the paper for the *American Journal of Psychology;* he warned only that he would defer a final decision until he could judge the consequences of an exposé of the medium Margery written by Walter Franklin Prince, which was about to appear in the journal.[82]

No doubt Boring was particularly open-minded towards psychical research at this time because of the interest aroused in Boston by the Margery case. He himself had had a few sittings with Margery. But he was by no means inclined to accept the field and had just taken the trouble to write a popular article showing why psychical phenomena were impossible or at least unverifiable on philosophical grounds.[83] Prince's savage critique of Margery may well have been accepted by the *American Journal of Psychology* at his behest.[84] It is therefore of considerable interest that Boring appears to have questioned, not so much the experimental results themselves, but their statistical treatment and the interpretation given them by Estabrooks. And most remarkable in the light of the traditional reaction of American psychologists to psychical research is of course his markedly positive attitude towards Estabrooks's paper; it is difficult to imagine Titchener or perhaps even Jastrow agreeing that research of this sort deserved publication.

Estabrooks reworked the paper in accordance with Boring's suggestions during the fall, but he showed the revised version first to Walter Franklin Prince, perhaps hoping for more ardent approval from the research officer of the Boston SPR. The consequences were enormous. The Boston society had not managed to win a large base of membership since breaking from the American SPR in 1925, and Prince had been carrying it virtually by himself. He was always anxious to find people besides himself who could contribute publishable materials, and when he saw Estabrooks's report, he evidently promised it immediate publication as a bulletin of the BSPR. Estabrooks found the prospect alluring, and when he sent the manuscript on to Boring in January 1927 he raised the possibility of dual publication. Boring replied with considerably more asperity than he had shown six months before. He agreed to publish the bulk of the manuscript in the *American Journal of Psychology*, but only if it were withdrawn from the Boston society, and he warned that it would be set in only 8-point type, "because the results do not seem to me to be very important, for the reason that they prove nothing."[85] Perhaps Boring's tone had something to do with Estabrooks's subsequent decision to give his paper to Prince to publish, or perhaps the temptation to have the paper reprinted uncut was instrumental; in any case, it appeared in February 1927 as Bulletin V of the Boston Society for Psychic Resarch. The opportunity for an article sympathetic to telepathy to appear in a major psychological journal had been declined.

If William McDougall had been in Cambridge during 1926/27, he would surely have encouraged his student to publish in the *American Journal of Psychology* instead; but McDougall, who was at this point in Italy, had decided in any case to abandon the Boston locale for good: during the late fall of 1926 he had announced his intention not to return to Harvard but to take a position at Duke University in North Carolina for 1927/28.[86] His motives for leaving, as expressed to the departmental chairman at Harvard, mingled his wife's poor health with his own advancing age,[87] but the strongest motive was probably the increasing discomfort and intellectual isolation that he felt within the Harvard department; the last straw (which he did not reveal to the chairman) had been his discovery that the university had kept him ignorant of a committee planning new directions for the course of Harvard psychology.[88] With McDougall's departure from Cambridge, the tradition he had initiated of applying the Hodgson Fund constructively to experimentation on psychical phenomena and making Harvard a center of psychical research came to an abrupt end.

Nothing McDougall had experienced at Harvard, however, had weakened his conviction that the universities were the proper home for serious psychical research. He made this plain in a deeply moving paper read to a Clark University symposium on psychical research held in the late fall of 1926. The symposium had been arranged the year before by McDougall, the magician Harry Houdini, and the Clark psychologist Carl Murchison. It was supported by the fund established at Clark in 1908 to subsidize psychical research, hitherto virtually unused.[89] The whole range of contemporary psychical research was represented:

LeRoi Crandon (Margery's husband), Walter Franklin Prince, and Joseph Jastrow attended from the United States (among others); Arthur Conan Doyle, F.C.S. Schiller, and Hans Driesch came from abroad; and additional papers were read from figures who could not be present—McDougall, Gardner Murphy, J. E. Coover, and Oliver Lodge. McDougall's paper, which dealt with "Psychical Research as a University Subject," was one of the few that transcended the frankly polemical. He made no secret of his personal opinion that the evidence for telepathy and survival was particularly strong and that substantial cases could also be made for clairvoyance and for some of the physical phenomena of mediumship, but it was not on these grounds that he argued for the inclusion of the field within the university. Whatever the outcome of a program of serious scientific inquiry into psychical research might be, whether psychical abilities be proven to exist or not, the very process of investigation would still be beneficial— providing an unmatched training in moral and intellectual discipline; revealing the potential and limitations of science as applied to the problem of mind; bringing some order and direction to a chaos of popular credulity and enthusiasm. "The whole civilized world increasingly becomes the scene of a confused welter of amateur investigation, of conflicting opinions, of bitter controversies, of sects and schools and parties, each confidently asserting its own views and scornfully accusing the others of error, and of woeful blindness or wilful deception."[90] McDougall politely avoided any more explicit reference to the work of the amateur societies that had previously been the centers for psychical research, but the memory of his own bitter experience with the ASPR must still have been clear in his mind: "Nowhere else [but in the universities] may we hope to find the calm critical temper of scientific enquiry sufficiently developed and sustained; to no other institutions or associations can we hopefully entrust the task of shedding the cold clear light of science upon this obscure and much troubled field of vague hopes and vaguer speculations."[91]

At just the moment when this credo was being delivered—early December 1926—McDougall was making known his decision to leave Harvard for North Carolina. It can scarcely be doubted that among his hopes for the future was to establish psychical research as a truly scientific subject at Duke University. The following summer he was joined at Duke by a biologist turned psychical researcher, J. B. Rhine, who would help bring that hope closer to reality.

CHAPTER FOUR
A Career in Psychical Research
—J.B. Rhine

I

To any student of the process by which psychical research moved from an amateur to a professional scientific footing the early career of Joseph Banks Rhine is bound to be of particular interest. It was around his person that the academic structure of the new field—parapsychology—began to take shape. Moreover, the paradigm for experimental work in the field was established by Rhine's 1934 book, *Extra-Sensory Perception,* which was the fruit of his first exploratory researches and readings. Rhine's early career is of interest not only for itself, however; it also exemplifies the way in which intellectual and social forces were at work upon many men in the 1920s, causing them to turn hopefully to psychical research as one field that might permit a reconciliation between science, value, and the autonomy of the human mind.

J. B. Rhine grew up in the communities of central Pennsylvania and northern Ohio, through which his family moved irregularly in search of farming opportunities at the turn of the century. His father had hoped to go to college but had never been able to do so, and he had similar hopes for his children. J. B. Rhine's own career was to be aimed at the Protestant ministry. In 1910, when he was 14, the family moved to Marshallville, Ohio, where he became friends with Louisa Weckesser, slightly older than he; the two found common ground in religious questions, for she was in the process of disengaging herself from her own family's strict Mennonite convictions. In 1915 the two went off to nearby colleges—Louisa to Wooster and Rhine to Ohio Northern University. The next year, Rhine too went to Wooster, attracted by its courses in religion, but he remained there less than a semester; like the young Gardner Murphy at Yale at almost the same time, Rhine was precipitated into an intellectual and religious crisis by his college studies—in particular the sciences—and abandoned school and his plans for a ministerial career in favor of travel to take stock of life. His restlessness coincided with the entry of the United States into World War I, and he enlisted with the Marines, emerging a sergeant in 1919.[1]

Over the next several years Rhine prepared systematically for a future in science. He was attracted by the idea of a life in forestry and so turned to botany as preparation for that career. Meanwhile, Louisa had followed her botany teacher from Wooster to the University of Chicago, where she received a B.S. in botany in 1919. She and Rhine were married in 1920, after her first year in the graduate program at Chicago (in plant physiology); he completed his undergraduate degree at the same school two years later and continued on to graduate work there (also in plant physiology) under the direction of Charles A. Shull. While J. B. studied the transport and metabolism of fats in plant seedlings, Louisa did research on the enzyme catalase. In 1923 both Rhines received graduate degrees—Louisa a Ph.D.,[2] her husband an M.S.—and that summer they took research positions at the Thompson Institute for Plant Research in Yonkers to work on various experimental problems in plant chemistry. Within a year, however, J. B. was offered and accepted an academic post in botany at the University of West Virginia in a department headed by Louisa's former teacher at Wooster, P. D. Strausbaugh. After his first year (1924/25), Rhine's future in botany was assured. He was well liked by his department, and Strausbaugh expressed great relief when Rhine turned down an offer from the University of Iowa.[3] He successfully defended his Ph.D. dissertation at Chicago in the summer of 1925; and he had a number of research papers at different stages of publication.[4] The scientific career towards which he had finally aimed his training was his.

Nevertheless, for several years the Rhines had been considering something more than orthodox scientific careers. When J. B. had abandoned thoughts of a religious vocation, he had at first found a surrogate in scientific mechanism. In the interval between his leaving Wooster and his arrival at Chicago he had read Ernst Haeckel's *The Riddle of the Universe*. His reaction to this argument for materialism was "a kind of worship, comparable to anything I had experienced in my period of response to the Christian religion." At Chicago, however, the Rhines read Arthur Mitchell's translation of Bergson's *Creative Evolution,* and Bergson triumphed over materialism. From this point on, their general outlook remained relatively unchanged. They would not turn away from science; whatever they would do would at least have its foundations in their background in science, for scientific method was an incomparably more effective tool for understanding than was mere philosophical speculation. Yet having wrestled so long with the problem of religious belief, they could not now simply abandon that problem, separately compartmentalizing their scientific research and their moral concerns. Scientific research into the material world, divorced from meaning and values, was incomplete; whatever their future work in science, it would have to lead them on to attack the fundamental questions with which philosophy and religion had traditionally toyed. As Rhine put it, they hoped "to do research on the rim of the great problem of the nature of life."

The Rhines were not alone in the early 1920s in feeling this tension between science and religion, which would become a national issue in the Scopes trial of 1925. At Chicago, Shailer Mathews, dean of the University's divinity school, organized a series of lectures on this very theme—many by sympathetic scientists of real stature, like Robert Millikan—which, however, the Rhines found inadequate to bridge the gulf between science and theology.[5] What struck them as far more exciting and as having real possibilities of being such a bridge was psychical research, brought to their attention first by the writings of Oliver Lodge and then, forcefully, in a lecture by Sir Arthur Conan Doyle in 1922. Doyle, who was making his first speaking tour of the United States, was then well into his crusade in support not merely of psychical research but of spiritualism as well; he was already becoming the despair of the "right wing" of the SPR. Of these internal divisions, of course, the Rhines knew nothing. They were moved by Doyle's sincerity and courage, and they found compelling the importance of spiritualism as Doyle presented it. Moreover, they were attracted by his argument that psychical phenomena could be subjected to scientific investigation and were impressed by the list of scientists Doyle could cite who had been convinced by their own careful inquiries. Doyle's own scientific creation, the famous Professor Challenger, would, in *The Land of Mist* (1925), become a prototype of the skeptical scientist convinced by the evidence, against his will, of the reality of a spirit world.

At the end of June 1923, therefore, with his M.S. degree freshly in hand and a research job at the Thompson Institute ahead of him, Rhine wrote similar letters to three people whom he knew to be involved with psychical research: Joseph Jastrow, William McDougall, and the new president of the American SPR, the Reverend Frederick Edwards:

> My situation, in brief, is this: I was, six years ago, a divinity student. I lost faith and went into science. My training has been largely physiological and plant chemistry, with an attempt at a broad scientific background....
>
> Having read Myers' "Human Personality" and some other impressive books on the subject I have been deeply impressed by these investigations and the tremendous importance they will be to the human race. If I can in any way further this research, I am sure it will be the best work of my life. I am fully prepared to devote my life, or rather, that part of it which is left after a livelihood is earned, to psychic investigation.
>
> I am writing to you now, at a time when I may make some choice of my future (as we use the word *choice*), to ask if there is any opportunity, known to you, for me to enter this field of research and leave the other behind? Do you know of fellowships or positions that would enable me to get some of the special training I would need and go on with investigational work?
>
> Can you give me the names of a number of reliable mediums in the city of Chicago?[6]

Rhine could not have chosen a more disparate group to ask for advice than these three men—an archskeptic, the ex-president of the ASPR, and the man who had just ousted him—and the letters he eventually received nicely reflected their different views. Jastrow was not one to encourage a young man to abandon experimental science for psychical research; he admitted only that for one "rightly [presumably skeptically] oriented" there was opportunity for important work, and he mentioned Coover and Troland as men to write to.[7] McDougall's reply, too, was discouraging if polite. The only fellowship he knew of in the field would be held by Gardner Murphy for at least the next two years, and he was pessimistic about prospects at Stanford.[8] Only Frederick Edwards was encouraging, which was natural enough in view of the ASPR's new populist orientation and its interest in establishing grass-roots research groups all over the country. He urged Rhine to take the job offered him and to earn a living at it while learning on his own to investigate mediums thoroughly and systematically, "and you will find that the time will come when there will be a demand for a man like you." [9]

Rhine followed essentially the course suggested by Edwards while at Yonkers in 1923/24. Edwards brought him into contact with the New York branch of the ASPR, just forming, with people like Titus Bull and Gertrude Tubby; he steered him away, understandably, from Walter Franklin Prince.[10] Rhine even attended one New York séance, at which he was introduced to Gardner Murphy. When the Rhines moved to Morgantown the next year, J. B. applied for membership in the ASPR; and in March 1925 he wrote to J. Malcolm Bird offering his services to the organization and inquiring anxiously about recent reports of division in the field.[11]

It was at just this time, of course, that McDougall, Prince, and other disaffected members of the ASPR had split off—over the issue of the society's generally unscientific credulity and in particular its stand, through Bird, favoring the authenticity of the medium Margery—to form the Boston Society for Psychic Research. Bird therefore must have seen the inquiry from a favorably disposed young scientist as heaven-sent. He replied to Rhine immediately, giving his own account of the split and offering him the opportunity to review the foreign periodical literature for the ASPR's *Journal*. Rhine had no reason to question Bird's good faith, of course, and he replied accordingly, saying "we feel certain that a big hindrance to progress has been removed by that Boston group. They are doing more good than they know"; and he expressed disappointment with McDougall's alignment. He also agreed to review German, French, Spanish, and Italian literature for the *Journal*, and he volunteered Louisa's services as well.[12]

Thus in the summer of 1925, unknown to his West Virginia friends, Rhine was in effect poised between two possible commitments, between a secure and straightforward future in botany and an admittedly insecure future in a field of potentially unlimited significance; a choice had to be made. In January 1926 he made the brave decision, writing privately to Bird that

> I have about decided to resign here this spring (in spite of assurance of raise in rank and salary) and go either to Chicago or Columbia to get thorough training in philosophy and psychology. While my botanical friends seem to expect me to make a successful botanist I am sure I cannot do so with such widely divided interests as I have at present. Moreover, the interest in psychic matters has greatly weakened the other, and I find myself drawn strongly to the point of giving my main attention to a study and interpretation (?) of these phenomena.[13]

He was ready to enlist—at least for a time—in psychical research.[14]

It was thus something of an acknowledgment of his concerns for Rhine, having made his decision to leave, to deliver two talks on psychical research to his West Virginia colleagues that spring—his first presentations in the field, covering the two principal areas of study. One, on telepathy, showed a familiarity both with the early English work on spontaneous cases reported in *Phantasms of the Living* and with more recent experimental research, Coover's report and Gardner Murphy's recent transatlantic experiments. The other, on mediums, from D. D. Home to the contemporary Franek Kluski, in particular vindicated Margery from "the misrepresentations she has been given by the riffraff of publicity seekers, such as Hoagland, Houdini, etc."[15] Rhine as yet had no very wide acquaintance with the literature of psychical research, and one should not expect these first talks to be marked by novel insights. Their real interest is in the evidence they offer of what was now the Rhines' principal concern, to evaluate the extent to which the empirical results of psychical research might provide support for eroding religious convictions. Rhine's draft for the introduction of his talk on telepathy is worth quoting at length:

> The supernormal has been the mainstay of our and other religions in the past; they owe their vitality pretty largely to that element. Today we are losing that element, we cannot but see it. Miracles, efficacy of prayer etc. are different—rather flat without any supernatural element—any definite spiritual world.
>
> Whether we want it to be different or not—it would be unpardonable for the scientific world today to overlook evidences of the supernormal in the world—if there are such. If there are any—these are the props and the only ones that can be rushed into the support of our collapsing orthodox theology. One *needs* only to look at the strenuous manner in which the physics of the *electron* and new *energy* and *matter* concepts are seized upon to see the *need* for evidence,—even possibilities and plausibilities, in the minds of the intelligent leaders of present day religion.
>
> Now that is only the *need*—Perhaps it will never be satisfied. Perhaps there is nothing supernormal. But if there is anything happening around us which has the appearance of not being explained on normal grounds, are we not bound to lay aside our prejudice if we have any and make an investigation? That is my excuse for bringing before you a subject which many of you consider debatable. However, among those who have studied it conscientiously for many years there seems to be no disagreement on the facts.

Organized psychical research can be dated, symbolically, from a conversation between Henry Sidgwick and his student F.W.H. Myers, one moonlit night in Cambridge about 1870, over the need to validate religious belief through the methods of empirical science.[16] The impulse that had moved the founders of psychical research had not lost its power in the intervening half-century.

To Chicago and Columbia as possible schools for retooling in philosophy and psychology Rhine soon added Harvard. William McDougall was a strong attraction, despite the Rhines' misgivings about his role in the new Boston society; the Hodgson fellowship was a further temptation, now that Murphy's tenure was over; but Margery's presence in Boston was even more of an enticement. They decided finally to attend at least the summer session in Cambridge and wrote immediately to the Crandons to ask for a sitting with Margery; an appointment was soon promised them for 1 July. They made no other preparation. There was no way that financial support could be assured, but the Rhines had been self-supporting from college on, and they were confident that they could manage for one year more even if Hodgson money were not forthcoming. Nor was there anyone in Boston of whom they felt able to ask advice, now that the Boston psychical researchers, in particular W. F. Prince and McDougall, had broken so sharply with the ASPR, to which the Rhines were attached. Nevertheless, on arriving in Boston that summer of 1926, one of the first things they did was to call at McDougall's home—only to find that he was leaving Harvard for a yearlong trip and was at that very moment en route to the pier with family and baggage. McDougall kept his taxi waiting for twenty minutes, long enough to give Rhine some hurried advice about whom to see at Harvard to apply for Hodgson money, and then was gone. The Rhines now no longer had the prospect of supervision and guidance as they began their studies in Cambridge.

There remained their eagerly anticipated sitting with Margery, in which they hoped to test their plans for a scientific study of claims for a spirit world. On the evening of 1 July the Rhines and six other sitters participated in a typical séance with Margery in the Crandons' home on Lime Street. How the other six reacted is unknown, but the Rhines were shocked to observe what were to their minds compelling signs that Margery's phenomena—primarily physical—were fraudulently produced. Rhine described what they had seen in an urgent and bitter statement, which they sent immediately to all the trustees of the ASPR:

> You may imagine our disappointment; I resigned my position at West Virginia University, pretty largely on the strength of my interest in Psychical Research, in order to change to abnormal Psychology and Philosophy. I came up here hoping to study the case because of the stand the A.S.P.R. has taken on it; I gave it the benefit of the doubt all the way along. And so when I found this poor bunch of tricks at the back of all these investigations and publications, it was a tremendous jolt; I am disgusted, not only with the case but with the attitude our Journal has taken on it, sponsoring it before the scientific world. The whole case is sure to crash some of these days and where will our reputation be then? We will be the laughing stock of the world for years to come![17]

The Rhines' disillusionment with Margery must have left them no little dismayed at the poor prospects for their future, but they did not abandon Boston or psychical research; their ties to their earlier life had already been broken.

The Margery fiasco was not without its positive effect upon Rhine's situation. For one thing, although his repudiation of Margery earned him the hostility of the left wing of psychical research, it gained him the favorable attention of the right. Rhine prepared an account of his sitting with Margery, intending it for the ASPR *Journal;* but when Bird and the Crandons began to accuse him of prejudice and treachery, the likelihood of its ever appearing there seemed remote. At the suggestion of Hudson Hoagland, the Harvard graduate student who had thought of writing his dissertation on Margery's mediumship before becoming disenchanted, Rhine submitted it instead to the *Journal of Abnormal and Social Psychology,* then edited by Morton Prince. The article was accepted, and it appeared in the January 1927 issue; Hoagland, who had urged Prince to publish it, stood the cost of mailing out extra reprints.[18] The Rhines' exposé attracted considerable publicity, enough to bring them immediately to the consciousness of psychical researchers generally and to stamp them there as sober, critical, professional researchers who belonged in the SPR / BSPR tradition. Walter Franklin Prince immediately became their ally in Boston, while ironically the man who had stimulated them to devote their lives to psychical research, Sir Arthur Conan Doyle, protested vigorously at their "colossal impertinence."[19] Margery herself remained as popular with her devoted supporters as ever.

Perhaps even more important to Rhine's future in the long run was the episode's effect upon his orientation towards psychic phenomena. For it helped lead him to treat as unsubstantiated and suspect what he labeled "the objective side of psychic phenomena"—that is, physical phenomena, such as table-tilting or the production of ectoplasm—and to concentrate his attention on mental phenomena, telepathy and clairvoyance, which were better suited to closely controlled testing and which seemed to him that summer to be "if anything, more assured in my mind."[20] This was a sharp restriction of the interests he had expressed at West Virginia. His new orientation provided another reason why he should have become allied with the "right wing" of psychical research. While it would be too much to say that the Rhines formulated concrete goals for an extended course of research during their year in Boston, there is some reason to think that the idea of studying telepathy and clairvoyance experimentally was at least planted in their minds then. That winter, indeed, they tried out several informal series of attempts at guessing playing cards, using friends and each other as subjects.

On the whole, the Rhines' first semester in Cambridge must have been a disappointing one. They found some minimal financial support from part-time jobs, but assistance from the Hodgson Fund was refused—in McDougall's absence, no one in the Harvard philosophy department was enthusiastic about subsidizing either an unknown plant physiologist or psychical research.[21] Nor was anyone left on campus doing actual experimental psychical research. Rhine managed to talk briefly with G. H. Estabrooks, in town for a day to collect his belongings, questioning him at length about his recent experiments, about whose success there had been much comment, and was considerably heartened by their meeting. Still, Estabrooks was now gone from Cambridge, teaching at Springfield College. In lieu of anything better, the Rhines' tentative plans were therefore for J. B. to work towards a Ph.D. in philosophy, taking a year of courses at Harvard and then finishing a dissertation during a year at Columbia and another year in Europe. To this end he enrolled in such classes as seemed to him to hold some promise of providing a broad understanding of man's nature—logic and philosophy with C. I. Lewis and Alfred North Whitehead, the psychology of sensation with E. G. Boring, and abnormal psychology with Morton Prince. The courses seemed promising enough at first, but they soon proved to have little relevance to Rhine's concerns. Whitehead was a charming person, but his philosophy was far too abstract to be useful. Morton Prince's course was "mostly junk on pseudo-psychic phenomena, but he is plainly taken in on the observations, even tho he emphatically denies there is anything more than the subconscious operative in such cases"; Rhine's increasing skepticism was affronted, and he began to skip Prince's classes. Of Boring's course Rhine wrote that it "has been thus far merely the *physiology* of sensation and I think will continue to be; the worst of it is, it is only part of the physiology. He is handsome, nice, etc. but I wanted psych."[22] What Rhine must have missed in Boring's course was the breadth of interest and imagination he had come to recognize as present in William McDougall's writings. At Yonkers the Rhines had read *Body and Mind*, a book whose sympathetic defense of animism had been of real value in confirming their tentative resolve to look further into psychical research. Then, before starting classes at Harvard, they had read McDougall's *Outline of Psychology* and *Outline of Abnormal Psychology;* the first work in particular reinforced their high opinion of his general approach, both for its anti-behaviorism and for its biological orientation. But McDougall himself, of course, was unavailable.

Finally, the Rhines were growing steadily more disillusioned with psychical research itself. They had taken it up under the impression that it had developed a body of evidence on the nature of man and life that deserved serious consideration, but they had not yet seen any convincing proof of supernormality in their studies. In the winter, they conducted a series of sittings with a medium, Mrs. Minnie M. Soule, for the Boston society; Prince had offered them this opportunity for experience and income. Mrs. Soule at least appeared to be honest, but she communicated nothing that could be confirmed: "She goes into trance and gives us a lot of mush about a lot of friends and relatives that we cannot find in our tree. We think it is bunk but we are getting $10 a week for doing it for the Boston Society for Psychic Research, and we will carry it on. Of course we may find it better later on, but we haven't any faith."[23]

In January 1927 they seem to have decided that their plans to merge psychical research with philosophy and science had been ill-conceived, and to clarify their intentions to themselves, they composed an assessment of their situation that shows just how close they were to breaking with the past and starting their lives over one more time:

> I came up here to study Psychology, Philosophy, and to see if there is anything to Psychic Research. Perhaps a real explanation of the enthusiasm we manifested was that we were escaping thereby from the narrow limitations of field to which we had affixed our interest. Naturally in our eagerness we held the subjects toward which we were directing ourselves with more hope than understand[ing] and yet there were certain well defined principles which claimed our deeper interest.

We hope[d] to find the solution for these somewhere in the general region of the three subjects mentioned. Our experiences of the half-year with Psychic Research have disappointed us leaving faint hope of genuineness.

Our experience with Philosophy has led us to expect little from its method that can be relied upon, much as it may comfort the soul (some souls). At any rate it cannot answer the problems that we are interested in.

Psychology remains, a subject of very great interest, a science whose method may be dependable and which may contribute to the solution of our problems, at least if anything may. But somehow we have changed. Having once cut loose from things we find it easier to change, and to change quickly, being constantly on the alert for leads and indications. We are finding out more about ourselves than we have ever been able to do before. I think too, we are tiring of chasing the Psychic rainbow or the Philosophic pot of gold.[24]

This did not necessarily mean that they had given up entirely the hope of finding any psychic phenomena to study. "There can hardly be the same end to all of it," Rhine wrote to Strausbaugh some weeks later, "especially the telepathy experiments in laboratory fashion. But we are going to see first. Then we have decided that we do not have much use for psychic research societies. It is a problem for the university, if it is a problem at all. And a university will take it up quickly enough if the phenomena are genuine. Harvard will, for one." But Rhine himself had no thought of staying at Harvard; he had already abandoned course work, and in any case he had just learned that William McDougall would not be returning to Cambridge, "having accepted one of the gilded offers of Duke University." Insofar as the Rhines had any plans, they were still to go on to Columbia the next year.[25]

It was E. G. Boring who indirectly caused the change in their plans that within six months would take the Rhines to McDougall's new department at Duke and that eventually resulted in a foothold there for parapsychology. None of this could have been foreseen by anyone, however. Boring had been approached that winter by John F. Thomas, a middle-aged public-school official in Detroit. Thomas had lost his wife suddenly in 1926, and her death had intensified his originally rather casual interest in psychical research: he had begun to participate in sittings with several mediums in the United States, including Mrs. Soule in Boston, and was planning to leave for London for still more sittings. He wrote to Boring in order to look into the possibility of eventually studying this mass of mediumistic material at Harvard, and Boring recommended to Thomas that he write to J. B. Rhine—Boring knew of Rhine's interests from his application for Hodgson Fund money, as well as from his recent article on Margery and the consequent furor. Thomas thereupon communicated with Rhine and suggested that they meet in Boston in April.[26]

The Rhines were sufficiently impressed by Thomas and his reports of sittings with Mrs. Soule and others to agree to help evaluate his records and prepare them for possible publication. It may not be easy to understand why, in the light of their hitherto uniformly disastrous experiences with mediums, the Rhines should now have committed themselves to such a project. The fact was, however, that they were impressed by Thomas's patent sincerity, his perplexity about the apparently veridical information he was being given, and his serious interest in studying the phenomenon of mediumship scientifically and systematically. And his evidence seemed good enough to convince them that the mediums *might* have communicated some evidential material to Thomas about his past life that the mediums themselves could not have known—that his data might require a super-normal (whether telepathic or survivalist) explanation.[27]

How the Rhines were to help Thomas was at first uncertain, for they could not begin work until they were settled again. But their plans to move to Columbia for 1927/28 were dependent on sufficient summer earnings, which grew less likely as the summer progressed. It was Thomas's optimistic thought that William McDougall might be willing to collaborate on his project that stirred Rhine in mid-August suddenly to wonder about working at Duke for a part of the coming year.[28] He had already arrived at the firm conclusion that scientific psychical research would have to be carried on in a university setting, and he respected McDougall's philosophical and psychological position enormously, though his adopted skepticism made him worry that McDougall "might be a little over-tolerant in Psychic Research."[29] Thomas pursued the idea eagerly. Within a week Thomas had received McDougall's agreement to a plan whereby the latter would supervise Rhine's work on Thomas's material at Duke for a semester, and Thomas so notified the Rhines, adding that he would be willing to pay their semester's expenses.[30] By the fourth week in September 1927 the Rhines were in Durham, North Carolina; it was to be a permanent home for them and for parapsychology.

II

Rhine's arrival at Duke, and the eventual association with McDougall that it brought about, was the first step in the professionalization of parapsychology. But at the outset Rhine had no thought that he would remain there permanently. Thoughout the summer of 1927 he had still had plans to work at Columbia, and it had only been John Thomas's intervention that had turned him south. When Rhine finally wrote McDougall, expressing both interest in Thomas's material and a desire to work in psychology, McDougall answered briefly that he could offer only sympathy and the resources of a new department, inquiring specifically whether Rhine hoped to do "purely S.P.R. work or general psychology."[31] During the ensuing year Rhine was in essence an honorary postdoctoral fellow at Duke: he audited some of McDougall's psychology courses, was given office space, and was allowed to order books on psychical research for the library, but he was being paid by Thomas to examine the sitting records critically and to help put them in order.

Beyond this project, the Rhines had discovered an area to explore on their own. Driving down to Durham from New York in September they had seen an account of a "mind-reading horse," Lady, who was supposed to have predicted the winner of the forthcoming Dempsey-Tunney fight. In December of their first year at Duke the Rhines arranged to go up to Richmond to observe Lady and returned with the tentative conclusion that her owner, Mrs. Claudia Fonda, could indeed communicate telepathically with her: Lady would answer unexpressed questions, directed at her mentally by Mrs. Fonda, by touching lettered or numbered blocks with her nose. They paid another visit to Richmond later in December, this time with William McDougall (who came away impressed), and made still other trips in January and April 1928. The Rhines were well aware of the studies that had already been made of apparently telepathic animals, *denkende Tiere,* and of how in most cases the animals could be shown to be guided by tiny movements and other cues given consciously or unconsciously by the experimenter. On the other hand, they knew too of some studies that seemed to have demonstrated communication to animals in situations where sensory cues were excluded.[32] They took care therefore to observe Lady's response to mental commands not only when Mrs. Fonda knew the answer Lady was to give but also when she did not, and from Lady's success in the latter cases as well as in the former they reasoned that the horse was not dependent upon cues from Mrs. Fonda.

Despite his earlier disappointments with Margery and mediums generally, Rhine felt that Lady presented a genuinely inexplicable situation, and he was anxious to report his findings before some other university "scooped" Duke.[33] McDougall acquiesced, recommending one further test "to satisfy the sceptics that there is [*sic*] no slight movements or signals of an unconscious order in our own bodies as we stand there 'willing' Lady."[34] In fact, the first public statement of the research came out in May, in a release from the news agency Science Service, whose managing editor, Watson Davis, had learned of McDougall's connection with it.[35] A formal paper was accepted by Morton Prince for the *Journal of Abnormal and Social Psychology* in August, and it appeared early the next year.[36] The paper was quite cautious in tone; it acknowledged the danger of unsuspected cues very frankly, reported carefully a series of experiments done under gradually tightened conditions, and concluded guardedly that "only the telepathic explanation, the transference of mental influence by an unknown process, . . . seems tenable in view of the results."

Rhine's interest in Lady's phenomena was ensured by his personal situation. John Thomas's material might be undoubtedly supernormal, but it could be explained as well by a telepathic as by a survivalist hypothesis; what was needed, Rhine recognized quite early, was some sort of experimental inquiry to distinguish between the two. He had not yet been given reason to hope that he could undertake this at Duke, however, and in the spring of 1928 he began to explore alternative possibilities for the following academic year, writing to J. E. Coover and Gardner Murphy, the only two men in the United States with some sort of academic connection with psychical research. Coover, characteristically, failed to reply, and Murphy wrote back discouragingly about the prospects for a career in the field. He himself, he wrote, had been quite unsuccessful at Columbia. What was needed was some sort of replicable experiment. "If anybody has the skill and good luck to get a definite controllable psychical phenomenon in the laboratory, so that anybody can come and see it and convince himself any time, then, I think, psychical research will reach its *1492* in the world of science. I think Brugmans could have done this if he had cared more about it, maybe Estabrooks too."[37]

This view agreed very well with Rhine's views as a winter observing Lady had helped to form them. Why would not replicable telepathic experiments with "dumb animals incapable of fraud" be just the necessary entering wedge, something that would justify the field and provide a further basis for the study of telepathy in man? Rhine wrote to Thomas: "[Gardner Murphy] thinks the time is not ripe. Thinks we need to get a case that will work for anybody in any laboratory first. Well, if we are not mistaken, the maturation process is progressing rapidly at any rate. For one thing I think Lady would fit the requirements, especially with a little more training."[38] Rhine had already been struck by conditions apparently necessary to successful telepathy in Lady's case as well as in cases reported in man—a relaxed, "trancelike" state and the absence of external distractions—which at least suggested the existence of a common psychological process.

It did not seem likely, however, that plans like these could be carried off at Duke. Rhine felt no assurance that he would be able to do independent work if he should stay. "Between the lines," he told Thomas, "I see that the Professor is the Big Baron in our research and that I should be something of a serf; and serfdom in research—And with his mind being more essentially philosophical in direction and mine trained more to laboratory procedure, perhaps we had better let well enough alone."[39] Thomas replied holding out the prospect of a job teaching at Detroit City College in the coming year; however, Thomas also raised the question whether he might not come to Duke and get a Ph.D. if McDougall invited Rhine to stay on. When in March Rhine received a firm offer from Duke for the year 1928/29 to do half-time research for McDougall and half-time precepting in psychology courses, he was torn, and he corresponded with both Thomas and Walter Franklin Prince as to the advisability of accepting. On the one hand, the work he would be doing for McDougall would not be in psychical research; he would be helping him continue his Lamarckian experiments. On the other hand, there was the possibility that given McDougall's longstanding convictions and Thomas's interest in working towards a Ph.D., the study of psychical research might eventually be institutionalized at Duke. Rhine finally decided to accept the Duke position, and he threw himself into a flurry of inquiry with reputedly telepathic or "talking" animals during the spring, while simultaneously continuing his work for John Thomas and beginning his initiation into the Lamarckian experimentation that would occupy much of his time for the next year and would continue to make claims on his attention for another three or four more.[40]

The Lamarckian experiments deserve extended consideration, both because they were Rhine's first experimental work with laboratory animals and because their methodology in certain respects anticipated that which he would employ in psychical research. McDougall had long felt that the Lamarckian hypothesis in biology was rejected principally because it lacked a mechanistic explanation, not because of any significant contrary evidence. He of course favored the Lamarckian position on theoretical grounds, since it would provide a biological analogue to his own hormic psychology. Consequently, upon his arrival at Harvard in 1920, he had begun a series of experiments in order to see whether empirical support for the inheritance of acquired characteristics could be provided. In their simplest and final form, the "tank experiment," he placed rats over and over again into a tank of water from which two gangways led out: one brightly lit and lightly electrified, the other unlit and not electrified. McDougall counted the number of times a group of rats was placed in the tank before they learned to escape by the dim gangway; he then inbred that generation and tested its descendants to see whether the new group of rats would learn more quickly to avoid the bright gangway. By the time he brought Rhine into the research, in 1928, he had tested thirteen generations and had seen their "errors" decrease markedly. This was the experiment, set up anew at Duke, on which Rhine collaborated from 1928 until 1933. Independently of McDougall, he too found a continuing decrease in the average number of errors committed by successive generations of rats.[41]

In many ways the Lamarckian work was an ideal introduction to research for an aspiring experimental parapsychologist. The two subjects were similar in a number of ways—it was not merely, as McDougall jested, that Lamarckianism and psychical research were equally dangerous to the reputation of the investigator.[42] The inheritance of acquired characteristics, like telepathy (or clairvoyance), was not something that could be precisely identified, mechanistically modeled; it was a true unknown, the evidence for whose existence lay only in behavior that transcended the presumed randomness of normality. This required McDougall—and Rhine—to come to grips with the problems of establishing behavioral "normality" in regard to the unknown factor and of measuring significant differences from such a standard of normality. As he got more deeply into the experiments, Rhine saw the need for statistical treatment of the data; in fact it may be that he saw this more pressingly than did McDougall himself, who evidently felt perturbed by the prospect of learning a complicated mathematics and beginning an experiment now ten years old all over again. At any rate,

Rhine's suggestions on this score were never followed; but he remained convinced of the necessity of statistical argument in certain areas of science.[43]

The Lamarckian work also made it clear to Rhine what one might look for in studying an unknown. McDougall had said that "in order to establish the reality of Lamarckian transmission it is not necessary to be able to define the nature of that which is transmitted";[44] this was the same positivist assertion that Charles Richet had made to spiritualists and psychologists alike. To those who understand twentieth-century positivism to imply the possibility of reducing all of science to the laws of physics, it will seem absurd to ascribe such a philosophy to McDougall. Yet this is not the only or indeed the most general sense of the term. More broadly characteristic of positivism has been its conviction that the concepts of the factual sciences must be reducible to common observational experience and that unobservables are at best useful fictions. McDougall would undoubtedly have accepted the latter, broader view (though not the former), arguing, however, that hormic psychology was a matter proven by experience. So, too, the fact of Lamarckian transmission could be demonstrated empirically, whether or not a mechanistic explanation should ever be given for it. It is further possible to devise experiments that will make more definite the conditions under which the unknown acts, though they may still say nothing about its essential nature. This was one of the tasks McDougall set Rhine, and it was in precisely this way that Rhine would eventually come to treat another unknown, extrasensory perception.

The Lamarckian work, beginning in the fall of 1928, coincided with Rhine's return to the study of Lady in an attempt to determine the factors conditioning her success at telepathy. Disconcertingly, when he visited her in December he saw immediately that Lady no longer displayed telepathic abilities, merely the knack of responding to physical cues given by Mrs. Fonda. He reported to McDougall that "it looks as if Lady has lost it, whatever it was, and has now a bag of tricks which anyone can see, and prevent."[45] An explanatory sequel to the first paper was promptly sent off to the *Journal of Abnormal and Social Psychology* and was published in October 1929.[46] The Rhines announced that the horse no longer entered the rather passive state in which it had had such marked success the previous year and that they themselves (as well as Mrs. Fonda) could now control Lady's actions by giving appropriate subtle cues; they concluded that a change must have occurred in the horse's abilities in the intervening months of regular public demonstration.

It has been all too easy for skeptics to declare that the Rhines' will to believe in psychic phenomena had let sensory cues go unobserved in the first tests and to deny that Lady was ever in the least telepathic. Nevertheless, this conclusion is not entirely satisfying. It neglects the fact that in 1927 the Rhines had still been emerging from a mood of skepticism and disillusionment with psychical research; it also overlooks the Rhines' alertness to the dangers posed by sensory cues and their conscious attempts to overcome them. The first set of experiments was by no means ironclad—the Rhines did not ever effectively rule out the possibility that cues might unconsciously have originated with themselves[47]—but similarly it is far from being as uncritical and casual as unfriendly critics might want to suppose.

Aside from working with Lady and continuing to organize and study John Thomas's mediumship records, Rhine pursued psychical research in 1928/29 by beginning systematically to read the literature of the field, which had not really been possible before he came to North Carolina. In Chicago, West Virginia, and Cambridge he had read whatever had come to hand—Lodge, Myers, W. F. Prince, Richet, the publications sent him to abstract for the ASPR—but he had never gone through the publications of the British SPR, although he had read in them. The dominant interest of the field in the early 1920s had been in mediums, and Rhine's familiarity with the English tradition of experimental work on mental phenomena was necessarily somewhat limited. But when McDougall's library finally arrived in Durham, and as the Duke collection began to grow, Rhine was at last able to read more closely in the English material, whose growing emphasis upon experimental telepathy helped solidify his own tendencies in that direction. He was particularly struck by Mrs. Sidgwick's 1928 review of Nea Walker's *The Bridge* (a study based on the "absent-sitter" technique of putting questions to a medium through an intermediary and absenting oneself from the sitting) and remarked on it to many friends.[48] The review insisted on the importance of studying telepathy experimentally, in order to understand how to exclude it from consideration when investigating a medium's claims to spirit communication, and reinforced Rhine's own similar convictions.[49]

By the spring of 1929, therefore, Rhine had come to command a far broader and deeper view of psychical research than he had had when he arrived at Duke. We are fortunate to have a summary of his perspective on the field in a "seminar" paper probably composed for delivery to a colloquium of psychologists and/or philosophers sometime that semester. The paper makes clear just how well acquainted the Rhines had become by mid-1929 with the assumptions and conclusions of European psychical research, showing (as they put it recently) "how the world looked before we had any experimental results."[50] The paper begins by presenting the "simpler" classes of phenomena, telepathy and clairvoyance. The former is well established from the work of Brugmans, Warcollier, and Estabrooks; it is a variable faculty that can be easily fatigued and depends upon passivity in the subject. There is some evidence that it is widespread, but effective in only a few individuals; there is also evidence tending to rule out an electromagnetic-wave explanation of the phenomena. Moreover, telepathy may involve, not communication through a medium, but direct mind-to-mind action, and hence would challenge "the defenders of the notion of entire-dependence-of-mind-upon-bodily-processes." Clairvoyance has been described in Ossowiecki's work and treated experimentally by Tischner and Ina Jephson; "her work yielded results comparable to those of Estabrooks on telepathy, and suggests that his may have been due to clairvoyance instead of telepathy.'' Indeed, Rhine went on, these two faculties may well be lumped together. The most economical hypothesis would attribute perceptive ability, whether of things or of other minds, to the mind of the sensitive and would draw no fundamental distinction between the two processes.

From these "relatively simple phenomena" Rhine went on to describe those of trance mediumship, beginning with the work of Mrs. Piper and passing on to the modern work with cross-correspondences and book tests. These and other tests seem to reveal some sort of agency independent of medium or sitter, something beyond telepathy and clairvoyance: "taking cases such as this as reported to us, we can hardly avoid a recognition of the spirit hypothesis as the most acceptable.'' The same is true for spontaneous physical phenomena such as occur in haunted houses and with poltergeist cases. For Rhine had at last become willing to accept the survivalist explanation of mediumistic phenomena as a useful working hypothesis. This willingness was no doubt aided by his growing familiarity with the impressive and solid literature of the field, but it was also partly due to his involvement with John Thomas's material, some of which at least he now felt was "genuinely evidential" and best explained by "the spirit interpretation."[51] He went on to remark that while the materialistic "French school" of psychical researchers had tried to explain all the phenomena by telepathy and clairvoyance in an agent or medium, there remained "the independent accomplishment of purposive actions by the communicator, after promising through a medium to produce them," which could only be ascribed to an external entity of some sort. Rhine concluded his paper with an acknowledgment of the problems that the spirit hypothesis entails—why are communications fragmentary only? why are they not straightforward in nature or addressed to persons concerned?—but treated these as problems to which answers would be forthcoming. If substantiated, the spirit hypothesis would quicken life with a greater interest, stimulate its higher standards and ideals, build up character and intellect, and, finally, fuse religion and evolution into a commonality.[52]

This seminar paper, which was limited to the structure and content of psychical research, illuminates only a portion of Rhine's hopes and ambitions in 1929. Another and at that moment perhaps even more important index to his thought is provided by a paper entitled "Experimental Religion," which he wrote at about this same time.[53] In this paper, Rhine made an attempt to resolve the problem that had been preoccupying him since his days at Wooster, that of finding a place for a spiritual element in a human nature increasingly comprehended by science. The weakness of religion, he argued, was that it rested on authority and inevitably collapsed before the modern spirit of investigation; consequently, to survive, religion would have to test its assumptions empirically in the spirit of modern science. This could scarcely be done by the bitter opponents or confirmed supporters of orthodox religion; nor by scientists professing religion, who would inevitably have compartmentalized the two subjects for good and all; nor by theologians who felt that science and religion must inevitably be in harmony—the best efforts of Shailer Mathews and H. N. Wieman had so far produced only literature. Instead, it was the open-minded scientists with no previous commitments who might hope to verify—or disprove—the claims of religion. Rhine called for an "Institute for Experimental Religion" to be founded, in which strict scientific method could test out the claims of supernatural power associated with religion: that is, test the efficacy of prayer, investigate the vital process in the organism for evidence of the soul, or study reports of apparitions or hauntings for proof of survival. Such an institute would be well worth funding, for even if the claims of religion for man's spiritual nature proved to be unsupportable, it would be better to *know* than to hope or to trust in authority. The paper never found its way into print, but it epitomized for Rhine deep convictions of long standing, and he often sent copies to like-minded friends and supporters in subsequent years.[54]

The academic year 1929/30 found Rhine with an intellectual framework within which to cast experimental research but with no time to do it, for McDougall had arranged for Rhine to be given a teaching appointment within the university. McDougall's arrival at Duke had involved the creation of psychology as a separate department out of the philosophy department, with which it had formerly been closely bound up (as was the common pattern in America), but his influence with the parent department was great enough to have Rhine appointed an assistant professor of philosophy and psychology, with the responsibility of teaching one course in philosophy fall and spring. One factor in the appointment was the real possibility that McDougall might soon be leaving Duke, and Rhine did not yet feel at home in academic psychology—his later assurance that experimental psychology would be the best framework for psychical research is quite markedly absent from a proposal he made to McDougall in early 1929 about courses he might teach during the coming year.

> I think I had better "make my nest" now rather than later, in the field of Philosophy, if I can and if you agree with me in this. While I am not appreciative of Philosophy as defined by many, yet I am devoted to it as defined in my own way—and what subject allows for so much personal latitude in defining it? . . . My disinclination to specialize in psychology would soon leave me unsheltered, whereas my own greatest interests coincide with a real need in the Phil. Dept., in a History of Science, and general science and orientation courses.[55]

In the fall of 1929 Rhine's independent teaching career began with this course he himself had proposed, a course surveying the history of science. Coincidentally, Gardner Murphy's *Historical Introduction to Modern Psychology* appeared in that same year. It is fascinating to find two future parapsychologists choosing to reflect upon the nature of the wider enterprise in which they hoped to join, even though psychical researchers had for nearly fifty years habitually referred to the history of science to reassure one another that in the end truth was sure to be recognized. Their two approaches to the subject, however, were rather different, for while Murphy dealt with the history of concepts and theories in a manner much closer to G. S. Brett than to Boring, Rhine approached science as Auguste Comte had, tracing out stages of scientific development from spirits to imponderables to laws and principles to processes and events and throughout paying very careful attention to the material and social factors conditioning scientific development. The spring semester involved Rhine in something rather different, a "Constructive Survey of Modern Science," also of his own devising, which was intended to correct the narrowness of students' overspecialization in one or two fields by providing a "synthetic working view" of the natural world.

In both these classes Rhine forcefully developed and refined the philosophy he had held since his arrival at Chicago ten years before. Science is a method of finding broad relationships in the universe through inference from data; it is a *certain* method, composed of observation, inference, and verification. History shows that to survive, science must be free, grounded in a democracy of thinking and education; autocracy has been a poor support, and the temple has been an enemy of science, for religion and science historically have been rival claimants to the true explanation of nature. Properly, religion should be a subheading of "science": the finding of personal, purposive forces in the universe by inference from experience. But if religion continues its traditional and authoritarian attitude, it is bound to persist in its current decline; while an unfettered science cannot fail to advance.

III

The spring and summer of 1930 was surely the decisive period for Rhine's future work; the linear progression of his research career begins, not in 1927, with his arrival at Duke, but three years later, when he had acquired the background necessary to plan out his own program of experimentation. There are signs that suggest that in the spring of 1930 Rhine was beginning to evolve a variety of ideas for projects to carry out. Animal telepathy and medium investigations, however, had now receded into the back of his mind. He applied to the University Research Committee in March for a four-hundred-dollar grant to make possible a survey of state orphanages and reform schools "in search of certain psychological types"[56]—no doubt (though he did not say so) to see whether any of these individuals might be psychically gifted. When it seemed that this proposal might not go through, he began to block out quite a different undertaking, an exhaustive history of scientific psychical research that would sort out truth from illusion.[57] Eventually the grant was made; but by this time Rhine had still other plans and was beginning to give serious consideration to the card-guessing experiments that would symbolize experimental parapsychology in most minds for years to come.

What made Rhine turn towards this sort of work at this particular moment is impossible to say. His participation in April in the SPR's attempted replication of the Jephson experiments may have had something to do with it,[58] or it may have been the stimulating news that G. H. Estabrooks (whose Harvard card-guessing experiments had become an often-cited model) would be coming to North Carolina to teach in the Duke summer school at Lake Junaluska.[59] Or it may be that Upton Sinclair's new book, *Mental Radio,* played a role. This was Sinclair's report of his wife's apparently largely successful attempts to reproduce telepathically (or clairvoyantly) drawings that he had made. His account is very much in the style of Warcollier or Tischner (he cites the latter's book at some length), and all the criticisms that Soal was then busy raising against Warcollier— impossibility of determining success, lack of controls—could have been applied to it. However, psychical researchers were in general not yet preoccupied with such difficulties. Sinclair gave William McDougall a copy of his manuscript in mid-1929, and McDougall responded with marked enthusiasm in an introduction to the book: "The experiments in telepathy, as reported in the pages of this book, were so remarkably successful as to rank among the very best hitherto reported."[60] Rhine would certainly have been impressed by the results as well as by McDougall's approval;[61] no doubt he would have been further affected by the generally favorable public response with which the book met when it was published in May 1930. At any rate, Rhine apparently gave over some of his 1929/30 classes to lectures on psychical research (John Thomas gave one) and experimental telepathy and led at least two of his students to try their own hand at such experiments. A rising junior, Charles E. Stuart, wrote enthusiastically that summer from his home in Rochester, New York, that he had carried out experiments à la Sinclair and that "mathematical probability would have to strain mightily to fit the successes"; Harvey Frick, a recent Duke graduate planning to enter the graduate school in philosophy that fall, carried on both picture- and card-guessing experiments—at close range and long-distance—that seemed to him to give encouraging results.[62]

In the summer of 1930 Rhine, too, began card-guessing work, and he continued it in the fall in collaboration with his colleagues in the psychology department. No doubt they were mindful of McDougall's support for the work, and while they themselves may have had no very deep interest or confidence in it, they did bring their special skills to bear upon the experimental situation. Helge Lundholm, who had come to Duke that fall to take over abnormal psychology, helped Rhine carry out experiments with subjects under hypnosis, trying to communicate with them telepathically; but these tests failed to produce strongly significant results. Karl Zener, who had been at Duke since 1928 and was concerned with experimental psychology, made far more important contributions to Rhine's methodology as it developed during the summer and fall.

> I began by giving "guessing contests" to some groups of children in summer recreation camps. The tests consisted simply in having each child guess the numeral (zero to nine) which was stamped on a card that I held concealed from him in my hand and

looked at. . . . From the thousand (approximately) trials thus made, no one individual stood out well enough to seem to warrant further investigation.

During the fall semester following my colleague, Dr. K. E. Zener, proposed that we try sealed envelope guessing tests on our own college classes. We accordingly prepared envelopes with numerals (or, in some classes, with letters of the alphabet) effectively concealed and sealed within. These were passed out to the students with instructions to guess the number (or letter) stamped inside, under certain conditions of quiet and relaxation. Of these trials 1,600 were carried out, also with quite insignificant results.[63]

In order to improve the target material, Rhine asked Zener (who had worked on the psychology of perception) to be responsible for designing a set of special cards with images that would be both easily distinguished and easily remembered. The result was what came to be known as the Zener cards, each bearing a circle, a rectangle, a plus sign, a star, or wavy lines, which Rhine and Zener used in decks of twenty-five (five of each kind). During the winter and spring of 1930/31 the two carried out 800 trials with these new cards and began to get positive results—207 of the cards were guessed correctly by the student subjects, instead of the expected 160. It was to experiments in clairvoyance with Zener cards that Rhine now began to direct his attention.

Rhine's attempts to bring his colleagues into collaboration in psychical research were not simply fortuitous but were in line with the hope he shared with McDougall of establishing the subject within an academic setting. In 1930, no doubt with this in mind, he had begun to introduce psychical research into psychology and philosophy courses; he had applied the professional expertise of his colleagues to psychical experimentation; and now he began to train his first graduate student. Harvey Frick had entered the graduate program that fall in the department of philosophy, of which Rhine was still formally a member, and had begun to work on a master's thesis on psychical research under Rhine's supervision. A. G. Widgery had been added to the philosophy department that year, however, and Widgery made it clear both that Frick's experimental work was not acceptable as philosophy and that Rhine would have to decide between philosophy and psychology.[64] By now Rhine had come to realize that his plans for psychical research would be better served at Duke in psychology than in philosophy, and with McDougall's approval, he began to teach psychology courses (the introductory survey and "Psychology of Reasoning") in the fall of 1931.

When Frick presented his M.A. thesis in May 1931, therefore, it was offered in psychology; W. I. Cranford (from philosophy), Rhine, and Helge Lundholm made up his committee. Rhine described the event to Walter Franklin Prince in the following terms:

We made a small step in the history of psychical research here in our department this year, by granting to a student a Master's degree on a thesis of purely psychical research material. His subject was "Extra-Sensory Cognition". . . , and his thesis consisted mainly of a historical review of experimental telepathy and clairvoyance of the non-mediumistic type, with reports on experiments he has himself made in telepathy. There are many weaknesses in it, it is far from complete, and the evidence produced by the author himself is not conclusive; but the dept. was satisfied with it. The M.A. requirements are of course not as high as we expect for ultimate publication.[65]

Rhine's judgment was dispassionate and accurate. The thesis was in large part a summary of the most prominent experiments in undifferentiated telepathy / clairvoyance over the last fifty years. Frick attempted little analysis of the material and no criticisms whatsoever of the methodologies employed, merely referring readers to the original papers; he himself took all reports at face value. He also included an account of the telepathic experiments he had made in the summer of 1930, reporting 15 of 108 as successful (though without either defining success or evaluating the significance of these results), and to this added a brief account of (unwitnessed) personal experiments in card-guessing, made "in order to render the results of experimentation mathematically calculable"[66]—yet while the results were identified as above theoretical and practical chance results, Frick made no attempt to calculate statistically his degree of success. For an M.A. paper by a first-year graduate student, however, Frick's thesis was by no means unusual or unacceptable. Rhine could take some satisfaction in it, despite its shortcomings as professional research, all the more since he felt the need to get psychical research established within the university. McDougall had long felt the same need and now shared Rhine's satisfaction; he wrote back from England on hearing of Frick's success, "My congratulations to Frick; it is indeed a real epoch marking event that Duke should give a degree for work in this field."[67]

In that same letter McDougall also congratulated Rhine on the mounting success of his experiments in clairvoyance, for the Zener cards had proved unexpectedly fruitful. A sophomore in the introductory psychology course, A. J. Linzmayer, proved to be able to guess the Zener cards at a rate markedly above what would be predicted by chance. In two weeks of experimental work at the end of the school year (May–June 1931) he guessed 404 of 1,500 cards correctly (300 would be expected by chance).[68] Rhine found Linzmayer's success particularly encouraging. It was not so much the demonstration of psychical ability that pleased him as it was the prospects that a consistently successful subject would offer for determining the laws and conditions governing the appearance of the phenomena. By 1931 English researchers were despairing of ever producing telepathic phenomena under sufficiently rigorous conditions and were concentrating all their efforts upon the task of demonstration. Rhine, however, did not share their obsession with strictness of conditions and in effect was ready to view much less skeptically the existence and replicability of the phenomena. This meant that he could proceed directly to look for regularities in Linzmayer's guesses; he commented on one or two such patterns in a letter to Prince early in July.[69] If (in the broad sense previously defined) we should understand a positivist psychology as one simply aiming to discover empirical laws relating human actions with other observables, divorced from any hypothetical causal or mechanical model of mind or mental activity, Rhine's approach must certainly conform. It was precisely laws and not explanations that he was looking for in the wealth of data now beginning to accumulate; a search for explanations could come later but was still premature.

During the year following the discovery of Linzmayer, Rhine recorded another thirty thousand guesses of Zener cards. Linzmayer himself came to Durham from his home in New Jersey twice, in October 1931 and March 1932, and in nearly two thousand trials continued to score above chance expectation, though at a gradually declining rate. Charles Stuart, who had graduated at the end of the fall semester 1931 and entered graduate school in psychology, had pursued his experimental work and during the year ran some 7,500 trials with himself as subject.[70] Finally, Rhine and J. G. Pratt (still another new graduate student in the department) came upon a divinity school student at Duke, Hubert Pearce, who proved to be another—and better—Linzmayer. He scored at about the same rate, with no indications of a decline in ability over time, and Rhine immediately discovered that Pearce's scoring appeared to be dependent upon much the same sort of physiological and psychological conditions.

By the spring of 1932, therefore, this program of experimentation had taken precedence over all the other projects Rhine had once had in mind. In late March he had applied to the research council for another two hundred dollars to carry on his studies (he did not mention that his direction of inquiry had shifted), which he explained were "not yet in a proper state for publication."[71] The grant did not appear, but of course Hubert Pearce did, and by mid-summer Rhine felt in a position to write up his results. McDougall had encouraged him to do so for some time and now wrote enthusiastically from England (where he was again spending the spring and summer) congratulating Rhine on his "striking results" and the article that could be expected.[72] In September Rhine sent a copy of the completed paper to Walter Franklin Prince in Boston, mentioning that he would be preparing a longer report that he hoped would be suitable for the Boston SPR to publish, even though his colleagues had suggested publishing it in the Duke monograph series on psychology.[73] Prince was utterly delighted by the paper Rhine had sent, and in any case he was avid for publishable material up to his own high standards. He had already hinted to Rhine that he should have submitted his work to the Boston society; now he pressed him to commit the future monograph to the society,[74] which Rhine eventually did.

IV

Presumably Rhine had thought to make his first statement in a psychology journal so as to prepare the way for the longer and less narrowly scientific discussion.[75] If this was his reasoning, however, it was disappointed. For some reason Rhine let his paper cool for nearly a year and did not submit it until June 1933[76]—again to the *Journal of Abnormal and Social Psychology*—by which time the longer and much more complete monograph was largely written. As a result, that monograph (under the title *Extra-Sensory Perception*) appeared well before the brief article, and it became the vehicle through which knowledge of the Duke work was first communicated to psychologist and psychical researcher alike (we shall shortly examine it in close detail). But the supplanted paper, "Extra-Sensory Perception of the Clairvoyant Type," is still of some interest because it shows very clearly what it was that Rhine was hoping to accomplish with his experiments in 1932.

Rhine saw the importance of the paper not simply as a presentation of data; supernormal phenomena had previously been reported time and again, as he well knew, always to be met with indifference or skepticism. But as he understood the problem, the skepticism had been due to the failure of psychical researchers to recognize any laws or relationships in the experimental evidence. If laws or curves (or, best of all, practical applications) could be demonstrated, the reaction of psychologists would, he insisted, be very different, and it was precisely such things that his research seemed to reveal—"natural relationships that would seem to be well within the reach of quantitative study."[77] The relationships he had in mind here were in part physiological—both Linzmayer and Pearce had lost their clairvoyant ability when given the drug sodium amytal—but the ones to which he devoted most attention were psychological. He commented upon work that seemed to reflect the volition of the subject, both conscious (Pearce could alternate high scores and low scores upon command) and unconscious (Linzmayer fell well below chance when forced to continue an experiment when inconvenient for him, which Rhine interpreted as indicating unconscious avoidance of success). And Rhine also particularly stressed the various performance curves he had found in different subjects: higher success on the earlier cards in each group of five guessed by Pearce, in a fashion reminiscent of the Jephson data; a "fatigue-curve," or decline in success, during intensively and tediously repeated experimentation by Harvey Frick; a loss of ability over time, seen in Linzmayer between 1931 and 1932.

> Whether it will be found that these curves (in general) are governed by motivational peculiarities, by the exhaustion of some function that is essential to extra-sensory perception, or by complex changes of the mental system as a whole, such as strain or distort or suppress this delicate process into the characteristic curve, in any case we will have to deal with interrelations of this extra-sensory function with the other processes of the psychic organization. We cannot escape this outcome.
>
> In a word, then, we have to face not only the fact of the actual existence of extra-sensory perception: we have to accept it as another normal process along with the rest—simply less understood than some.[78]

Any doubts that these ideas expressed in the 1934 paper were uppermost in Rhine's mind in 1932 should be dispelled by his joyful remark to Charles Stuart in July 1932: "I am quite enthusiastic about this drug treatment since it promises to give us the laws we so badly need in order that we can make the world of psychology safe for the psychic researcher."[79]

Why, then, did Rhine not proceed immediately to publish these results for which he had such hopes? No doubt a busy schedule of new courses in 1932/33 made it difficult for him to give the manuscript the necessary last-minute polishing; Rhine now was teaching psychology courses for the first time—the survey course plus a course on suggestion and hypnotism in the fall, courses on personality and on psychical research in the spring. In the last-named he took great pride: it was "simply a series of lectures, one a week, with a few demonstration[s]," he wrote to Upton Sinclair, "but it seems to me to have some historical importance as initiating the subject into the university routine."[80] It attracted twenty-two students, a considerable number for an upper-level psychology course. But Rhine became ill in the spring, and the course did not go as well as he had hoped[81]—and his illness also prevented work on the paper.

There was clearly more behind the delay than this, however. When he submitted the article to Henry Moore, now the editor of the *Journal,* Rhine explained, "I wrote this paper [last summer], and left it to lie a year in order to feel still more secure in its soundness."[82] In the paper itself he attributed the hiatus to "the interests of conservatism."[83] Apparently during 1932, as Rhine was making the transition from a philosophical to a psychological orientation and future, he was becoming conscious of the importance of meeting the expectations of psychologists and growing cautious about the expression of his conclusions. Significantly, it was during just these months that he was renaming the set of phenomena he had chosen to study. Rhine had introduced the term "extra-sensory cognition" for Harvey Frick's thesis the previous year, a broad term chosen "to throw the closed-minded critic off the scent."[84] In the summer of 1932, however, he told Upton Sinclair that he was studying "extra-sensory modes of perception."[85] To Prince he wrote in September 1932: "I am using the term "extra-sensory perception" just now, in order to make it sound as normal as may be. I want psychologists to recognize that it is a branch of perception, however different it may be from other branches."[86] "Cognition" was too broad for his purposes now; his experiments had been in *perception,* after all, studying a subject's reaction to an external object. And certainly perception was a subject that psychologists were accustomed to think of as an experimental field.

Something specific that may have alerted Rhine to the difficulties inherent in introducing psychical research into experimental psychology was the odd reception that John Thomas's Ph.D. dissertation had met from the Duke psychology department in that very spring of 1932. Thomas had taken a B.A. in 1898 and an M.A. (in education) in 1915 from the University of Michigan; he had formally been a Ph.D. candidate at Duke since 1930 and had spent parts of four years in Durham working with Rhine with McDougall's sanction. In the dissertation that he submitted—"An Evaluative Study of the Mental Content of Certain Trance Phenomena"—he set out in very full detail the sittings he had had with the

English medium Mrs. Osborne Leonard and others over the last eight years and, by judging the applicability to him of the information the medium had communicated, attempted to decide whether she had in trance revealed some supernormal source of knowledge. Whether telepathy-cum-clairvoyance or the spirit hypothesis was the explanation was initially left as an open question, at Rhine's suggestion, although Thomas had by now become personally convinced that his dead wife was indeed communicating with him through the various mediums.

It is no longer possible to be sure, but evidently two difficulties were raised by Thomas's committee in his final orals in May 1932, which resulted in their refusal to accept the dissertation. First, it was felt—in particular by a new member of the department, Donald Adams—that Thomas had not adequately proven statistically that chance coincidence might be eliminated as an explanation of the medium's successes: Were her remarks not so vague that they could apply to almost anyone?[87] Second (and far more remarkably), the committee apparently was not content to have Thomas merely say cautiously that the evidence indicated Mrs. Leonard to have supernormal means of acquiring information; instead, they wanted him to go ahead and argue the survivalist explanation. Rhine, who had confidently expected the final oral to be merely a *pro forma* exercise,[88] was understandably disconcerted by the department's rejection of the dissertation, particularly its objection to the more cautious approach towards supernormality. To Rhine, a policy of conservatism would have been tactically preferable, as he wrote that summer to W. F. Prince. Prince agreed with Rhine, but apparently most others, including McDougall, did not.[89] So Thomas set about rewriting his conclusion in order to show the strength of the survivalist interpretation, to which in any event he inclined.

Thomas had also to devise a convincing way of proving that an abnormally high degree of coincidence existed between Mrs. Leonard's comments and his own past life. Here he settled on a variation of the technique described by H. F. Saltmarsh in the SPR *Proceedings* for 1929, which involved surveying a more or less random sample of individuals to see how applicable they felt the same information was to them.[90] There was obvious scope for error here—applicability or success remains a wholly subjective criterion, and the problem of the "random" sample is difficult to solve—but the only other technique that had been employed for determining abnormal levels of success, Richet's technique of estimating the probability of each of a series of successes and multiplying the resultant fractions together to get a probability for the whole, was still more subjective and arbitrary, and Rhine encouraged Thomas to employ the former. During 1932/33, then, Thomas sent a questionnaire to a group of sixty-odd individuals (including W. F. Prince and Rhine), asking them how well their experiences checked with Mrs. Leonard's statements; no one found nearly so high a degree of correspondence as had Thomas. Putting this material and a new conclusion into his work, Thomas successfully defended his revised dissertation before a new committee—McDougall, Rhine, Adams, Lundholm, and two outside members of the Duke faculty—at the end of May 1933. Rhine wrote in relief to Prince, "He passed without any question and, I think, with a good margin."[91]

The Rhines themselves had become much less confident than Thomas that his evidence proved post-mortem survival; indeed, close associates and friends though they had become, they were temporarily estranged over this issue. Rhine continued to feel that telepathy and clairvoyance had to be thoroughly understood before it could be decided whether in the residue of mental phenomena after their removal there remained anything unmistakably supporting personal post-mortem survival. He continued to feel, too, that psychical phenomena would probably prove to be consonant with the rest of natural science, even though they might force a radical reformulation of that science. Rhine discussed these exploratory ideas in an illuminating letter to McDougall in the summer of 1931.

> If we can "treat" the psychical as physical in its production of effects in the physical world, the energy ambiguity I always felt was there has disappeared. One can really mean energy in precisely the same usage that physicists mean it. And to me that is relief. To connect mind up in hypothetical energy chains with the alimentary canal, the green leaf and the sun, is to me almost a necessity, to escape the need of a divine hand to supply it. But I do fully appreciate the stand you have always taken in opposing the "reduction" of mind and mental energy and mental process to something so simple as 19th century mechanics, even when these are properly elevated to adequate powers of complexity. The psychical is not reducible to what we call physics to-day, I feel as strongly as you do. I have only been inclined to ask, But what know we of the physics of to-morrow and its boundaries?[92]

This is not to say, however, that Rhine's concern for the deeper and ultimately more meaningful issue of post-mortem survival had been dissipated or lost. He had not discarded the possibility that it might eventually prove feasible to discover scientific proof of survival, as is clear from an exchange between him and H. F. Saltmarsh at just this time. Saltmarsh had read a paper entitled "Is Proof of Survival Possible?" to the SPR in October 1931; it was published in the *Proceedings* early the next year.[93] It was of course evoked by the continuing debates of the 1920s within European psychical research over the merits of the survival hypothesis, and Saltmarsh decided, regretfully, that the debate was without possible solution, "that no logical proof of survival is at present possible from the evidence studied by psychical research." His argument was, in brief, that in order to verify a communication from a supposed spirit there must be available either physical evidence in support of the communication or confirmation from a living mind; hence in principle it would always be possible to put forward telepathy or clairvoyance on the part of the medium as an equally conceivable explanation for her acquisition of the information. Whichever explanation a student of the phenomena will choose must depend on the opinions he has already formed as to their relative likelihood. Therefore, Saltmarsh concluded, "conviction will be, and seems likely to remain, completely subjective and dependent on individual idiosyncrasy."

Rhine responded to Saltmarsh's paper at just about the time Thomas was being told by his committee to argue the case for survival, for his letter to the SPR was sent off to London in the spring of 1932; however, it expressed thoughts that clearly had been maturing in his mind for some time.[94] It was based on an idea already set out in his 1929 seminar paper. Rhine's carefully argued thesis was that the case for survival could be strengthened by recognizing in a medium's communications not simply factual information or personal attributes characteristic of the supposed spirit but evidence of *purposive behavior* whose motivation "cannot plausibly be ascribed to any of the living or the dead except the 'spirit' identified." He illustrated his argument with a variety of examples. Suppose an individual has a dream in which an agitated man (later identified from the dreamer's description) begs her to rescue his dog, left to starve by its master's sudden death. Such a purposive act is not only appropriate, it is almost *peculiar* to the supposed communicator. We will be at a loss to explain why this one mistreated animal out of thousands should have been singled out by the percipient if we do not turn to the "plausible supposition, that the mind of the dead owner of the dog is in some way still alive."[95] Here, of course, Rhine was reiterating the argument of selectivity against the purely telepathic/clairvoyant hypothesis, though now it was clothed in a purposive psychology.

Rhine's concern went deeper than this, however. Scientific method, he believed, was a universal, no less applicable to post-mortem survival than to other, more established subjects of study. Saltmarsh's belief that the survival hypothesis alone was fated always to be a matter of "individual idiosyncrasy" could not be right. "This is scientific heresy of the most utterly ruinous type," Rhine exclaimed to Thomas,[96] and busy as he was with experimentation and the preparation of the manuscript of *Extra-Sensory Perception* for Walter Franklin Prince, he found the time to dash off a further riposte to Saltmarsh that makes his inner convictions still clearer.

> Mr Saltmarsh states that in this matter, "Apodeictic proof is quite impossible, conviction or plausibility is subjective." I agree. *In quite the same way* (though not the same degree) the remark applies to the question of Lamarckian inheritance, the planetesimal hypothesis, localization of function in cortical areas of the brain, or the corpuscular theory of light. But the fact that conviction is, and must be, subjective does not mean that psychical research must forever be a matter of individual prejudice and idiosyncrasy. Biologists will determine the value of the Lamarckian hypothesis, not by "individual idiosyncrasy" but by improved methods and experiments, and accumulated results of varied character and condition. And psychic research students will determine the value of the survival hypothesis by the same general means. If Mr Saltmarsh were to persuade me of his essential rightness, I shall despair for science, not only in the field we are discussing, but in every field in which such subjective factors as "antecedent acceptability," "plausibility," "individual idiosyncrasy," and the like are possibly present in any degree. And in what field are they not?[97]

Rhine was far more thoroughly committed than was the English school to treating psychical research as methodologically if not yet substantively a normal aspect of scientific inquiry.

V

During the year that intervened between Rhine's submission of the clairvoyance paper to the *Journal of Abnormal and Social Psychology* and his completion of the manuscript of *Extra-Sensory Perception* his research developed several new features. For one thing, he began to make an experimental distinction between telepathy and clairvoyance. The conceptual distinction between the two had been well established for some time; telepathy involved supernormal knowledge of the contents of another mind, while clairvoyance involved supernormal knowledge of the material world. But in general earlier experimenters had not recognized the sort of conditions this distinction forced the scientist to place upon a test for one or the other phenomenon. A test for telepathy should properly attempt to communicate mental content, and no possible material referent should exist, while a test for clairvoyance should have as its target a physical object or situation unknown to anyone; otherwise a successful experiment can yield no results clearly attributable to one rather than the other. Nevertheless, the earliest experimenters—those of the SPR, say, or Warcollier—would typically look at a drawing or picture and then try to communicate it; and if the percipient were successful, they would call the phenomenon telepathy rather than the clairvoyance it might logically have been. Of course, as we have seen, most researchers before the 1920s were in any case disinclined to admit the existence of *both* phenomena, and this inevitably made them insensitive to the methodological problems their definition involved. Only in the early 1920s did the SPR begin to take seriously the idea that clairvoyance might be as real an ability as telepathy, and Ina Jephson's experiment in clairvoyance (conceived, she herself said, when she realized that "telepathy" might actually be masking results obtained by clairvoyance) was a natural consequence, designed to test for clairvoyance exclusive of telepathy. And it was her work, as Rhine acknowledged, that helped to stimulate his own interest in clairvoyance.

Rhine's early work with Linzmayer and Pearce had included tests both for pure clairvoyance and for undifferentiated extra-sensory perception (or "ESP" as he came to refer to it). The former were of two principal types, "BT" and "DT," as Rhine labeled them. In the first of these tests, the subject called the top card of a deck of Zener cards "before touching"; the card was then removed and the call recorded, and the calling continued through the deck of twenty-five in this way. In the other type of test, the subject called "down through" the deck, and his twenty-five calls were recorded as they were made, but the deck of cards remained untouched until the end of the run. Beyond these, Rhine had also run some series in which he looked at the card that the subject was to guess, and in these cases of course either telepathy or clairvoyance could have explained success.

It was apparently in the spring of 1932 that Rhine began to think seriously about distinguishing clairvoyance from telepathy experimentally. In February of that year Walter Franklin Prince asked Rhine to help him evaluate some "telepathic" material. Prince had actually had relatively little direct experience of narrowly experimental psychical research upon mental phenomena, although he was closely familiar with its literature. One important exception to this was his 1921 involvement in Gustav Pagenstecher's investigation of a case of psychometry—a percipient's ability to describe settings, events, or personalities from the past history of a material object—which employed innovative techniques and elicited very strikingly positive evidence.[98] But in general Prince had expended his efforts on the study and exposure of physical mediums and (after his move to Boston) on summarizing and appraising the published reports of other psychical researchers.[99] Now he had read and been impressed by the telepathic experiments reported in Upton Sinclair's *Mental Radio* and was busy preparing a bulletin for the Boston SPR that would discuss them against the background of earlier research on telepathy.[100] He asked Rhine to help judge telepathic experiments carried out in 1885 by Max Dessoir, in which (as in the Sinclair work) the agent looked at a drawing and attempted to communicate the image mentally to a percipient.[101] During March and April Rhine several times raised the question with Prince whether these successes might not just as reasonably be attributed to clairvoyance as to telepathy,[102] and eventually Prince agreed, suggesting only that telepathy and clairvoyance might both be aspects of the same function, or themselves be equivalent to psychometry.[103]

A few months later, in the fall of 1932, Rhine began systematically to test Hubert Pearce for "PT," or "pure telepathy": "The agent and percipient were in the same room. The percipient sat with closed eyes, waiting for the uniform tap of a telegraph key as a signal. At the time of the signal the agent would be holding in consciousness the image of one of the symbols of the Zener cards, but actually had no cards present. The choice of order of images was planned by the agent for each 5 trials, avoiding any natural expected order, and varying, repeating, in 'random' fashion from one five to another." Rhine coupled these experiments in telepathy with experiments designed to test for clairvoyance in the same subject conducted under the same conditions and at the same time, and he found, to his satisfaction, that Pearce tended to make comparable scores in these two areas under such circumstances—thus confirming the likelihood of a relationship (or an identity) between the two.[104]

Another important feature of Rhine's research during the academic year 1932/33 was a significantly broader base of experimental subjects. Five more Duke students proved to have essentially the Linzmayer/Pearce gift, if not quite to the same degree. Of all Rhine's claims in *Extra-Sensory Perception,* his announcement that he had had no trouble in finding many consistently good performers in card-guessing would cause the most astonishment and even incredulity, not merely among the general public but among committed psychical researchers. After decades of disappointing inquiry, investigators as different in attitude as S. G. Soal and Gardner Murphy were quite willing to agree that clear-cut demonstrable telepathic or clairvoyant ability was an exceptionally rare phenomenon—yet Rhine seemed to have no trouble laying his hands on half a dozen good subjects in the relatively limited population of a small university community. In one way or another, the new subjects all had been associated with Rhine or with psychology at Duke before their discovery: Sara Ownbey and George Zirkle as graduate students; May Frances Turner, T. Coleman Cooper, and June Bailey as undergraduates. Rhine stressed further common elements in their background: artistic interests, sociability, a family history of "mildly clairvoyant" powers. During 1932/33 the five had some 16,675 trials at clairvoyance (BT and DT), averaging 7.6 per 25 guesses correct, and 10,275 at telepathy, averaging 9.6 per 25 guesses correct.[105]

The two graduate students, Sara Ownbey and George Zirkle (who married at the end of the summer of 1933), supervised much of the group's work. In effect, Rhine was beginning to assemble the nucleus of a research unit, in a manner familiar to psychologists and scientists generally. While he oversaw a part of the experimentation personally, in the main he tended to follow the pattern, already established with Stuart and Pratt, of delegating different aspects of the experimental program to graduate students in the department. Most of his subjects were volunteers, but in a few instances, in order to secure the continued interest and participation of his more important subjects in a monotonous and long-drawn-out research program, he arranged to pay them for the time they gave to the research—fifty cents an hour, drawn from the research fund granted him by the university. As analyzed by him, the experimentation done with these five subjects produced results wholly comparable to those obtained with Pearce and Linzmayer: scoring rate above chance to about the same degree; correspondence in scoring rates for both telepathy and clairvoyance; decline in success with fatigue and illness on the part of the percipient; depression of scoring upon administration of sodium amytal, subsequently restored to normal by caffeine; effect of psychological factors (for example, skepticism about a new experimental technique) upon a percipient's scoring.[106]

During the summer of 1933, Rhine's students began to explore one further new problem at his direction, that of the effect of distance upon ESP, and managed to obtain some striking results in time for inclusion in the manuscript he was preparing. By "new" is of course meant merely new in Rhine's experimental program; the SPR's earliest studies of veridical hallucinations had strongly suggested that not even the diameter of the earth need be a barrier to spontaneous telepathy, and subsequently a number of apparently successful cases of experimental telepathy at long distance had been reported. One impressive and often quoted case was that of Usher and Burt, in 1910, involving the transmission of a dinner menu from Prague to Paris;[107] and of course the Paris-New York program of experimental telepathy directed by René Warcollier and Gardner Murphy in the mid-1920s seemed also to succeed no less well than experiments conducted within New York City or within Paris.[108] But the eternally vexing problems of inconsistency and of difficulty in evaluating success affected these works like all others. The research reported by Rhine, on the other hand, yielded apparently clear-cut and consistent results. Clairvoyant and telepathic guessing, he found in 1932/33, seemed if anything to *improve* with increased distance between agent and percipient. In one clairvoyant series in July 1933 George Zirkle (with Sara Ownbey as agent) had had 51 successes in 100 trials at close range but 160 in 250 trials at 30 feet.[109] In the same month, Sara Ownbey and May Frances Turner carried out a telepathic series between Lake Junaluska and Durham, a distance of over 250 miles, in which the percipient, Miss Turner, scored 19, 16, 16, 7, 7, 8, 6, and 2 in successive blocks of 25 guesses—Rhine particularly noted that Miss Turner had never scored so high under any other conditions.[110] Finally, at the very moment Rhine was drafting his manuscript, J. G. Pratt was conducting a successful distance experiment in clairvoyance with Hubert Pearce, still enjoying the exercise of his faculty.

> It is in many respects the best yet and answers several important questions at once.... Pratt picks up, in a room in the Physics Building of Duke University, every minute during the running period a card taken from a cut and shuffled pack that lies on the table before him, and puts it face down on top of a book. He does not look at its face. At the beginning of the same minute, Pearce, in the Duke Library, over 100 yards away, tries to perceive the card then "exposed" by Pratt. He has succeeded, magnificently, in doing so. At first he failed, as he nearly always does with a new condition procedure. But the runs mount as he goes, as follows: 3, 8, 5, 9, 10, 12, 11, 12, 11, 13, 13, 12. The total 300 at that distance average 9.9 per 25.... Then the cards were taken to the Duke Medical Building, with over 250 yards between cards and percipient. Again there was the low-scoring adjustment period at first. This lasted over more runs this time but was followed by good scoring, which is now going on daily at this distance.[111]

In this aspect, as in several others, Rhine's reported results were unique in degree if not in kind of success.

Rhine had already begun to study still other types of supernormal mental faculties, but he did not feel ready to discuss them and did not refer to them in the work he was preparing. The completed manuscript of *Extra-Sensory Perception*—sent off to Walter Franklin Prince in October 1933 and published by the Boston Society for Psychic Research in an edition of nine hundred copies in April 1934—was of such a scope and of such promise as to revolutionize psychical research and to make its title literally a household phrase.

CHAPTER FIVE
Extra-Sensory Perception and Contemporary Psychical Research

I

Rhine's first work at Duke, the publication of *Extra-Sensory Perception* in particular, has long been felt by psychical researchers or parapsychologists to have somehow transformed their field.[1] Their view is no doubt in part a retrospective response to the fact that Rhine was the first to institutionalize psychical research within the university, thereby giving it an academic respectability that it had not enjoyed before. More than this, however, there soon arose a feeling that psychical research had been altered in intellectual content and structure by the monograph, which described a program of experimental investigation on a scale rare in psychical research and reported results of overwhelming statistical significance. Whately Carington (he had changed his name from Whately Smith in 1933) expressed this view very simply in 1945 when he said that "the modern phase of experimental work . . . may be regarded as beginning with the publication in 1934 of Dr. J. B. Rhine's book, *Extra-Sensory Perception*."[2] Most parapsychologists since who have reflected upon the history of their field have in one way or another expressed the same feeling.[3] But such history has tended to be narrative rather than analytical; it has been particularly sketchy about the precise relationship of Rhine's work to earlier psychical research. What *was* that relationship? and in what sense was Rhine's work transformingly new?

These are questions very much worth raising, and we might approach them by proposing the answer that *Extra-Sensory Perception* provided something of a paradigm for psychical researchers. In using that notoriously loose and perhaps overworked word, we mean to suggest that Rhine's initial work very soon came to give focus and definition to a field that had previously been an inchoate mass of conflicting assumptions, techniques, and goals.[4] This is of course just what Charles Richet had hoped his *Traité de métapsychique* would do back in 1922; as we have already seen, he was unsuccessful in his attempt to establish common ground among psychical researchers. There are several reasons why Rhine should have had the success that Richet did not. One very important reason is that in the intervening twelve years many of the disagreements within psychical research had lost their force and intensity. The most divisive issue of the early 1920s—whether the survivalist hypothesis could be accepted as a starting point for scientific investigation or whether it was necessary to adopt an essentially positivist attitude, remaining skeptical of all hypotheses—had been tacitly decided by 1930. Cautious survivalists like the Sidgwick group at the SPR, caught between an increasingly noisy and unscientific spiritualism on the left and a fiercely tough-minded skepticism on the right, had tended to put aside the survival question, in the expectation that research conclusively proving simple supernormality would eventually call it forth again. On the other hand, the committed survivalists, like Margery's supporters, were being edged out of the research community: by January 1933 Harry Price contemptuously described the ASPR to C.E.M. Joad as having "sunk to the level of a fifth-rate spiritualist society,"[5] and very few researchers took it any more seriously. The debate over mental versus physical phenomena had been settled, by default, by the disappearance of all convincing physical mediums. And the suspicion of professional science, of mathematics and of laboratory methods, had started to die out as new researchers—Estabrooks and Ina Jephson, Whately (Smith) Carington and S. G. Soal—began to think that restricted experimental techniques need not necessarily inhibit psychical phenomena. Rhine's book appeared at the right time to profit from this easing of disagreements.

Before proceeding further, it is important to be clear as to the structure and aims of *Extra-Sensory Perception*. It began with a general introduction composed with the members of the Boston SPR in mind. In this Rhine first presented a complicated classification of psychical phenomena so as to make clear the restriction of his own concerns to "corporeal parapsychical" material—mental or subjective phenomena, excluding spiritualism—and suggested some of the ramifications of this subject into other scientific fields. From this he passed to an account of previous experimental research into mental phenomena, a critical survey of the existing literature in which he reviewed the area under investigation, summarized the achievements and conclusions of previous experimenters, and concluded with a rather pessimistic judgment of the various theories so far put forward to explain the phenomena. The survey reveals an impressive knowledge of the literature of this particular aspect of psychical research, on which Rhine had concentrated his reading once he had arrived at Duke. The materials to which he had paid most attention were Prince's bulletins from the Boston SPR and the experimental reports in the *Proceedings* and the *Journal* of the British society. From their reviews (and other sources) he was able to stay informed of the occasional experimentation published in the ASPR *Journal* or the *Revue métapsychique;* what he knew of Warcollier's work, for example, came in this indirect fashion.[6] Because this historical survey discussed only the experimental reports on mental phenomena, it presented a picture of a straightforward if sporadic tradition of research. The complex questions that had so dominated psychical research in the 1920s never made an appearance, for Rhine had decided at the outset of his work that they were irrelevant to a truly scientific research program.

The second major section of Rhine's book described "The Experimental Results." In it he explained the techniques that had been used by him and his colleagues to permit statistical study of an isolable instance of extra-sensory perception and then described the historical course of research with his major and minor subjects, identifying the Zener cards and defining the different ways in which the cards had been guessed by the subject: GESP, general extra-sensory perception, when the experimenter-agent, hidden from the percipient, looked at each card; PC, pure clairvoyance, when the agent held or touched the card being called but did not know what suit it was; or DT, "down through," when the agent simply shuffled the cards and left the deck as a whole stacked in front of the percipient, who was to guess the cards in order (without touching them) going down through the deck. The results obtained by each subject were then given, series by series: Pearce, for example, was reported to have called 3,049 hits in 8,075 trials (averaging 9.4 per 25 Zener cards instead of the 5 to be expected by chance) by pure clairvoyance in one such series, a record whose likelihood Rhine calculated as something under 10^{-320}.[7] This section also recorded some attempts to study these phenomena under varying conditions: when the subject had been given sedatives or stimulants, for example, or when agent and percipient were separated by considerable distances—of 250 yards and even 250 miles. Again, Rhine repeatedly reported results of extraordinarily high statistical significance.

The experiments were interpreted as a whole in the concluding section, "Explanation and Discussion." It is to this section that one might naturally turn in trying to understand the book's success within parapsychology, expecting some significantly new element of understanding to be presented. But while to the layman Rhine's conclusions, his structuring of the subject matter, might have seemed novel, to someone who had carefully studied the work of earlier psychical researchers they would have had a comfortably familiar ring. Certainly Rhine himself had no thought of claiming the establishment of a new science. Indeed, he was insistent in underlining his dependence upon the accomplishments of such individuals as Charles Richet and Eleanor Sidgwick. The conclusions of *Extra-Sensory Perception,* in fact, confirm those of a fifty-year tradition of psychical research, though they are built up from an original program of successful experimentation. The monograph's effect upon parapsychologists arose in part from the fact that it developed in coherent fashion all the principles of psychical research that two generations of students had laboriously uncovered, bit by bit, but had not yet properly drawn together. This judgment is of enough interest to warrant study in detail of Rhine's conclusions, including an exploration of their relation to those of his predecessors; at the same time we can make clear the paradigm that served to guide so much parapsychological research in the 1930s and 1940s. But we should first lay some stress upon the conclusions at which we have arrived. Historians and philosophers today would agree in the main that no scientist can ever divorce his observations from a certain framework of expectation created by his previous experience. It seems clear to us, similarly, that Rhine's methods and generalizations in this book of 1934 were to some extent prepared for him by his concentrated study of the experimental literature of the previous decades. In turn, *Extra-Sensory Perception* was received with enthusiasm by many experimental psychical researchers because—naturally enough—it agreed with what they were already prepared to believe and because it promised what had so long been lacking, the replicable experiment.

We must remember at the outset that psychical research had never been a strictly experimental enterprise. The founders of the SPR had planned to use all the methods of science in their investigations, including experimentation, but the difficulty of developing a repeatable demonstration led them to concentrate on what we have called the "natural-historical," or anecdotal, approach and on the collection and analysis of spontaneous phenomena. Many psychical researchers, as we have seen, became convinced by their own experience that laboratory methods were almost always inherently sterile and were at best one of many routes to truth. Still, in the first years of organized psychical research a variety of experimental techniques were explored, among them the one that Rhine would eventually adopt, card-guessing. Charles Richet, in the 1870s, was one of the earliest to undertake experimental investigations into mental phenomena, particularly clairvoyance. The leaders of the SPR encouraged members to undertake their own researches to try to demonstrate communication of thoughts, and for more than a decade such researches were widely carried on and regularly reported in the society's *Proceedings* and *Journal*. However, there was no consistent pattern to these early experiments; tastes, colors, commands, pictures, imaginary scenes, playing cards, and numbers were all objects chosen for experimentation. In the late 1880s reports of experiments with playing cards were particularly frequent, the percipient trying to identify the suit of the card drawn, the number of pips on it, or perhaps the very card itself.

These first experimenters were certainly not blind to the fact that simple entities like numbers and cards held particular advantages in that their successful or partial transmission could be more satisfactorily measured. Richet in 1884 had worked with cards (among other objects) and observed in qualitative terms that, given a large number of attempted communications, a pronounced variance between expected and achieved successes would indicate that something besides accident was at work.[8] Richet's discussion immediately led Oliver Lodge and Edmund Gurney to try to derive an algorithm that would permit an experimenter to measure the relative improbability of different sorts of success (for example, at guessing suits, colors, cards);[9] and F. Y. Edgeworth did the same in two much more sophisticated papers (perhaps too difficult for their immediate audience) in the 1885 and 1886 SPR *Proceedings*.[10] Gurney, reviewing Richet's work in the *Proceedings,* had urged the interested public to attempt such experiments themselves so as to make possible a general accumulation of data. Two persons, he suggested, should take part, one drawing a card and looking at it steadily while the other tried to guess it. He added a number of cogent experimental precautions, advising that the total number of guesses to be made in any run should be specified before beginning and that the percipient should not be allowed to learn any results until each run had been completed.[11]

Yet, whether because the task seemed too mechanical and boring or not closely enough connected with original, spontaneous thoughts[12] or because the difficulties involved in applying the calculus of probability seemed scarcely worthwhile,[13] reports of card-guessing became quite rare in the 1890s, and there were only infrequent accounts of attempts of this sort in the literature for several decades. Instead, the attention of psychical researchers shifted to the study of several impressive mediums and to allied investigations that seemed to bear more directly on post-mortem survival. Those who reported card-guessing experiments tended at most to compare their results with what would be achieved by chance; they did not try to come to grips with the statistical question of just how improbable their results had been.[14] Richet himself did not attempt to extend the statistical side of his researches, although he continued to carry on experiments that would permit him to evaluate success as compared with chance. Nevertheless, Richet long insisted upon the strengths of the statistical methods that he had helped to introduce. In his *Traité de métapsychique,* Richet provided a widely ranging survey of the English and Continental researches, in the course of which he described earlier card-guessing experiments and repeated his conviction that even though such methods were not at all dramatic, they were precise and convincing.[15]

It is natural enough that Richet should have sounded somewhat apologetic about this view. Ever since the 1890s what experimental work was reported on thought-transference had centered largely upon attempts to convey vivid, complex thoughts that would be laden with meaning and emotion for the agent communicating the thought and might thereby presumably be better received. When successful, such experiments seemed vastly impressive. The feats carried off by Gilbert Murray that attracted so much attention were of this sort. Friends would agree on an event or a scene from literature in his absence; then he would join them, take the hand of one of their number, and with remarkable accuracy identify the imagined scene. The following exchange made up a part of the game of 23 November 1910:

> MRS. ARNOLD TOYNBEE (agent): "Out of *l'Espion.* Evsei finding a fly in his ink."
> PROFESSOR MURRAY: "This is a book—a Russian book. It's *l'Espion.* It's a scene I don't remember at all. I get an impression of the boy squashing a fly, but I can't remember it at all. I confuse it with Joseph Vance."[16]

By this time most amateur investigators were seeking dramatic, overwhelming, essentially qualitative "proof" of thought-transference rather than mere statistical evidence of its presence.

This issue was not merely one of personal preference or style. Since psychical phenomena were apparently incapable of consistent or repeated demonstration, any single occurrence of telepathy or clairvoyance might equally well be attributed to the chance concurrence of natural causes. For all psychical researchers, therefore, a major question was how to exclude chance as an explanation for the phenomenon in question. To some it seemed that what should be presented was a qualitatively convincing case, one resting upon the exceptionally rare individuals who on occasion can produce results of incalculable but obviously enormous improbability, like Stefan Ossowiecki; others tried more or less seriously to demonstrate a high degree of improbability in mathematical terms. Each approach presents difficulties. If the evidence advanced in support of psychical abilities is so sparse, it will never seem convincing, however conclusive the individual report may appear; as Charles Richet pointed out, such isolated occurrences will always be easily assigned to chance. On the other hand, a more broadly based statistical approach is hard to design and to carry through and leaves the uninitiated reader wondering how to evaluate the results. If it is argued that the odds of a supposedly telepathic event having occurred by chance are only one in five hundred, for example, has anything really been proved?

This issue was brought into the open in the mid-1920s as a result of Ina Jephson's attempt to design a card-guessing experiment that would be susceptible of statistical evaluation. Her initial appeals for card-guessing data had come early to the attention of René Sudre, in France, and had stung him to attack her projected experiment. In early 1928, before her results were published, he begged investigators not to perform collective experiments of this sort. His argument, set out in the journal of the American SPR, was that a single definitive experiment was needed; that the use of statistical reasoning was unconvincing and beyond the competence of most psychical researchers.[17] Miss Jephson replied briefly to his criticisms in her report, but he repeated them the next year in somewhat greater detail, adding a complaint against her practice of evaluating cases of partial success and insisting that only precise identification of a guessed card as to both suit and number should be scored as success. Vastly more convincing than her reported results, he insisted, would have been the report of a single case in which five consecutive cards were identified on five separate occasions.[18]

To this, Miss Jephson responded more fully, arguing that what constitutes proof in science is a repeatable effect, in this case a continuing statistical regularity, rather than a unique event. "Scientific conviction, after all, is only the name we give to the effect of the prolonged persuasion due to repeatedly observed facts or events."[19] R. A. Fisher independently explained the importance of "statistical significance" for anyone investigating biological phenomena, pointing out that this criterion prevents the scientist from being deceived by circumstances that he cannot control.[20] Miss Jephson's position—that high improbability determined statistically might provide psychical research with something like a replicable phenomenon—was shared by a number of the newly active experimenters in the SPR, like Whately Carington and S. G. Soal. Moreover, it had been invariably adopted by academic psychologists taking up experimental psychical research—Coover, Troland, and Estabrooks—for, after all, experimentation that does not give consistent results must be repeated over and over and studied en masse.

Rhine's work was very much in this latter, experimental-mathematical tradition. What was novel about his experimental program was not so much the idea of card-guessing statistically analyzed as it was the design of the particular cards that he made the targets for his subjects, a contribution shared by his Duke colleague Karl Zener. It was with these cards bearing five simple geometric designs that Rhine began to record his first signal experimental successes, in the spring of 1931, with the Duke undergraduate A. J. Linzmayer.[21]

The particular advantages that the Zener cards promised the experimenter are obvious. Earlier researchers using playing cards had recognized that individual preferences for particular suits or ranks might distort chance expectations, but rather than devise a new experimental situation, they had chosen to try to appraise the weight of public preference attaching to different cards.[22] Zener and Rhine had intended the new designs to offer neutral, equally familiar targets while precluding the habitual associations of playing cards. Considerations from sensory perception directed them towards designs that would be distinct, easily differentiated, and easily remembered.[23] Beside this, the Zener cards made it possible for the experimenter to employ a relatively simple statistical technique in evaluating his results, a technique that had already been employed by George Estabrooks (and, presented more obscurely, by J. E. Coover). In Estabrooks's experiments, students were asked to guess either the color or the suit of a playing card (not the card itself), making p equal to .50 or .25; he proceeded to calculate the standard deviation ($\sqrt{npq}$) and, explicitly assuming that the binomial expansion of his numbers of trials approximated a normal distribution curve, the probable error (.67450). Using this last figure, he was able to obtain from published tables the probability that his results were merely due to the normal distribution.[24] Rhine used the same technique, though of course for him p was equal to .20. Since he did not explain his method of calculation as fully as had Estabrooks, he left his nonmathematical readers somewhat confused: he did not make explicit how the probable error was to be computed, nor did he make explicit the assumptions on which his statistics rested—for example, that of normal distribution.[25] In this latter respect, however, he was not unusual, even among psychologists, who were not yet fully conscious of the subtleties underlying statistical reasoning and who ordinarily used it, as Rhine did, as a straightforward tool. In any case, it must be remembered that *Extra-Sensory Perception* had been written rapidly, as a preliminary report on certain exploratory experiments that had proved to be markedly successful; in drafting the monograph, Rhine had not thought of being either definitive or exhaustive.

The exploratory character of the reported research was responsible for something with which some psychical researchers were later to find fault, an occasional seeming looseness in Rhine's experimental conditions. Originally Zener and Rhine had handed out to their classes packs of cards sealed in opaque envelopes, but in order to canvass more possible subjects and to enable tests to be done more rapidly, Rhine (who had begun to carry on the experimentation alone) soon came to work in a variety of circumstances that did not always so rigidly exclude sensory contact. Conditions were excellent in many of the experiments carried on over considerable distances in 1933; they were almost equally good in the DT work that Rhine began in 1932. They were obviously less tight, in varying degrees, in other situations: when the cards were hidden by a screen eleven to twenty-four inches high; when they were hidden merely by a book or by the hand; when the subject held the card in his own hand, looking at its back, as he made his guess.[26] Roughly comparable scores were obtained under all these conditions, and there was no prima facie reason to question the adequacy of any of the safeguards, *except*—for a skeptic—the positive results obtained; but successful psychical researchers may always expect skepticism.

The method that Rhine adopted for presenting his results tended to strengthen the impression that his experiments might not always have been rigorously planned and supervised. The fact that his report was historical rather than analytical in form meant that clear, detailed statements of experimental conditions were scarcely ever to be found in his account of the research. For example, when Rhine summarized the results of a series, it was not always apparent whether the percipient had been told of his success or failure after every five guesses or after every twenty-five; in some cases Rhine used one method, in some cases the other (identified as BT-5 or BT-25, respectively). The reader could not always readily determine what sort of screening techniques (if any) had been used to prevent the percipient from consciously or unconsciously using sensory cues in identifying the cards, nor whether the series had been witnessed by an obvious observer—or by anyone at all. Indeed, by 1934 Rhine might in some cases have found it difficult to say what specific conditions had originally obtained, for his methods of recording data in this preliminary work had not always been consistent. In the foreword that William McDougall contributed to *Extra-Sensory Perception,* he wrote that "the experimenters have been at special pains from the beginning to exclude by the conditions maintained, any possibility of deception, conscious or unconscious";[27] but the statement was somewhat stronger than was justified by the informal, exploratory nature of the initial experimentation.

Still, shortcomings such as these of technique and presentation are entirely unremarkable in a first report in any normal-scientific field. Wholly similar weaknesses are to be found in the reports of the experimental psychologists like Estabrooks and Troland who had earlier studied psychical phenomena: had Troland claimed marked extra-chance results, his casual description of the informal experimental situation he employed would have been sure to attract sharp criticism. Rhine's spectacular success ensured that his work would undergo searching scrutiny, from psychical researchers perhaps more than from psychologists. The English psychical-research tradition had developed its techniques in the study of mediums, many of whom were extremely subtle frauds, and as a result it had come to be extremely demanding, almost obsessive, about the need to exclude the possibility of a normal explanation for the transient phenomena being studied. The small group of experiment-minded psychical researchers like Soal and Carington, then taking shape, took for granted the need for such precautions in their own research—precautions far exceeding those taken by the ordinary experimental scientist, who can presume that his phenomena may be reproduced and studied again and again. The approach taken by Rhine in his report thus reflects his original normal-scientific training and to some degree sets him off from the SPR. However, this does not alter the fact that the experimental situation he reported, like the statistical treatment he gave his results, was entirely in keeping with lines of investigation long familiar in psychical research. SPR members might come to urge Rhine to tighten up future experimentation, but they would not perceive his work as something fundamentally misconceived or unorthodox.

The conclusions that Rhine eventually developed in part 3 of *Extra-Sensory Perception,* his generalizations about the nature and occurrence of experimentally observed psychical phenomena, show (as does his experimental approach) very strong resemblances to the ideas of earlier researchers. Sometimes they were in fact based directly upon that earlier work, which had suggested confirmatory experiments for the Duke laboratory to try out; at other times, it seems, Rhine's interpretation of his experimental data was reinforced by the views already present in the literature—including reports not only of experiments but of mediumistic and spontaneous cases, in which he had also read widely. This use of the literature, as we have already observed, is typical of scientific research, which ordinarily originates in an attempt to confirm and extend the judgments of others. Even so, at the deeper level of synthesis Rhine's work was indeed original, for it offered psychical researchers a model study that *integrated* all these earlier scattered conclusions on the basis of an unparalleled series of experimental observations. We can best appreciate this originality by considering individually some of Rhine's most important conclusions in the light of their antecedents or of anticipations in the earlier literature of psychical research and by recognizing how they mesh together in the "Explanation and Discussion," with which *Extra-Sensory Perception* concluded.

The first elements of this discussion were based on Rhine's desire to distinguish experimentally between clairvoyant perception (in which the percipient was asked to identify a physical object) and telepathic perception (in which the percipient was to pick out a mental image or impression). Conceptually, the distinction had already been made. It had been well recognized in the early days of the SPR that there were two possible different modes of "acquiring supernormal knowledge," as Mrs. Sidgwick put it in a paper of 1892. In this paper, she attempted to examine the evidence for "clairvoyance," which she defined at the outset as "a faculty of acquiring supernormally, but not by reading the minds of people present, a knowledge of facts such as we normally acquire by the use of our senses," thus explicitly excluding cases of "thought-transference" or telepathy.[28] However, while Mrs. Sidgwick was interested in the possibility of moving on from the telepathic phenomena she considered to have been established to a technique for demonstrating clairvoyance, she was not seriously concerned with the need for procedures that would separate the two types of psychical phenomena:

> Cases like this [i.e., of thought-transference] show that though, if a faculty of independent clairvoyance exists, it may doubtless be exercised in the presence of persons cognisant of the facts clairvoyantly known, it can hardly under these circumstances prove its existence. But though the evidential reason for dividing off these cases from clairvoyance proper is clear, I am not prepared to say that the line so drawn has much scientific value. It is undeniable that such evidence as we have of clairvoyant perception of things at a distance is often very much mixed up with evidence of similar perceptions possibly due to thought-transference from persons present, and this suggests the possibility that clairvoyant perception of distant scenes is facilitated when it can be led up to by thought-transference from those present.[29]

The English experimenters of the 1880s and 1890s were no more impressed than Mrs. Sidgwick and made no consistent attempt to segregate the two classes of phenomena. Even though they may have been conscious of the distinction, they seem not to have been alert to the possibility that one type of phenomenon might *interfere* with the experimental identification of the other.

Rhine was to develop the conceptual distinction between telepathy and clairvoyance into two different experimental situations that would rigidly separate instances of the one from the other. This experimental differentiation also had its antecedent—in the work of Ina Jephson, as Rhine himself acknowledged.[30] In her 1928 paper Miss Jephson had described her own realization that one type of perception could mask the other—ironically, she suspected that clairvoyance might actually be the cause of apparent instances of telepathy—and had devised her experimental technique so as to exclude telepathy.[31] Rhine's early researches complemented experiments of this type with others designed to test for telepathy alone by asking the percipient to guess cards chosen mentally by the agent and noted down only after the guess had been made. It was in his systematic attempt to distinguish experimentally between these different psychical phenomena that Rhine was doing something new.

While Rhine was led to segregate instances of telepathy from instances of clairvoyance, he was nevertheless convinced by his evidence that the two were products of a common underlying process—and, moreover, that the so-called percipient was in fact the active entity in these psychical phenomena, whose concentrated attention was required to produce the effect. The feeling that thought-transference or clairvoyance had to be consciously willed went back, again, far into the history of the SPR; in 1890 Oliver Lodge was commenting that thought-transference without conscious agency had never yet received experimental confirmation.[32] But Rhine's judgment that the percipient played the active role had not always been shared by earlier psychical researchers. The Society for Psychical Research had originated out of an attempt to investigate (or, for some of its members, to validate) the claims of spiritualism, which perhaps inevitably led them to consider thought-transference as the most fundamental psychical phenomenon; they evolved the concept of agent and percipient as, respectively, the active and passive figures in telepathy[33] and, as we have seen, gave second place to the problem of clairvoyance. In France, however, Charles Richet maintained a much more critical attitude towards spiritualist explanations and perhaps as a result treated telepathy as simply "a special and frequent case of cryptesthesia" (as he preferred to call clairvoyance).[34] Other Continental researchers went on to stress explicitly the part played by the percipient.[35] The English researchers came to share this view at least in part. There are signs that in the late 1890s telepathy was being construed as conceivably the positive act of a percipient,[36] and Gilbert Murray's achievements twenty years later established the "percipient's" importance for many in Great Britain. Ina Jephson's card-guessing experiments—and Rhine's—are in a sense consequences of this shift in values that now made clairvoyance (in which there is no "agent") the type-phenomenon for ESP.

The study of the presumed general mental function behind the two classes of phenomena led Rhine to enunciate a number of general rules applicable to the manifestation of ESP, bearing on the psychological and physiological state of the percipient. Once again, Rhine himself repeatedly called attention to their enunciation in the earlier literature of psychical research. But they had never been brought together as they were in *Extra-Sensory Perception,* in a synthetic model defining conditions of experimentation for subsequent exploration.

By far the most heavily emphasized point in the book concerning the experimental context was the need for abstraction and attention on the part of the percipient and the importance of techniques that would maintain these at a high level. Rhine cited Gilbert Murray's comments about his personal experience: "The least disturbance of our customary method, change of time or place, presence of strangers, controversy or especially noise is apt to make things go wrong";[37] and similar judgments go back to the very first decade of psychical research, even though then the emphasis was more on the circumstances of the "agent" in telepathic experiments than on those of the "percipient."[38] Then, as interpretation shifted in the period after 1900, the same conditions came to be recognized as essential to a percipient's success.

Linked to the emphasis upon maintaining the attention of the experimental subject was the desire to maintain his interest at a high level. The early work on card-guessing had served to convince many that experimental situations had to avoid monotony if the subject was to display continuing psychical ability. Mrs.

Verrall offered this as one explanation why Gilbert Murray was able to achieve his striking successes: "[His] experience is entirely consistent with observations in other telepathic experiments; the great difficulty is always to keep alive the interest of the percipient, and prevent the deadly boredom which comes from thinking of cards, or numbers, or diagrams, even in the case of persons whose scientific curiosity induces them to multiply experiments capable of 'statistical' estimate as to success or failure."[39] Rhine accepted this view completely; but committed to card-guessing as he was, he had to develop ways to keep the subject's mood light and relaxed—by joking with him, challenging him to better his score, varying the experimental situation in trivial ways—while encouraging him and building up his confidence, never expressing doubt or hostility.[40] An easy, casual atmosphere became one hallmark of experiments at the Duke laboratory.

Early psychical researchers usually explained a falling off of a subject's success over a long period of time—not merely in guessing cards but in telepathic transmission of images as well—as reflecting presumed lapses of attention or of concentration or attacks of boredom.[41] It was Ina Jephson who first raised the possibility that these indications of varying attention might be observable over short spans of time, for example, within the scope of a single run of card guesses. Her experiments with herself and others seemed initially to suggest a relatively higher proportion of successful clairvoyant guesses at the beginning of each run of guesses, and she attributed this "fatigue-curve" to the increasing tediousness of the situation.[42] Rhine was thus prepared for the discovery of both long-term decline of success and short-term curves of operation in his experiments. Even before he had begun to concentrate upon work with the Zener cards, in mid-1931, he had encouraged a graduate student in the Duke psychology department to look for the decline effect in guessing suits of playing cards and had been informed of apparent success;[43] subsequently Rhine identified the decline effect in the work of his principal subjects, Linzmayer and Stuart, although he attempted to show experimentally that the decline was the result of weakening motivation and not of physical fatigue. At the same time, Rhine pointed out the likelihood of finding other curves, "curves of operation" or "attention-curves"; for example, a subject might be expected consistently to give more attention to the first or last call in a run.[44] This notion, probably suggested by Rhine's reinterpretation in psychological terms of Miss Jephson's results,[45] was confirmed by results of experiments with most of the Duke subjects, although certain individuals appeared to display characteristic and quite different curves of attention.

In the areas just discussed, it is clear that *Extra-Sensory Perception* shared the attitudes of previous psychical researchers as to the proper psychological conditions of the experimental situation. There remains a final question that it is natural for the experimenter to raise: is it possible and fruitful to inquire of the percipient what he thinks is taking place at the moment when he is achieving clairvoyant or telepathic perception? Psychical research had not adopted a general consensus on this. In the first few decades of experimentation, perhaps encouraged by the tendencies of the dominant school of psychology, Wundtian introspectionism, researchers had often tried to identify the nature of the process taking place within the subject's mind: whether the communication was mediated by words or by images, whether accompanying "thrills" were noted in successful instances, and so forth.[46] Increasingly, however, researchers in the 1920s came to mistrust the subject given to self-conscious examination or to an analytical interest in the experiment and his own thought processes. On this issue *Extra-Sensory Perception* expressed a certain ambivalence. Rhine did try to discover how his subjects received their correct guesses, in terms of the feeling and imagery involved, but he was not remarkably successful and indeed arrived at the conclusion that the percipient "knows but cannot tell 'how he knows'; there is no analysis possible apparently."[47] In any case, he did not devote much attention to the problem, conscious that a "subjective exploration" was difficult to evaluate and might well make a subject self-conscious and cause his scores to fall off.[48] On the whole, *Extra-Sensory Perception* showed much more serious and consistent interest in investigating the external, objective psychological and physiological conditions that accompany psychical phenomena than it did in exploring the internal, subjective content of the percipient mind.

The mathematical and experimental techniques that were essential to J. B. Rhine's first research, and his basic interpretative model, were (as we have seen) closely related to the psychical research of the previous half-century. So, too, were his attempts to appraise possible explanatory theories, to evaluate his phenomena and generalizations in the light of accepted scientific knowledge. The parallel is obvious, for example, in his consideration of extra-sensory perception against the canons of contemporary physics: Rhine concluded that his experimental results were totally inexplicable by modern physical theory, since this would have to suppose some sort of radiation hypothesis or waves being emitted both from telepathic "agents" and from Zener cards and picked up selectively by the percipient. One of Rhine's arguments against such a theory arose out of the parapsychological experiments that he had had carried out over considerable distances. During 1933 he compared Hubert Pearce's card-guessing success at 2, 10, and 30 feet; then in 1934 at 100 and 250 yards. At the same time, two assistants in his research, Sara Ownbey and May Frances Turner, carried on an experiment in pure telepathy over 250 miles.[49] On the assumption that radiant energy was linking cards or agent with the percipient in such cases, Rhine reasoned that success should have followed the inverse square law, that is, it should have decreased with increasing distance—and yet in all cases the percipient performed at least as well at a distance as at close range. He did not balk at drawing the radical conclusion—"bold" was his word—that "at this point we are, then, it seems, faced with the need of another order of energy, not radiant," and presumably hitherto unknown to science.[50]

While this may have been a radical conclusion, it was a response to a long recognized difficulty. The early investigators of thought-transference, the leaders of the British Society for Psychical Research, had included a considerable number of physical scientists, who naturally tended to look into possible physical explanations for the phenomena they were observing. William Crookes, the discoverer of the cathode-ray tube and the element thallium, was the principal defender of the "radiation hypothesis." His presidential address to the SPR in 1896, delivered in the excitement over Röntgens's discovery of x-rays the previous year, included the suggestion that thought might be transmitted by similar radiation.[51] The idea of thought waves had certainly, and not unnaturally, been raised before; in 1894 A. J. Balfour (Mrs. Sidgwick's brother), in his own presidential address, had cited the analogy between gravity and telepathy as instances of "action at a distance" but had pointed out that in the latter case the phenomena did not seem to conform to any of the accustomed ways in which forces were known to act from one part of space upon another and certainly did not seem to obey the inverse square law.[52] The SPR had of course amassed numerous accounts of spontaneous veridical phenomena witnessed hundreds or thousands of miles away from their source, and what experimental studies had been carried out along the same line did not indicate a significant diminution of physical effect with increasing distance. Crookes, referring to these points two years later, commented that "these are weighty objections, but not, I think, insurmountable."[53] Few other members of the society shared his confidence, however. Myers, in 1901, pointed out further difficulties that would arise in trying to accommodate a wave theory to clairvoyance and to precognition;[54] and W. F. Barrett, in 1920, once again emphasized the problems posed for a radiation theory by its unremitted action over thousands of miles to a specific target rather than in every direction.[55] Rhine's conclusions were thus again a confirmation of those of earlier students of psychical phenomena, except that his concern for clairvoyance led him to lay particular stress upon the difficulties of imagining how Zener cards could emit radiation that could be picked up by individual minds.[56]

Rhine's suggestions as to the biological implications were, naturally enough, somewhat more fully developed than his predecessors', for his graduate training had been in biology, while earlier psychical research had tended on the whole to attract attention from fewer biologists than physicists (although Charles Richet, A. R. Wallace, and Hans Driesch were certainly eminent defenders). In particular, physiological experimentation and speculation had been limited. It now seems odd, for example, that in the fifty years before the publication of *Extra-Sensory Perception* only one experiment was published that seriously assessed the variations of an individual's psychical power when placed under the influence of drugs (although such experiments had long been proposed): this was the Groningen research carried out by Brugmans in 1919, in which the student being given telepathic commands was observed under the influence of both alcohol (which improved his scoring) and bromide (which depressed it). The publication of this suggestive work apparently did not lead to further tests of a similar nature anywhere else.[57]

Far more common than physiological experimentation had been speculation into the relationship between psychical abilities and mankind considered as a species. Was it the evolutionary product of natural selection? and was it present in all mankind, or only in a few isolated individuals? One might have expected Alfred Russel Wallace to take the lead in such speculations, but he held to the belief that the evolutionary process could in no way explain psychical phenomena.[58] F.W.H. Myers was among the first to try seriously to interpret telepathy in the light of natural selection, suggesting, in 1886, that it might be akin "to certain forms of sensibility which we perceive, in some rare examples, to be possible to the human organism, but which have not, apparently, been valuable enough to our ancestors to get themselves established among recognised human faculties."[59] In this view, telepathy was a feature of biological life that had never been fully developed but had remained subliminal rather than become "supraliminal."[60] Variations on this idea were subsequently taken up; for example, A. J. Balfour wrote:

> Is it not in itself likely that here and there we should come across rudimentary beginnings of such senses [i.e., psychical powers]; beginnings never developed and probably never to be developed by the operation of selection; mere by-products of the great evolutionary machine, never destined to be turned to any useful account? And it may be . . . that in these cases of the individuals thus abnormally endowed, we really have come across faculties which, had it been worth Nature's while, had they been of any value or purpose in the struggle for existence, might have been normally developed, and thus become the common possession of the whole human race.[61]

Evolutionary interpretations of this sort could help explain the depressing fact that psychic power was neither widely nor easily demonstrable in man.

Rhine's book pursued both these lines of investigation, through physiology and through evolution. He developed Brugmans's research, for example, by administering both sodium amytal and caffeine to his subjects and determined that the former caused their rate of scoring to fall off, while the latter caused it to increase; this he took to be consistent with the Dutch work as indicating a need for "a certain degree of integration of the nervous system."[62] This conclusion, taken in conjunction with the apparent need for alertness in subjects, led him tentatively to separate experimental successes from the spontaneous occurrences that often took place in dissociated conditions, such as drowsiness or sleep. Given the fact that sensory phenomena remained stable when, under administration of sodium amytal, extra-sensory perception disappeared, Rhine also advanced the suggestion that ESP might be a higher development than ordinary sensory perception, a later stage in mental evolution. The consistent if slight success that Rhine had had with virtually all the seventy-seven students the Duke laboratory had tested, together with the quite marked ability that extensive testing had revealed in many of the department's graduate students, led him to suspect further that "for aught that may be said to the contrary, E.S.P. may be as widely distributed a natural capacity in the species as is the highest mode of cognition, reasoning."[63] The fact that his major subjects all averaged from six to ten successes in twenty-five guesses suggested a "species level" corresponding to "native species endowment," though individuals could certainly transcend this level on occasion. Rhine's unprecedented experimental success had thus brought him to break away from earlier views in this respect.

While *Extra-Sensory Perception* was open to the implications of physics and biology for psychical research, it was particularly sensitive to possible cross-relations with psychology. For several years Rhine had been immersed in all the activities of a psychology department, and the experience he had gained was responsible for the particular thoughtfulness with which he tried to make psychological sense of his experiments. Earlier workers had certainly shared his view that psychical phenomena were closely akin to psychological phenomena. Rhine's own adoption of the German term "Parapsychologie" to denote the field—the prefix indicating, not that it was *outside* psychology, but that it was "'beside' psychology in the older and narrower conception"—emphasized the connection.[64] But in the previous fifty years no serious attempt had been made to integrate the study of psychical phenomena into the domain of psychology per se. Most of the earlier investigators were not in any remote sense professional psychologists, even though they might have developed a great interest in and knowledge of psychology, as F.W.H. Myers did. Those interested in psychical research who *were* psychologists—like William James and William McDougall—had little personal concern for the experimental aspect of the subject. Coover and Troland, on the other hand, who were competent experimentalists, had no interest in undertaking a sustained investigation of the subject, and Gardner Murphy, who once had had such an interest, was ill. In 1933 Rhine was very much alone in such a commitment.

What seemed most important to Rhine as a psychologist was the atmosphere of the parapsychological experiment: the conditions of stress and calm, attention and relaxation, novelty and boredom, which seemed to affect the manifestation of telepathy and clairvoyance. Such observations were not new with Rhine, for comments on each of these factors can be found in the literature before Rhine; but he gave them order and at least the outline of a psychological framework, by attempting to relate psychical phenomena to other known and testable mental features: perception, cognition, motivation, personality, and learning. His preliminary conclusion was that since ESP ability could decline but apparently could not be learned and developed, on further inquiry it would probably prove to be just as much an innate capacity as was sensory perception.[65] In addition, Rhine regularly preceded his accounts of his subjects' performances with personality profiles, and in his general summary, "The Psychological Conditions," Rhine made some tentative essays at correlating personality traits and ESP abilities.

Imaginativeness, artistic ability, and sociability were among those traits that he thought might be linked to ESP performance. But he made no claim to have established these connections conclusively; the relations of ESP to personality, as well as to age, race, and other variables, remained another subject for future research.[66] The psychological framework of *Extra-Sensory Perception* was not yet capable of explaining the material being reported; rather, it was being used to suggest possible ways to carry out further studies of extra-sensory perception, ways in which a more fully developed parapsychology might eventually be tied into experimental psychology. Many of these research suggestions were indeed followed up (and are still being pursued) by parapsychologists working with Rhine and elsewhere.

If Rhine went beyond many of his predecessors in suggesting parallels with experimental psychology, he did not share the interest in the more speculative areas of psychology that some of them had manifested; in particular, he gave short shrift to neurological or psychodynamical speculations. Rhine did admit that ESP "depends upon the higher functions of the nervous system," but beyond this he did not go; he gave even less attention—really none at all—to a possible psychodynamical model that would relate it to unconscious mental processes.

In both these areas of speculation, Rhine's attitude differed from that of some of the earlier psychical researchers, in particular F.W.H. Myers. Myers, an avid follower of the latest developments in psychology during the 1880s and 1890s, had been interested in both neurological and psychodynamical speculations: the former in connection with the attempt to devise a localization-of-brain-function model to explain "automatic writing," the latter in connection with his developing a generalized conception of the unconscious—the "subliminal self," as he called it—to account for the host of strange mental states that were intriguing psychologists and psychiatrists in the last decades of the nineteenth century: hypnosis, hysteria, multiple personalities, and psychical phenomena.[67] Myers was in close touch with the French investigators, especially Janet, and he was one of the first men to publish an appreciative account of Freud and Breuer's work on hysteria.[68] Rhine put aside grandiose speculations like those of Myers—fascinating but not readily testable—in favor of a program to demonstrate experimentally and statistically the existence of psychical phenomena and their variation with change in conditions and over time.

II

J. B. Rhine was by no means unknown to the SPR in 1934, even though he had not yet become a member. Their coverage of the Margery case had first called him to their attention—favorably, on the whole, since on principle so many of them suspected physical mediums. They would surely have heard of his work at Duke through William McDougall, who spent the summers of 1931-33 in England and was in contact with psychical researchers while there. They would also have had good reports of him from Walter Franklin Prince, who went to England in the summer of 1930 to accept the presidency of the society. Rhine's exchange with Saltmarsh, emphasizing scientific proof while expressing sympathy towards the survival hypothesis, had also struck the proper tone for the SPR establishment. Hence Rhine must have appeared to the SPR, not at all as an outsider, but as one of the few men in America who might be doing psychical research up to serious standards.

Extra-Sensory Perception had also had its advance notices in England. Prince, who had been much impressed with the brief paper that Rhine had shown him in September 1932, was overjoyed at the prospect of publishing the longer monograph. Even before he saw it, he was writing about it in glowing terms to all his correspondents in the SPR, of both the left and the right wings.[69] His letter to the perennially skeptical E. J. Dingwall was perhaps not entirely tactful: "... When the long paper embodying a big series of experiments for telepathy conducted by members of the teaching staff of an American University is printed by the Boston Society next Fall you must be prepared to say why and how they were not under scientific control, and just why they are not worth respectful attention. Perhaps it will be because the experiments were in an *American* university. Can any good come out of Nazareth—otherwise out of this crude country?" It is not difficult to imagine in what spirit Dingwall must have awaited the book. But in general there would have been a climate of expectancy and sympathy when copies began to arrive in England in May 1934.

Most English psychical researchers in fact greeted *Extra-Sensory Perception* warmly, but it is easy to discern two different levels in their enthusiasm, depending on orientation. The traditionalist center of the SPR, including the leaders of the society, welcomed it as a piece of research that should provide a basis for still wider experimentation—no doubt silently hoping that the study of mental phenomena would eventually lead back to the survival hypothesis.[70] In some measure, however, they shared the feeling, widespread in the earlier days of the society, that by limiting oneself to narrowly experimental research one ran the risk of artificiality in technique and restrictiveness in conclusions. These individuals were undoubtedly pleased by the success of Rhine's experiments, but they retained a very broad conception of the proper methods and goals of their field and were not themselves tempted to do tests with Zener cards.

The fullest and generally most appreciative English responses to Rhine's monograph came, in fact, from the tiny group of experiment-oriented psychical researchers, most notably Whately (Smith) Carington and H. F. Saltmarsh. As it happened, Carington and Saltmarsh were strikingly close to Rhine not merely in their interest in experimental psychical research but in their philosophy, their vision of man and nature. Carington had recently begun to speak out on the importance of psychical research for the destruction of a sterile and inadequate materialism. In a paper considering "Positive Implications of Telepathy," read to the SPR in February 1933, Carington argued that admitting telepathy would overthrow materialism and "make the world safe for survival."[71] In analyzing the contradiction between telepathy and materialism, he went far beyond what Rhine was to do in *Extra-Sensory Perception,* claiming that telepathic communication could not possibly be viewed as analogous to communication by electromagnetic radiation, since (1) no bodily organ even suggested that it might be able to generate the necessary radiation; (2) telepathy did not obey the inverse square law; and (3) a radiation theory of telepathy would involve the fantastic requirement that "some process of encoding and decoding goes on—*automatically, unconsciously and with respect to an unknown code.*" These considerations encouraged Carington to argue instead that telepathy supported the contention of "Mysticism in the large" that there is in humanity an underlying community of consciousness. He developed these same ideas at much greater length in *The Death of Materialism* (1933), a work that Rhine found to be largely congruent with his own views when he read it the following year.[72]

Yet Saltmarsh was still closer to Rhine than was Carington, in that his position was more humanistic. It was Saltmarsh who reviewed *The Death of Materialism* for the SPR, and his generally favorable discussion at one point touched on the inherent conflict between human ideals and, not simply materialism, but scientific determinism. He regretted openly that Carington, while rejecting materialism, proclaimed himself still a determinist, and he insisted on the need for a system that would assure man of his free will.[73] What Saltmarsh hoped psychical research would yield was nothing metaphysical, not access to a cosmic consciousness nor proof of post-mortem survival (it will be remembered that he and Rhine were debating the latter point in 1933), but the establishment of a solid basis for a science of ethics. Such a science would reveal the secrets of human nature, would make plain the proper purpose of mankind, and teach men to gain control over themselves. To limit man to the logical and rational is to make free will an illusion and ethics "a mere branch of descriptive sociology. The world is reduced to a very dismal affair of soulless mechanisms & morality, a sort of clockwork model of the real living thing." It is psychical research that holds promise of confirming the reality of a suprarational order, which will allow man to recognize his freedom. In this personal philosophy, outlined in a letter to Rhine in September 1934,[74] Rhine would have recognized a spirit very like McDougall's, and like his own.

Carington and Saltmarsh reacted to Rhine's experimental demonstration of telepathy very much as they had to Ina Jephson, with praise mingled with constructive criticism. Carington described the book to Prince as "an extraordinarily important piece of work—quite the most important that has ever been done in this field";[75] in a separate letter to Rhine, in which he announced that he had undertaken similar experiments, Carington suggested slight emendations in the statistical techniques used and proposed for investigation other psychological questions, concerning guessing preferences and the need for introspective data.[76] Saltmarsh reacted in much the same way, writing immediately to Rhine to agree that he had definitively established the reality of ESP and had gone far towards defining the laws of its operation.[77] The letter that Saltmarsh wrote to *The Listener* in August 1934, enlarging upon the very favorable review it had given *Extra-Sensory Perception*,[78] insisted that "Dr. Rhine has done the most important piece of work so far accomplished in psychical research" and called attention to experiments that might be performed to test out certain questions his data raised—for example, why did gifted subjects appear to have a fixed maximum level (rate) of success in guessing?[79] Both Carington and Saltmarsh continued to correspond regularly with Rhine during 1934-35, discussing projects and techniques for further investigation and keeping Rhine informed of the work and opinions of the English. Rhine's book clearly exercised a strong incitement to research upon these exponents of the experimental approach.

The spiritualist left of the SPR ignored *Extra-Sensory Perception*, but the skeptical right seized upon it with no less vigor than the center. E. J. Dingwall read the book intently, only to find in it the same sort of wishful thinking and slipshod technique that to his mind had recently come to characterize the SPR; as Dingwall put it in a letter to Harry Price in which he complained of the worldwide decline in psychical research, "at the moment this Rhine business and the Carington stuff here is the thing to go for."[80] If (as appears likely) it was Dingwall who reviewed *Extra-Sensory Perception* for *Nature* in September, he did his part by disposing of it in a vein more ironical than sympathetic: "[Rhine] appears to have been fortunate in obtaining so many subjects who were able to demonstrate their remarkable gifts, although it is possible that other investigators have been more critical in their requirements, for Dr. Rhine, it would seem, was convinced of the reality of E.S.P. before he started his own experiments, mentioning such suspicious tests as those formerly given by the Creery Sisters [who eventually confessed to using a code in some experiments in the 1880s] as contributing to his own conviction." He complained that Rhine had poorly described the cards and their shuffling, hinting by implication that this might vitiate Rhine's results. Even so, he concluded by calling for "a repetition of the experiments under much more stringent conditions," since "if confirmed, the phenomena present problems of great interest and complexity."[81]

S. G. Soal's attitude towards Rhine was somewhat less clear-cut, because his skepticism was to a degree less obsessive than Dingwall's. To be sure, he found *Extra-Sensory Perception* incredible, particularly in view of his own laborious experiments on both telepathy and clairvoyance, which had given unequivocally negative results. A study of a very different sort carried out in early 1934, his investigation of the stage telepathist "Marion," had convinced Soal that Marion's remarkable feats were made possible by subconscious interpretation of subtle sensory indicia and were not truly telepathic—negative results again.[82] In a private letter to Walter Franklin Prince in July 1934, Soal raised just those objections to ESP that would soon appear in Dingwall's review: unsatisfactory description of method, inadequate randomization of cards, bias towards positive results. In the same letter, he enclosed a list of particular criticisms of Rhine's procedure that he agreed could be sent on (anonymously) to Duke. He insisted that in psychical research the onus of proof was on the experimenter and that all normal explanations would have to be rigidly excluded before the supernormal could be entertained. Consequently, he felt, far more stringent precautions were required in preparing, shuffling, and presenting cards to the subject than Rhine appeared to have taken; and a committee of skeptical scientists would be far preferable to Rhine alone, or Pratt alone, as witness to the phenomena.[83]

Nevertheless, in spite of his expressions of scorn for Rhine's technique, Soal could not resist the temptation to try out Rhine's general approach, for it offered precisely the sort of experimental situation that he had long believed would be necessary if mental phenomena were ever to be demonstrated conclusively. In his letter to Prince, Soal conceded that he would retract all his suspicions if he could obtain similar results by following the Duke procedure under more thoroughly controlled conditions, and later that month he began to plan out an ambitious attempt to replicate Rhine's work just as he had "replicated" Ina Jephson's. The new replication could not be done under the wing of the SPR, of course, for Soal had been estranged from the society since 1932; instead, it was carried on under the auspices of Harry Price, with whom Soal had found common ground in distrust of the SPR. He proposed a total of 100,000 guesses from the 134 participants in his last experiment and gave Price the task of finding absolutely opaque cards identical in size and design.[84]

Meanwhile, Soal had decided to allow Rhine to see his detailed criticisms and wrote to Prince, Rhine's mentor for so long, to tell him so. But Prince had died on 7 August after an operation for intestinal cancer, and his secretary at the Boston SPR sent both critique and covering letter on to Duke. When Rhine sent a bland acknowledgment of the critique directly to Soal, mildly questioning his outspoken skepticism, Soal was slightly at a loss; he replied in some embarrassment, trying to emphasize the difference between scientific skepticism and blind hostility, and promising that he himself would give two years or more to an experimental test of Rhine's conclusions. He insisted that he would guard against an atmosphere of distrust or suspicion—that he was not like Dingwall, who, Soal claimed, made it obvious to everyone that he did not *want* to observe anything inexplicable.[85] Perhaps this last was disingenuousness on Soal's part. But he was perfectly serious about the experimentation; by Christmas 1934 it had begun.[86] For skeptics no less than for sympathizers, *Extra-Sensory Perception* had proved a powerful stimulus to research.

If private correspondence among this "invisible college" of psychical researchers had not survived, we would know nothing of this first English response to *Extra-Sensory Perception,* for it was only in January 1935 that a review of the book appeared in the *Proceedings* of the SPR.[87] It had been assigned to Robert

S. G. Soal, supervising a card experiment with Frederick Marion at the National Laboratory for Psychical Research in early 1934. Soal is in the center, Harry Price on the far right, and Marion is between them. (Courtesy of the Harry Price Library.)

H. Thouless, on the face of it an odd choice, in that it passed over such figures as Soal and Ina Jephson, who might have seemed better placed to evaluate Rhine's work; indeed, initially Miss Jephson was supposed to do the review. Thouless was forty and had been in charge of the psychology department at Glasgow for eight years. His research had lain entirely in "orthodox" subjects—social psychology, psychology of religion, hypnosis—and hitherto his only serious involvement with the field of psychical research had been his participation in an investigation of the claims of psychical ability made for the hypnotist Wallenius.[88] Thouless was not convinced of the reality of psychical phenomena; apparently he was asked to review Rhine's work simply because of his professional status as a psychologist who belonged to the society. His review was read at a meeting of the society on 30 January 1935 and was energetically debated afterwards by the Salters, Whately Carington, Soal, and Dingwall, among others.[89] It is not hard to imagine the lines along which the discussion must have run.

Broadly speaking, Thouless's reaction to *Extra-Sensory Perception* was that of the "experimentalists" in the SPR like Whately Carington. His review failed almost entirely to comment upon the theoretical or speculative elements in *Extra-Sensory Perception,* and we can see why these features would not have excited comment from an audience of English psychical researchers, for the book's conclusions of course do not differ in any significant respect from those laboriously hammered out by the English tradition; indeed, they are often built upon them. The distinction between telepathy and clairvoyance, on which Rhine placed considerable emphasis, was mentioned by Thouless, but in the context of the Duke techniques for *demonstrating* the distinction; the conceptual differentiation between the two was by this time well established in England,[90] but no one there had yet attempted to separate the two experimentally. The English had once found it difficult to accept fully the possibility of clairvoyance, to be sure, but Ina Jephson's experiments had led at least some of them to consider it more seriously, and hence Rhine's claim to have demonstrated clairvoyance seemed by no means outrageous. E. R. Dodds, professor of Greek at the University of Birmingham, described the effect of Rhine's ideas on his thinking in these terms: "About clairvoyance, again, in view of the relatively restricted amount of unambiguous experimental evidence for it, I did not care to say much or commit myself very definitely [in a recent article]. The paper was written before your book appeared. Since then I have been looking into the case for clairvoyance with renewed interest. One of the leading English psychical researchers observed to me the other day that if there were no 'snag' in it, *Extra-Sensory Perception* was the most important contribution to psychical research since the war; and with this judgement I fully concur."[91]

Thouless's lack of surprise on this point, then, is not remarkable. Instead, what is notable about Thouless's review is its mixture of admiration and criticism for Rhine's report. No doubt Thouless's reaction to *Extra-Sensory Perception* was in this respect conditioned in part by his training in psychology. As a psychologist, he found many of the features of Rhine's experimental technique highly stimulating. He singled out Rhine's introduction of the Zener cards as "one of the most important changes that Rhine makes in method," commenting that they were much more easily "imaged";[92] he spoke approvingly of Rhine's habit of using preselected subjects; and he discussed at some length the conclusions to be drawn from some experiments in which subjects had been asked to identify the target card incorrectly. But his training and exchanges with psychical researchers had also sensitized him to the weaknesses in Rhine's description of the experimental conditions, and his review brought into the open the reservations that many members of the SPR had come to feel about the Duke work. Whately Carington had written Rhine on New Year's Day of 1935 to report that English psychical researchers were now of the opinion that while "chance" was completely excluded as a possible explanation for the Duke results, "leakage" was not. "Speaking frankly and brutally, it is clear that your work will not carry conviction to the critical mind until and unless you have systematically closed and rivetted and clinched the door of leakage till not the minutest crack remains."[93] Thouless's review four weeks later raised just this issue. While he accepted Rhine's calculations as unquestionably proving that something besides chance had given him his results, he pointed out that Rhine had sometimes naïvely misapplied his statistics; furthermore, he complained that in any case Rhine's emphasis on high anti-chance probabilities had led him to neglect the reporting of the precautions he had taken against possible sources of experimental error.[94] Beyond objecting that "it is generally quite impossible to discover for any particular experiment what the experimental conditions were," Thouless identified a number of specific respects in which Rhine's stated procedure did not exclude the possible use of normal means of cognition. The BT-5 method, for example, permitted the use of inference in guessing the last cards in a series of twenty-five and, worse yet, allowed subjects to come to recognize some cards from the back. Some of Rhine's experimental results—such as successful DT guessing—could only be explained by assuming clairvoyance, to be sure, but Thouless was unwilling to let those occasional experiments stand warrant for the remainder. For while the English psychical researchers were quite ready to admit extra-sensory perception as an extremely rare mental trait, they were quite unprepared for Rhine's claim that it was a widespread human power.

Indeed, it cannot be emphasized too strongly that successes on the scale reported by Rhine were totally unprecedented in psychical research in 1934. Soal's very recent work had sealed a growing conviction that psychical powers, if they existed, could only be the gift of an exceedingly small number of individuals or would only be demonstrable under irreplicable circumstances, or perhaps both. Now Rhine's work seemed to suggest that his simple techniques could be used to reveal "measurable telepathic and clairvoyant capacity in as many as one in three or four persons."[95] As Harry Price pointed out in a letter to *The Listener*, stimulated by its review of *Extra-Sensory Perception*, "the greatest phenomenon revealed by the book [is that] whereas for many years psychists have been scouring the world for first class percipients, and failing to find them, Dr. Rhine discovered a batch of them in his own university—even among his own students."[96] It was bewilderment at this feature of the Rhine work, not a conviction that it was necessarily fundamentally unsound, that now led Thouless to call for renewed experimentation under more careful testing conditions.

One other thing encouraged Thouless to think that Rhine's research might perhaps be worth following up. In the historical introduction to his book, Rhine had commented upon certain oversights in J. E. Coover's elaborate study (supposedly compellingly negative) that, when corrected, in fact offered data supportive of the view that extra-sensory perception actually existed.[97] Coover had compared the card-guessing performance of subjects in situations in which the experimenter saw the target card, a situation permitting telepathy, with the performances of subjects in situations in which the card was not known to the experimenter, treating the latter results as a control; but as Rhine pointed out, he had not allowed for the fact that this "control" group might be using clairvoyance to identify the cards, and indeed their guesses did seem to show signs of extra-chance success. Considering both groups of guesses together yielded a strong probability against chance being the sole operant in these experiments. Thouless now confirmed this suggestion and pursued it, devoting nearly half his review of Rhine's book to its résumé and elaboration, "since it is a common opinion that Coover's results were entirely negative and show nothing but chance distribution."[98] He determined more precisely the statistical unlikeliness of Coover's results as two hundred to one against chance—Coover had claimed "that various statistical treatments of the data fail to reveal any cause beyond chance operating"![99] Thouless criticized Coover sharply for this, as well as for having adopted an absurdly high level of statistical significance ($p = .0000221$); he concluded that "about six of Coover's hundred subjects had measurable power of exceeding chance expectation in the guessing of playing cards."[100] This revelation that the piece of experimental research that had seemed to be perhaps the single most damaging attack upon psychical abilities actually contained unrecognized evidence in favor of such powers was in its own way even more encouraging than Rhine's own experimental results.[101]

Hence on balance Thouless was still tentatively prepared to accept some of the claims of *Extra-Sensory Perception*. Granted, he wrote, weakness in the procedures and in the manner of reporting them would give ammunition to the skeptical, but "at least we may say that Dr Rhine has shifted the burden of proof on to those who deny that extra-sensory perception is a fairly common capacity."[102] Rhine replied to Thouless's review in a letter published in the SPR *Proceedings* in December 1935, insisting that by emphasizing his unusual degree of success, it had misrepresented his intentions: "The 'novelty' of the Duke work lies in its experimental separation for the first time of telepathy and clairvoyance, . . . [which] is much more novel than the number of subjects found."[103] Privately, however, he admitted the generally constructive tone of Thouless's remarks, describing them as "very exhaustive and a bit grudging, but on the whole a valuable contribution and one I appreciate."[104]

The text of Thouless's review of *Extra-Sensory Perception* in the SPR *Proceedings* was followed immediately by an appeal from the secretary of the society, W. H. Salter, speaking for its council, asking for members' participation in "further experiments (*a*) with assessable material, (*b*) under strict conditions, and (*c*) in such a form as to distinguish between the different types of extra-sensory perceptions."[105] It is not merely Salter's unforced use of that last phrase that reveals the influence of Rhine's book; his note makes it clear that it was "the high percentage of success obtained by Dr. Rhine" that had led the council to pursue the problem formally. The individual singled out to direct the new experiments was G.N.M. Tyrrell, who a few months before had been inspired by Rhine's book to renew his own researches. In 1921 Tyrrell had carried out several series of varied experiments with a young woman, Gertrude Johnson, who in the course of the work succeeded repeatedly in guessing the denominations of the first half-dozen in a freshly shuffled deck of cards.[106] He "had no opportunity to experiment at length" with Miss Johnson until 1934, when

Rhine's book stimulated him to return to experimental psychical research. During the fall of that year, Tyrrell devised a special apparatus for demonstrating generalized extra-sensory perception that did not involve card-guessing, although Rhine's methods inspired certain features of its form. Working again with Miss Johnson, Tyrrell again obtained some success, reported in the spring of 1935,[107] and it was because of this independent interest that Tyrrell had been chosen by the council as the man to conduct the new series of experiments. We have already seen that S. G. Soal had independently begun a similar attempt at replication. It is in fact quite evident that the publication of *Extra-Sensory Perception* was responsible for the revival in England of interest in carrying on large-scale, rigorous, quantifiable experimentation into psychical phenomena. The doyenne of English psychical research, Eleanor Sidgwick, no doubt represented the cautious hopes of many in her congratulatory letter to Rhine of March 1935: "I am reading what you write on the subject with great interest feeling that it has opened to us what is likely, for the moment at least, to be our most fruitful line of psychical investigation. It is a pity that good subjects seem to be so comparatively rare, though you seem to be in this respect very fortunate."[108] It is easy to imagine that Mrs. Sidgwick, at ninety, must have looked with some satisfaction at this work conforming in many respects to the traditions of the society that she had helped to found over fifty years before.

On the Continent, as well as in England, *Extra-Sensory Perception* was received by psychical researchers with general approval. However, the immediate consequences for research were far less marked in Europe, precisely because truly experimental psychical research had not become well established there. We may see this, for example, in Germany, where after the death of Schrenck-Notzing, Hans Driesch had become the most eminent representative of psychical research—primarily because of his reputation as a biologist and a philosopher—but a representative of philosophical rather than experimental activity. Driesch had begun his career as an embryologist in the 1880s, and his experiments in this field led him increasingly to try to explain philosophically the fundamental principles behind biological development. By 1895 he had become a vitalist, and he subsequently began to attempt to interpret organic growth in terms of the activity of a nonmechanistic agent that he labeled an "entelechy"; twenty years later he had completed the transition from experimental biology to philosophy and had become professor of philosophy at Cologne in 1919 before moving to Leipzig in 1921. By this time he had already been attracted to psychical research by its congruence with the implications of his own vitalistic biology. Driesch did not himself undertake any experimental work in the field, although he acted as an observer in a number of mediumistic investigations. His continuing concern as a philosopher was rather to evolve a general methodological and theoretical framework to direct and organize psychical research.[109]

Driesch had offered his final perspective on the field in his *Parapsychologie* (1932). Unlike Richet's *Traité* of a decade before, Driesch's work did not pretend to be a textbook; it was presented as a student's guide, to be of help in grappling with the methodological and theoretical difficulties inherent in psychical research. Driesch showed far more concern for the possibilities of deception than had Richet and in effect allied himself with the critical approach of the leadership of the SPR (he had been president in 1926/27): he insisted that no "paraphysical" phenomena were well enough established to warrant study and that "parapsychical," or mental, phenomena alone should receive attention. In this latter class he distinguished five types—psychometry, clairvoyance, prophecy, telepathy, and thought-reading—provisionally accepting each as a fundamental and irreducible phenomenon. Driesch separated the latter two because of considerations of varying agency in thought-transference. "In *telepathy* [*Telepathie*] the agent gives his mental content actively, whether unconsciously or consciously, while the percipient 'receives' in a purely passive manner. In *thought-reading* [*Gedankenlesen*], on the contrary, the percipient is active, he wants, even if only unconsciously, to receive; the agent, who gives his knowledge, his mental content..., here plays a wholly passive part."[110] It was perhaps unusual to make this distinction so sharply—the English tradition had been to treat the two classes as one, first the "agent" and later the "percipient" being understood to have the active role—but it enabled Driesch to distinguish between spontaneous and experimental results.[111] In any case, clairvoyance and thought-transmission together laid the phenomenal basis for the general theory of parapsychology that Driesch permitted himself to sketch out at the conclusion of his book. Between them, they implied the existence of a superpersonal, nonspatial "mental field," in which living personal minds could directly exchange knowledge. Indeed, Driesch argued, there was sufficient evidence that discarnate minds could exchange knowledge with living ones to justify taking seriously the spiritist hypothesis of post-mortem survival (which, however, he preferred to label "monadism"). In any case, his "mental field" provided a uniform theoretical framework for these and all other parapsychological phenomena.

Driesch was sent a copy of *Extra-Sensory Perception* in April 1934, and he quickly replied to express his pleasure with the work. "It is, in fact, a very important book," he wrote, "one of the best we possess in this field of science. For it is absolutely convincing." He went on, however, in such a way as to make it clear that his interest in the theoretical aspects of parapsychology was still paramount. Rhine's emphasis upon the similar success enjoyed by his subjects at telepathy and clairvoyance and his suggestion that the two classes of phenomena might therefore be reducible to a single one led Driesch to urge upon Rhine a sharper distinction between the two. Driesch argued that it was more appropriate to view telepathy as related to clairvoyance in the same way that seeing was related to hearing—"there is something equally present in both cases, namely paranormal *sensation* and normal *sensation*"—and Rhine very largely acquiesced in this. Significantly, Driesch also insisted upon the importance of distinguishing the difference in appearance between the experimental phenomena Rhine had dealt with and the spontaneous and mediumistic phenomena in the literature,[112] for Driesch, like many of the leaders of the SPR, did not share Rhine's complete commitment to a purely experimental program of research.

Nevertheless, like them, Driesch unquestionably felt that Rhine's work could be of real significance for psychical research. He subsequently took it upon himself to translate Rhine's second book for a German audience and contributed to this edition a foreword that suggested where he felt its significance lay. While he insisted that the study of mediums and spontaneous phenomena carried on by such researchers as the SPR and Osty should not be neglected in favor of the narrowly experimental procedure here described—"both one *and* the other is the only correct approach"—he could not help recognizing that the experimental demonstration of extra-sensory perception provided a strong, explicit repudiation of the materialistic world view that he himself had so long opposed.[113]

The more strictly experimental position had at least one committed defender in Germany, and on him *Extra-Sensory Perception* was expectably an incentive to further research. Hans Bender, a young psychologist at the Psychology Laboratory in Bonn, had begun his own experiments on clairvoyance in 1932, knowing relatively little of the psychical-research literature. It will be more convenient to look at Bender's career later,[114] but here we should at least say that he saw Rhine's work as supporting his own researches and as suggesting areas for further study. In 1935 Bender prepared a review of Rhine's book for the *Zeitschrift für Psychologie,* describing it warmly as "a work which, despite the heretofore and still controversial character of its theoretical conclusions, emphasized by the author, is still capable of freeing future works in this area from proving the justifiability of such investigations and of leading them directly into numerous problems that it raises for inquiry."[115] For Bender, Rhine's book marked a watershed in psychical research: at last, it seemed, a technique had been devised for bringing telepathy and clairvoyance under experimental control.

The French reaction to *Extra-Sensory Perception* was equally enthusiastic; the work was interpreted as fulfilling a longstanding tradition of research. René Sudre, who had earlier combined strong support for experimental psychical research with scorn for the English statistics of card-guessing and had encouraged researchers to pursue instead the one irrefutable experiment with selected subjects that would transcend "fractions of success," now hastened to assert that Rhine's data provided just the sort of statistically unimpeachable conclusions that he had been calling for. "I have no hesitation in saying," he announced, "that this is the first indisputable demonstration of the reality of clairvoyance by the statistical method."[116] The book was reviewed for the *Revue métapsychique* in July/August 1934 by César de Vesme, who had been associated with the field for forty years and had been deeply disturbed by Soal's negative reports of 1931.[117] Quoting Rhine's conclusions at some length, he too insisted upon their traditional character—they were "more or less conformable" to the results obtained by other investigators, from F.W.H. Myers to Eugène Osty. "The difference that exists between Dr. Rhine's 'Extra-Sensory Perception' and Dr. Osty's 'Connaissance Paranormale,' " de Vesme concluded, "is very small."[118]

What René Warcollier's first reaction to the book was we do not know. But we may find some indication in Gardner Murphy's response, given the very close affinity between the two men in their approach to psychical research. Murphy had recovered quite suddenly from a decade of ill health early in 1934, just when *Extra-Sensory Perception* appeared, and his immediate feeling was evidently one of unalloyed delight.[119] His more considered judgments, published in a review in the *Journal of General Psychology* later that year, stressed Rhine's successful mastery of the experimental problems as the one thing that distinguished *Extra-Sensory Perception* from previous work:

> The *finesse* of the experimental method is perhaps not equal to that of the method used by the Gröningen experimenters . . ., nor is the psychological analysis of the conditions attending success as thorough and illuminating as that of Warcollier (articles since 1921 in the *Revue Métapsychique*); but at our present stage in the history of the subject these are not our chief needs. Our first need is to get the phenomena *under control,* with a group of subjects willing to work for a long time with a standard method upon which planned systematic variations can be imposed, the results being uniformly compared against chance expectation. Dr. Rhine has achieved all these things.[120]

Murphy, of course, had long been sure that successful experimentation was entirely feasible and was essential to the development of psychical research, while the English were on the whole still far from convinced on either point. Hence Murphy, more unreservedly than the English, was willing to take Rhine's approach as paradigmatic for future research, and whereas in England *Extra-Sensory Perception* led researchers to try to duplicate its results, in the United States it produced attempts at articulating its consequences experimentally.

Surely both these responses owe a great deal to the entirely "normal" character of Rhine's methods and conclusions. The preexisting tradition of psychical research was bound to take Rhine's work with particular seriousness because it conformed so well to everything that that tradition felt it had learned. The book brought together the conclusions of a great number of different researchers of the past representing a remarkably wide range of fields—dowsing, mediumship, experimentation of all sorts, parlor games—and it presented these conclusions, found piecemeal in the earlier literature, as all derivable from a single model experimental situation, a situation that itself had strong elements of familiarity. No wonder Rhine's book led quickly to a variety of further research. Situations like this, when one piece of work serves to give orientation or direction to a field, to cause a diversity of available theory and data to coalesce suddenly into a coherent whole, are common in the history of science; Benjamin Franklin's letters of the 1740s on electricity provide an often cited example. *Extra-Sensory Perception* had a similar effect upon psychical research. It united into a single system conclusions that had previously been enunciated in relative isolation; it offered a simple experimental technique for their demonstration; it suggested areas wherein the system could be tested or expanded; and above all, it seemed to promise success.

CHAPTER SIX
Parapsychology in Its Public Aspect

I

It seems unlikely that parapsychology could ever have been given a measure of toleration by the American scientific community without the institutional and popular support that it soon enjoyed. National publicity and popular enthusiasm might render the Duke enterprise somewhat suspect to scientific orthodoxy, but on the scale to which it grew it forced the orthodox to examine this new subject and to treat its claims seriously. Often the examination was openly intended to find the errors in parapsychology and to expose them so that a naïve public would no longer be deceived; but at least there *was* an examination. One of the most difficult tasks for any unorthodox claimant to scientific status is to get the scientific mainstream to study its evidence with any care, and the enormous welling up of publicity that occurred between 1934 and 1938 to some extent accomplished this for parapsychology, as we will see. No less vital, however, was the institutional and financial backing that made it possible for a few would-be parapsychologists to hope for careers in the field. The prospect of professional opportunities was essential if young researchers were to be attracted to and remain with the discipline. But of most importance to the future of the new science was Rhine's own position at Duke, which from the first assured it at least a limited academic respectability.

Duke University had come into existence only a few years before Rhine's arrival, as a refoundation of a Methodist school, Trinity College, established in Durham, North Carolina, almost a century earlier. Trinity had long been a principal beneficiary of the philanthropy of the brothers Ben and James B. Duke, who had created the American Tobacco Company in the 1890s and subsequently extended the family fortunes by investing in the production of electrical power for the industrialization of the Carolinas. The new institution was not a dream of the Dukes, however, but of William Preston Few, president of Trinity College from 1910 on. It was Few, ambitious for his school and for education in the South, who in 1924 persuaded James B. Duke to endow a university with Trinity College at its core, and indeed it was Few's idea that it should be named for the Dukes. Born in rural South Carolina and educated in Methodist schools there, Few had done doctoral work in English at Harvard and had come to teach at Trinity College in 1896. Although not a gregarious or outwardly forceful person, Few was an extremely effective leader for the school, for he was dominated by his concern for its health and future development. And his very unobtrusiveness made him well suited to guide a denominational college towards academic excellence in the face of possible church pressure and southern public opinion.[1]

Of course, Few himself believed it possible—and essential—to incorporate Christian values into academic life, not only at a liberal arts college but at a research university. It was not merely that Christian instruction should be taught in universities—as the Duke motto, "Religio et eruditio," might imply—although the new school did include a graduate "School of Religious Training." Few felt strongly that religious values should not just coexist with academic life, but actively enrich it. Reflecting on his career and the development of the university, he explained that "too many people in the universities and out have come to assume that freedom in religion means freedom from religion; and to reassert a strong moral note will not always be easy for universities in our day. It would, none the less, seem to me to be perfectly clear that information, training, learning, scientific research, intellectual culture—any or all these alone will not be sufficient to save the world of our troubled day. The world needs spiritual regeneration, and our University halls ought to echo with the voice of moral authority."[2]

Few's convictions determined the way in which psychology developed at Duke. At Trinity College there had been no psychology department as such; instead, students were offered occasional courses in psychology within the departments of Biblical literature and religious studies, philosophy, and education. In 1924, however, Few took steps to secure the appointment of a professor of psychology, who would be a member of the philosophy department. The successful candidate was George I. Mount, an older man who had already taught at Iowa State and Dubuque. He had been promoted by his graduate mentor at Iowa, Carl E. Seashore, as "from a family of long and highly respected standing in the Methodist Church, and... an active church worker."[3] Still, it was not *only* Methodist virtue that Duke was looking for; the chairman of the philosophy department, W. I. Cranford, expressed the hope to Mount that he would "be able to give a good deal of time and attention to the experimental work [in which] practically nothing has ever been done here."[4]

Mount stayed at Duke for only one year, after which he moved on to the University of Southern California, and in March 1926 President Few initiated another search for a "first-rate psychologist" by getting in touch with William McDougall at Harvard and asking him to suggest possible candidates, either an outstanding mature psychologist or a very promising young one.[5] It is understandable why, of all the psychologists in America, Few should have made inquiries first of McDougall, for Few dreaded the new behaviorist psychology, of which McDougall was then the most prominent critic. McDougall began by commenting favorably upon W. S. Taylor's qualifications for the position, adding, however, the reservation that "when here, he had given adherence to that foolish and pernicious perversion of psychology which goes by the name 'Behaviorism' of the extreme kind," and Few quickly replied that he had "dropped the man at once." Next, McDougall sent a list of rising young psychologists, including E. G. Boring, W. S. Hunter, E. C. Tolman, and K. S. Lashley—describing all except Boring as "infected with the behaviorist virus."[6]

A few days later McDougall, who had been feeling increasingly dissatisfied with Harvard and Boston, announced his own availability for the Duke post.[7] Few immediately invited him to Durham for an interview and began to ask for opinions on the appointment. Characteristically, among those Few chose to consult was the dean of Duke's School of Religion, E. D. Soper, and Soper in turn asked a colleague in the Yale Divinity School, Luther Allan Weigle, to make inquiries for him. Weigle replied that McDougall would probably figure on most psychologists' list of the ten most notable men in their field. He added, however, that

> he has not taken the place of leadership in American psychology that he undoubtedly expected to take, and one of my friends believes that he has lost his chance for that leadership. It is not because of his attitude towards behaviorism. Watsonian behaviorism represents only one wing of American Psychology and is a passing fad. . . . Strictly speaking, McDougall is not an experimental psychologist. He is rather a speculative psychologist, belonging to the school of Ward and his followers. . . . The American psychologists do not regard his work as scientific, in the strict sense of the term.[8]

Weigle's advice was to proceed slowly. Few, however, hastened to make McDougall an offer in July, and, after some hesitation, McDougall accepted. As we have seen, McDougall came to Duke for good in the summer of 1927.[9]

Why was President Few so eager to bring McDougall to Duke? He did not seem likely to build up the sort of dynamic psychology department that Few's referees automatically supposed a new university would want; and he had little interest in the problems that were coming to occupy more and more of the attention of American psychologists. Still, there was no question of McDougall's stature. Duke would undoubtedly be acquiring an eminence of international reputation, even if that reputation had been won outside the main lines of academic psychology as it was coming to be taught in America. As McDougall pointed out to Few, his training had been such as to provide him with an incomparable breadth of experience and competence.[10] And at a mere fifty-six, he was still a productive and creative mind; he was still actively publishing, and he continued to do so at Duke at a remarkable rate.

To Few, however, none of these assets was as important as McDougall's indomitable opposition to behaviorist psychology and, more generally, to materialistic philosophy. This far outweighed his possible limitations as a leader of experimental research. Few summarized McDougall's qualifications in just such terms in announcing the appointment to the student newspaper, the *Chronicle:* "Above all, Professor McDougall is recognized as the outstanding leader of the group of psychologists who have refused to accept the mechanistic view of mind and life which certain current schools of psychology have maintained. And incidentally, for the reinforcement on philosophical grounds of the theories of education, of personality, of life,—the essential foundations upon which this institution through all its history has rested—there is, I think, no abler spokesman than Professor McDougall.'"[11] McDougall did not disappoint Few's characterization. In his Duke years his customary antimaterialism became pointedly allied to a defense of religious values; moreover he continued to invoke and defend psychical research, which had always been a prominent thread in the fabric of his antimaterialism, in the books he published from Durham. In the first of them, for example, *Modern Materialism and Emergent Evolution,* McDougall appealed to telepathic phenomena as indicating "a mode of manifestation to one person of the life activities of another . . . which is an exception to the generalization that the life of other beings is manifested to each one of us only through physical media and impressions of the senses."[12] In turn, Few came to appreciate the potential importance of psychical research as a buttress of moral life.

During his first years at Duke, McDougall built up a departmental staff from former students and disciples. In 1928 Karl Zener, a specialist in psychophysical and Pavlovian experimentation who had recently finished his dissertation at Harvard and was teaching at Princeton, joined the department. Two years later, a somewhat older psychologist, Helge Lundholm, arrived. Though Lundholm had not been one of McDougall's students, he had come into contact with McDougall while a research fellow at Harvard in the mid-1920s, and of all the members of the department it was he who most closely identified himself with McDougall's approach to psychology. In 1931 another former student of McDougall's, Donald K. Adams (specializing in the psychology of learning), arrived from a two-year postdoctoral stay in Berlin; and in the same year, J. B. Rhine shifted to psychology from the philosophy department. A graduate program was well established by now: the first Duke M. A. in psychology had been awarded in 1928, and the first Ph.D. in 1930. The department temporarily stabilized at six members in 1934 with the appointment of William Stern. Stern, a sixty-five-year-old émigré from Germany, never became deeply involved in departmental affairs and died in 1938.

As one might expect of a staff personally assembled by McDougall, the members of his department were all sympathetic to his approach to psychology and indeed showed what might broadly be termed "McDougallian" thought in their own work. Thus one of Zener's modifications of classical Pavlovian conditioned-reflex work was to show that the subject animals did not serve merely as inert test objects in these tests, but played an active, motivated role; Donald Adams sought to emphasize in his learning theory that the learning process was holistic in nature and could not simply be analyzed into atomistic stimulus-response reinforcements. Their espousal of Gestalt psychology and the field theory of Kurt Lewin was likewise in line with their general sympathy for psychological holism and for the active role of individual consciousness. Given their psychological orientation and their obvious personal admiration for McDougall, it is not surprising that the department should initially have shown no hostility to J. B. Rhine's pursuit of psychical research, so warmly supported by their chairman. Zener, Lundholm, and Adams all helped Rhine in his early work, as witnesses and as advisors, and their assistance continued in some measure well after the publication of *Extra-Sensory Perception* in 1934. Adams, probably the least sympathetic of the three, was still willing to serve as an occasional test subject as late as 1936. Lundholm showed a continued willingness to confront some of the psychological implications of psychical phenomena.[13] Karl Zener of course helped Rhine to design the original test cards, although he later expressed some dismay when they were referred to as "Zener cards" in Rhine's monograph, so that in the English edition (1935) Rhine changed their name to "E.S.P. cards."[14] The granting in 1930 of an M.A. to Harvey Frick for his experiments and, even more significantly, the granting of a Ph.D. in 1933 to John Thomas for his dissertation testing the spiritist hypothesis certainly showed that the department was willing to accept the pursuit of psychical research in at least a peripheral role.

Yet to describe the members of the department as wholly favorable to the field would be going too far. Respect for McDougall and sympathy for his psychological orientation undoubtedly made them more open-minded than many American psychologists, but they still felt that telepathy and clairvoyance were at best subjects of marginal relevance or importance to psychology proper. This became a matter of real concern in 1934, when the younger Duke psychologists began to perceive Rhine's eager and successful prosecution of psychical research as tending to overshadow the normal research and training program of a psychology department. His attractive personality and the fascination of his experimental results made him inevitably a focus of student interest, which the department was powerless to affect. Their frustration finally burst out in the spring of 1934— before *Extra-Sensory Perception* had even appeared—in a private protest sent to McDougall, who was spending his customary half-year in England. "We share with you an interest in psychical research," they wrote, "and we agree with you that psychical research should have a place in a university, but we do not like to have such research attain a dominance which would exclude the investigation of psychological problems in general."[15] They accused Rhine of trying to influence students to do their dissertations on parapsychological topics; of promising a scholarship to a good research subject; of building up his own situation at the expense of the rest of the staff; and in general of spreading the view that an active interest in psychical research was a necessary condition for success in the Duke psychology graduate program.

McDougall could understand the professional apprehensions of his younger colleagues and suggested cautiously to Rhine (without mentioning their letter) that he consider limiting his activities.

William McDougall and J. B. Rhine in the early 1930s. (Courtesy of the Duke University Archives.)

> It would be a great pity if our endeavour to introduce P[sychical] R[esearch] into the circle of university studies should go awry through excess of zeal leading students to neglect all the more orthodox parts of psychology in favour of P. R. I am sure the danger is a real one & that in any case we have to take precautions against it. It was with that in mind that I have been inclined to advocate confining lectures & all work in P. R. to graduate students, & even possibly (tho' here I am quite uncertain & open to guidance) to men students. The fact that our colleagues seem to observe indications of need for greater precautions than we have taken, strengthens this inclination in me, & I would wish you to consider it carefully.[16]

But when Rhine found out about his colleagues' surreptitious complaint later that summer, he was deeply offended. He felt that their sensitivity to being associated with an activity that their professional peers would find suspect had led them to view his actions in the worst possible light, and he began to look for some way to establish his own independence of action at Duke. From mid-1934 on, Rhine looked forward to establishing psychical research in a University-based "institute" distinct from the psychology department.

To found an autonomous research institute for the scientific study of psychical phenomena had been an ambition of Rhine's for some time—a prototype had been the "Institute for Experimental Religion" that he outlined in 1929. He was himself temperamentally drawn to independent research and, besides, had always been concerned that academic orthodoxy might impose severe constraints upon psychical investigation—a concern that now seemed borne out by his colleagues' campaign against him. Furthermore, McDougall too had once expressed thoughts along these lines. In 1930, at the behest of President Few, the Duke faculty had met repeatedly to explore what directions the growing university should follow; among the possibilities discussed was the creation of an interdisciplinary institute for the social sciences.[17] McDougall contributed to this discussion his own hopes for an "Institute for Research in the Sciences of Man and Society," giving psychology a particular eminence therein and adding that "a very special branch, hitherto neglected by all universities, namely Psychical Research, should be represented in a sub-department."[18] Although nothing came of these social-science proposals, the idea remained alive in McDougall's mind, and he wholeheartedly supported Rhine in his project when eventually it began to be implemented.

Funds, of course, were the first requisite for an independent research unit. Here Rhine was fortunate in a visit to Duke in April 1934 by the English medium Eileen Garrett for tests of her abilities in telepathy and clairvoyance. Mrs. Garrett's trip to North Carolina had been sponsored by Frances Bolton, the wife of a wealthy Cleveland industrialist who was herself independently wealthy and a willing philanthropist.[19] Through Mrs. Garrett, Rhine soon came into correspondence with Mrs. Bolton, discussing with her their common interest in psychical research. Mrs. Bolton made it clear that her principal concern was with the question, Do we survive after death? Rhine by now had seen the wisdom of deferring this question; he explained to her that while the survival problem was his own ultimate research goal, he felt it necessary to work slowly towards that end, systematically and scientifically, from topics more immediately susceptible of experimental investigation, like telepathy and clairvoyance.[20] He also described his departmental difficulties at Duke and his hopes for a research institute and eventually asked whether she would be willing to help finance such an institute: "I have estimated that, if Duke gives us suitable space—as I am assured it will—and furnishes us with the necessary capital equipment, special library, office materials, full-time secretary, janitor, etc. to a value of about $5,000. per year, that with $25,000. annuity 'for 10 or 15 years,' we could establish a splendid institute amply manned and equipped to enable us to push toward the main goal, and to give attention to the side interests and lateral spread at the same time." Towards this program Mrs. Bolton agreed to contribute ten thousand dollars per year for three to five years, beginning in September 1935.[21]

One factor that clearly helped Mrs. Bolton to her decision was William McDougall's firm support. McDougall traveled with Rhine to Washington in February 1935 to discuss the details of her gift with her, and it was he who drew up the memorandum of their discussion.[22] While the document does not mention the word *institute,* the money was clearly to be used for what Rhine had been dreaming of, a research organization within the university with ties to the psychology department but as little encumbered with normal faculty responsibilities and as financially independent as possible. One fourth of the annual gift would supplement Rhine's salary; he was to retain his appointment as associate professor in the psychology department but would be required to teach no more than one course per semester. Another fourth was to be used to pay a full-time research assistant, thus in effect creating a second position in parapsychology at Duke. The remainder would subsidize travel and publication, the services of special subjects, general expenses for apparatus, and so forth. Duke would be requested to supply Rhine with secretarial staff and office and laboratory space. Four trustees were named to administer the gift under the title, "William McDougall Research Fund." Rhine was to find other financial backers, but Mrs. Bolton's support was certainly the most crucial for the establishment of parapsychology at Duke.

At the same time that he was securing the Bolton gift, Rhine was also negotiating for the creation of graduate fellowships in parapsychology. He had suggested to Elwood Worcester, of the Boston SPR, that two such fellowships be set up in memory of Walter Franklin Prince, who had died in August 1934—one for a graduate student at Duke and one for the support of an assistant research officer of the Boston society, who could work part-time in the Harvard psychology laboratory.[23] On Worcester's advice, Rhine approached Mrs. William (Ellen A.) Wood, the wealthy Boston matron who had been a friend of the great Mrs. Piper and whose generosity had made possible the establishment of the Boston SPR in 1925. Again he was successful: Mrs. Wood agreed to sponsor two fellowships of one thousand dollars annually for five years—*both* fellowships to be held at Duke. They were to be known as the Walter Franklin Prince Fellowships in Psychology and were to be integrated with the proposed institute when it came into being.[24] By the fall of 1935 both gifts had been accepted by the university, and Rhine now had very considerable support for his research activities. His work held out to these thoughtful patrons the hope of a continuation and development of the tradition of William James and Walter Franklin Prince; in the slightly improved economic conditions of the mid-1930s, they were once more willing to invest their money in rigorous investigations that might lead science to the final question of life after death.

It is instructive here to understand how Rhine's financial support compared with the monies available to other scholars and researchers at Duke. These were of course the days before heavy government support had given birth to Big Science. Most funding for scientific research was provided by foundations or by universities themselves, and in this respect Duke had a comparatively undistinguished record, even for these lean years. For 1934/35 the research council of the university had proposed a budget of twelve thousand dollars for the research needs of the entire college of arts and sciences, with an additional eight thousand dollars to be utilized for the salaries of research assistants.[25] By the next year, the budgetary request for research had risen to twenty thousand dollars, with the hope that the utilization of these funds might attract a matching sum from sources outside the university. In fact, Duke spent just over forty-six thousand dollars on research in 1935/36.[26]

Out of research council funds—apparently their only source of research funding—the members of the psychology department received a total of $595 in 1934/35 and $1,650 in 1935/36; and $350 of the latter amount went to Rhine.[27] Thus, even if the Bolton bequest is considered in isolation, and even discounting the quarter of it that went to supplement Rhine's income and the quarter meant for a full-time research assistant, Rhine still had five thousand dollars for "research," compared with the thirteen hundred dollars available for all the other members of the psychology department together. And this money of his was more than a tenth of what the whole university—including the medical school—spent on research in 1935/36! The Walter Franklin Prince Fellowships, too, were as generous as anything in the university. Duke provided forty-eight stipends of $350 to $650 to graduate students during 1935/36 (none of which went to students in psychology), and the only grant comparable to the Prince fellowships was the Angier B. Duke Memorial Fellowship, which also paid $1,000 per year.[28] These figures show that by 1936 Rhine had secured funds that were sizable for any social or behavioral scientist and huge by Duke standards.

The private funding and the national publicity that Rhine was beginning to attract to Duke were welcomed eagerly by President Few and the university administration. The publicity was two-edged, as we will see, but Few seems to have believed that it was "worth millions" to Duke, and he encouraged Rhine in his contacts with the press. Yet Few saw much more than financial advantage in Rhine's work, which appeared to him a vindication of his sense of the university's purpose. Rhine kept him informed of his research, and Few praised it warmly, suggesting that he might someday write something to "top James' Varieties of Religious Experience."[29] Mrs. Few, who was deeply interested in psychic matters, was also consistently enthusiastic about ESP research. Moreover, as Rhine discovered to his pleasure, two of the Duke trustees were convinced dowsers, and they too backed his work.[30] This firm administrative support of ESP had many tangible consequences: provision of a secretary to handle Rhine's correspondence; the establishment of special laboratory quarters for Rhine's staff when the psychology department moved into a new building in the summer of 1935; and steady promotion and raises in salary for Rhine himself.[31]

With independent financial support and the backing of Few and McDougall, Rhine was able to establish the nucleus of a research team at Duke with no interference from his colleagues. This happened in a wholly conventional way, similar to that in which any research specialty comes into existence. Students, both undergraduate and graduate, were drawn into participation as subjects or testers in Rhine's experiments (a process that had begun well before 1934), and eventually some decided to commit themselves at least partially to parapsychology. By the summer of 1934 this nucleus comprised J. G. Pratt, Charles E.

Stuart, and Sara and George Zirkle. All were graduate students in psychology; all but Pratt had originally been successful test subjects in Rhine's early work; and none but Stuart was formally concentrating on parapsychology as his major graduate field within the department. Their common commitment to the work was the product of a dedication to Rhine and, no doubt, to the continuing fascination of the subject matter. George Zirkle eventually took his degree and with Sara left Duke for a position in administration at Hanover College (Indiana), where they carried on some experimentation briefly but did not remain actively involved in the field; Gaither Pratt and Charles Stuart, however, stayed at Duke and, with Rhine, became the fixed core of the program in parapsychology. Stuart died in 1947; Pratt remained with Rhine until 1963.

Pratt, like Rhine and Murphy, had experienced an early crisis of conscience over traditional Christianity, though it did not lead him quite so immediately into psychical research.[32] He had come to Duke out of rural central North Carolina in 1927, intending to realize childhood plans of entering the Methodist ministry, and had gone on to the Duke School of Religion in 1931 after obtaining an excellent undergraduate record. But by this time he had become consumed by doubt that traditional religion, particularly the southern Methodism of his home, could have any meaning for him. During his first year in the School of Religion he turned to Rhine, with whom he had taken several courses, to discuss his doubts. It was partly at Rhine's suggestion that Pratt transferred to the graduate school in psychology, hoping that a science of human nature would prove more successful than religion at meeting the needs of mankind. At this point Pratt had no intention of trying to make psychical research a career. While he knew Rhine well and in fact assisted him in ESP experiments from the spring of 1933 on, in order to earn money for his education, it was to psychology proper that he was drawn; he soon became Donald Adams's student and carried out research in the psychology of learning first for his A.M. degree (1933) and then for a Ph.D. (1936).[33] Yet during these same years he continued his involvement with the parapsychological experimentation—the distance work with Hubert Pearce in 1933; the study of Eileen Garrett in 1934—a division of interest that was one of the spurs to the department's complaint against Rhine. In the fall of 1935 he moved to New York temporarily to work with Gardner Murphy on parapsychology while completing his dissertation, and it was only in the course of the next year that Pratt made the decision to commit himself entirely to parapsychology.

Charles Stuart was apparently drawn to parapsychology for other reasons—not to fulfill some inner psychological need, but to satisfy a simple fascination with the experimental successes he had enjoyed as an undergraduate research subject in 1930, in some of Rhine's earliest work. He did some experiments of his own that summer, before his junior year, and wrote back excitedly to Rhine about his results. He continued to assist in Rhine's research, completing his B.A. with a major in mathematics and a minor in philosophy, and passed almost inevitably into the graduate program, taking his Ph.D. under Rhine in 1941.[34]

The deeper concerns that had driven Rhine, Murphy, and even Pratt into parapsychology were by no means the only reasons why a bright young student might find the field attractive, especially as prospects of careers in parapsychology began to appear. In the course of the next several years, a number of other graduate students attached themselves to Rhine's program: Burke Smith, who had received his bachelor's degree in psychology from Duke (1934); Margaret Pegram, who as a Guilford College (North Carolina) undergraduate had been introduced to *Extra-Sensory Perception* by one of her teachers; and Joseph Woodruff, from Tarkio College (Missouri), where his psychology professor had been a Duke Ph.D. (1932).[35]

Rhine did his best to ally the Duke group with the existing tradition of psychical research in American universities, but with only limited success; in no sense was there yet established a true community of interested and active scholars in the field. J. E. Coover, the one American academic with an actual university position in psychical research, had long maintained a disbelieving inactivity and had contributed nothing to the field (apart from a few debunking lectures and quasi-popular articles) since his massive publication of 1917. Rhine sent Coover a copy of *Extra-Sensory Perception* and asked for his opinions and advice but found it inordinately difficult to extract so much as an acknowledgment from him.[36] George Estabrooks was a second individual with whom Rhine tried to establish communications, but it soon became clear that Estabrooks had deliberately chosen to remove himself from psychical research. He had gone from Harvard to Springfield College to Colgate, where in 1934 he was an associate professor of psychology and director of placement. At Springfield he had tried unsuccessfully to duplicate his Harvard successes at telepathic and clairvoyant guessing and had grown skeptical about the first work, though he never actually repudiated it. But he had not involved himself with it at Colgate and now, though he exchanged friendly and encouraging letters with Rhine, politely declined Rhine's invitation to return to experimental research.[37]

The enthusiastic reaction elicited when Rhine approached Gardner Murphy, at Columbia, more than made up for these other disappointing responses, for Murphy's eager enlistment in Rhine's plans led to the establishment of another solid base for parapsychology in New York City. In 1934 Murphy possessed a certain symbolic leadership as the one academic psychologist who was manifestly sympathetic to psychical research, even though since holding Hodgson money in 1925 he had been prevented by a debilitating illness from devoting much attention to the subject. What energy he had had he had used to compose and see through the press three books that gave him an immediate reputation among psychologists: *An Historical Introduction to Modern Psychology* (1929), *An Outline of Abnormal Psychology* (1929), and *General Psychology* (1933). The illness had just been brought under control—by means of a medically unorthodox diet—when *Extra-Sensory Perception* appeared. Thus Rhine's approach to Murphy in May 1934 came at a fortunate moment. In addition to inquiring cautiously whether Murphy might be willing to review the monograph for a psychological journal, he also asked whether it might be possible for his two principal assistants, Pratt and Stuart, to get further training with Murphy in New York. Murphy's reply was an offer of unreserved alliance. Not only did he agree to write the review, he enthusiastically endorsed Rhine's plans for his aides, and with breath-taking generosity offered to defray their costs by a personal contribution of one thousand dollars from the royalties on his books![38] In September of the next year, Pratt took advantage of the proposal and went to New York to begin what turned out to be a two-year association with Murphy.

Over the next several years Murphy developed his renewed involvement with parapsychology in a number of directions. He was slow to take up experimentation himself because his experience a decade before had convinced him that he lacked the gift of discovering good subjects. Instead, he gave over his attention to advising and supervising younger investigators in their work, at first Pratt and then others in the New York area who began to take up ESP research. As an advisor Murphy proved to offer sharp contrasts with Rhine. Murphy met Rhine's enthusiasms with caution and restraint and complemented the Duke concentration upon the mathematics of card-guessing with his own interest in qualitative target materials and the psychodynamics of ESP guessing. Their differences often led to private exasperation on both sides, but Murphy always remained a firm public supporter of the Duke research. Indeed, by 1937 he had begun (like Rhine at Duke) to encourage undergraduate and graduate students at Columbia to undertake parapsychological experiments, and even to try to make contact with the still rabidly spiritualist New York-based ASPR so as to be able to use their facilities. Murphy's deepening involvement with parapsychology meant a great deal to the new field, for his growing stature in psychology (he was elected to the American Psychological Association (APA) council of directors in 1937, and he missed election to the presidency by only eleven votes in 1939) forced his professional colleagues to take it slightly more seriously. Equally importantly, his insistence eventually led to the opening up of the Hodgson Fund to parapsychologists.[39] By the end of the decade, Gardner Murphy, with Rhine, was in a position of real (not merely symbolic) and independent leadership, even if he and Columbia were far less in the public eye than Duke.

It is wholly natural for a new scientific specialty to coalesce in the laboratory of a researcher who has discovered an inexplicable effect; natural, too, that it should soon be reinforced by other researchers who have been looking for analogous effects. One further way in which a truly new specialty may grow is by attracting individuals from marginally related, well-established disciplines, generally young scientists hopeful of making a career for themselves. In the case of parapsychology, the established field with the closest relationship was psychology itself, and a few young psychologists did in fact think of moving into the new area. Their move was of course made difficult by the labels of unprofessionalism and unorthodoxy that clung to the heir of psychical research. The reactions of professional psychologists to ESP were by no means simple and will be discussed in detail in later chapters. Generally speaking, however, they were of three classes: (1) actively disbelieving or hostile; (2) neutral or incurious, more concerned to pursue their own research, well under way, than to enter a wholly new area of unproven fruitfulness, however provocative; and (3) interested enough to carry out investigations of their own, whether skeptical or sympathetic in spirit. From this third class of psychologists Rhine received a number of letters of more or less open-minded inquiry.

Two of these contacts led to additions to the membership of the new discipline. C. R. Carpenter had been a Duke undergraduate psychology major; but since he had graduated in 1928, he had not known Rhine well. He had gone to Stanford for graduate work (in animal behavior) and was finishing his dissertation when *Extra-Sensory Perception* came out; he had had the chance of discussing it with Coover, who, he subsequently wrote, had expressed "much interest."[40] In the fall of 1934 he wrote to Rhine from Bard College (New York), where he was teaching psychology, and a few months later was engrossed in conducting an ambitious series of ESP experiments. Lucien Warner, five years older than Carpenter, was another psychologist with a strong biological orientation. When he first made contact with Rhine in December 1934 he had not only a Columbia Ph.D. in psychology (1927) but publications in animal behavior and genetics; his interest in psychical research went back to graduate school, when he had been discouraged from pursuing it by his department. He now had no academic position, although he had earlier taught at both New York University and Pomona College, and he tried during the next half-dozen years to make parapsychology something of a new career.

Yet neither Carpenter nor Warner remained permanently in parapsychology: the former chose in 1938 to devote himself instead to primatology (where he did distinguished work), while the latter left science entirely for a time for publishing before returning to psychology. What influenced both in their decision was the problem of a career. Pratt, his Ph.D. completed, could hope for a position in Rhine's laboratory; Carpenter and Warner had to find a future elsewhere. Carpenter, already launched on the professional track at Bard, was extremely sensitive to the effect his parapsychological research might have upon his reputation among other psychologists. "Dr. Rhine," he wrote, "frankly, I am undecided relative to publishing the results of my experimental work on E.S.P. I have recently discussed the experiments with several of my psychologists friends at Yale and judging from their opinions, I feel that I cannot afford to publish my results. If I were permanently located in a desirable university, the matter would be different. Also, the fact that I am young without an established scientific reputation, affects the decision."[41] Warner, older and with no secure situation, tried for several years to find a way to earn a living in parapsychology but could not. There could scarcely be a clearer illustration of the difficulties involved in establishing the new science over any sort of broad academic front. To attract the serious, capable investigators essential to its development, it had not merely to offer intellectual appeal and excitement but also to hold some promise of a professional future in order to make up for that certain lack of respectability that made parapsychological involvement difficult to reconcile with a career in the more orthodox sciences. Other psychologists besides Carpenter and Warner began a commitment to ESP and parapsychology in the later 1930s—Bernard Riess of Hunter and Dorothy Martin of Colorado most fruitfully—but always a little diffidently.

The emergent community of parapsychologists was not, however, at first limited merely to former psychologists or even more generally to trained scientists, and in this respect the new field first grew in a manner quite unlike most other scientific specialties. The psychical-research tradition out of which it emerged had never insisted upon scientific credentials over demonstrable experimental rigor and skill; indeed, it tended to be almost as suspicious of the professional scientist as the latter was of the unqualified amateur researcher. The British SPR could pride itself upon its program of cautious "scientific" investigation, even though its membership was open to all and it was guided by individuals many of whom had had no formal training in any branch of science. Such suspicions of academic orthodoxy were just as strong or stronger in American psychical research in the 1920s, and they were initially shared by J. B. Rhine.

Hence Rhine was from the first openly encouraging to the amateur enthusiasts for psychic studies who wrote in to Duke by the hundreds and even thousands in the aftermath of the publicity given to *Extra-Sensory Perception*. Within less than a year he had built up a network of individuals throughout the United States who had learned of ESP through the press and were anxious to make contributions to its study. What was most noteworthy about this group was its heterogeneity, much more typical of the traditional psychical-research societies than of a developing academic discipline. It included two engineers, each of whom led a group of psychical researchers in his hometown (Grand Rapids, Michigan, and Schenectady, New York); an elementary-school teacher in Sarasota, Florida; a furniture and machine manufacturer in Mt. Airy, North Carolina; a pediatrician in Montgomery, Alabama; a distinguished psychiatrist in Los Angeles; and a professor of philosophy at the University of Pittsburgh. There was also a small cluster of interested workers associated with the General Science Course taught at New York University, headed by the science writer E. E. Free. Most of these people had been interested in psychic matters before they heard of Rhine's work, and they saw their association with him as capable of furthering that interest in a systematic fashion. For despite their varied backgrounds, they were a serious and intelligent group—some, such as the philosopher (Oliver Reiser), the manufacturer (O. K. Merritt), and the pediatrician (C. Hilton Rice), strike the reader of their letters as uncommonly perceptive and purposeful.

At the onset, many of Rhine's fullest discussions about his work were with the nonacademics, in particular O. K. Merritt and the engineers, E. P. Gibson and G. E. Buck. He found Merritt especially useful for his inventiveness, which provided Rhine with new experimental approaches (the matching technique) and new mechanical contrivances (the first mechanical shuffler employed at Duke).[42] Perhaps this greater intimacy on Rhine's part with the nonpsychologists was a reflection of his original ambivalence towards the academic community.[43] Yet his respect for their insights did not blind him to the potentially superior expertise of his psychologist contacts. And in a sense this expectation was borne out by a comparison between the approaches of the two groups. For the amateurs, ESP research generally tended to remain concentrated upon the need to demonstrate spectacular abilities, not to understand the phenomena; the psychologists, on the other hand, adapted their activity much more easily to the needs of a sustained program of research. During the 1930s, therefore, the amateur, untrained element lost its original prominence within the new parapsychological community.

II

In mid-1936, with a research group at Duke already in existence and a number of individuals elsewhere taking up promising lines of investigation, Rhine began to think earnestly about finding outlets for the publication of parapsychological reports. The publications of the psychical-research societies were out of the question. No one associated with Rhine could yet hope to publish through the American SPR, still obsessed with Margery, nor would he want to. The Boston society approached scientific standards much more closely, but it had no formal journal, and W. F. Prince's death in 1934 had left it without an editor for its bulletins; in any case, there were of necessity only occasional publications, since each had to be subsidized by a new private donation. The British SPR had a high reputation and issued its proceedings regularly, but its traditionally broad approach to the field did not suggest that it could be counted on to publish the quantities of experimental papers on ESP that Rhine expected. Besides, the publications of any psychical-research society were bound to circulate almost exclusively within the world of the extrascientific, amateur investigator, and it was from precisely this world that Rhine was attempting to remove parapsychology.[44]

The professional psychological journals remained as obvious possibilities. From almost the beginning of his involvement in psychical research Rhine had had good success with these journals. The *Journal of Abnormal and Social Psychology* had proven particularly receptive, for its editor, Henry Moore, shared the belief widespread in the teens and twenties that psychical research fell naturally within abnormal psychology. Rhine had had no fewer than four papers accepted there without hesitation—his exposure of Margery, two papers on Lady, and one on ESP clairvoyance—and none rejected. *Character and Personality,* which was published by the Duke University Press and had Karl Zener as associate editor, had printed his account of Mrs. Garrett's ESP work. Still, there was no way to be sure that this openness towards parapsychology would continue, given the dependence of editorial policy upon individual attitude.

Rhine had had one particularly worrisome experience in 1935. In May, through his friend Hudson Hoagland, he had submitted to Carl Murchison at Clark University a manuscript bearing on some of R. R. Willoughby's recent criticisms, for the *Journal of General Psychology*. Hoagland initially reported that Murchison "looks with favor on the idea," but within two weeks Rhine had a letter of rejection from Murchison, who told him that his article was unsuitable because it was propagandistic in form. "It is your logic that I dislike," insisted Murchison, "and not your experimental data," and he hinted that he might be willing to accept a revision.[45] Rhine thanked Murchison somewhat sarcastically and offered the paper to "more tolerant quarters" (*Character and Personality*) instead. His conviction that Murchison's rejection was due to his bias was reinforced by Hoagland's subsequent assessment of the situation at Clark: "Your paper did not strike me as propagandistic in form. I am sorry that Murchison did not wish to use it. . . . I, personally, think he is rather embarrassed at the thought that he may open his journals to psychic research matters in general if he once starts a precedent."[46]

Still more disturbingly, when Karl Zener showed Rhine's article to Charles Spearman, who was just succeeding Robert Saudek as editor of *Character and Personality,* he received a second rejection—for quite different reasons: Spearman wrote that the paper was little more than a restatement of the results reported in Rhine's book and so did not merit publication; "further I must confess some qualms in respect to the subject." Zener expressed to Rhine his own distress at this decision and his disagreement with it, but he did not feel he should try to override a new editor at the beginning of his tenure.[47] The paper was returned to Rhine once more, and apparently it never appeared.

Under these circumstances, Rhine could be forgiven for concluding that some parapsychological papers, even if publishable in the opinion of neutral observers, might never be allowed to see print in psychological journals. A week after his rejection by Spearman he approached E. G. Boring to see whether C. R. Carpenter's work might be publishable in the *American Journal of Psychology;* again he received a guardedly discouraging reply.[48] In any event, whether or not editors generally would let prejudice interfere with their judgment of a paper, the anticipated rapid growth of the new field could be expected soon to yield more articles than existing psychology journals would be able to print. Carpenter, Warner, and Pratt all had papers well along by June 1936.

What Rhine therefore began to turn over in his mind was the possibility of establishing a new scientific journal, linked to Duke, that would assure parapsychologists of a place to publish research of acceptable quality. Such a journal would have a further advantage, he felt: it would automatically give the field professional visibility and, more than that, would force other scientists to admit that there was something in parapsychology to be taken seriously. As Rhine explained to Gardner Murphy that June: "A regular periodical would do a great deal, I believe, to standardize work in the field, on the principle that nothing succeeds like success; that is, in the aspect of recognition. It would considerably stabilize and command respect for parapsychology to have a scientific journal in the hands of academic people in line with the best scientific publication and publishing only first quality experimental material."[49] He proposed the regular publication of such a journal to McDougall a week later. McDougall replied encouragingly, though he questioned whether subscriptions would warrant publishing it on a regular basis.[50] With the promise of a thousand-dollar annual subsidy for ten years from a private person, Mrs. Alice B. Crunden of New York (who had already made sizable contributions to the McDougall Research Fund), however, Rhine could proceed to implement the plan.[51] In the fall, therefore, he began to encourage his younger associates to think about publishing their work in the new journal rather than in a purely psychological one; he also began to solicit papers from psychologists who had expressed interest in ESP or had notified him that they were undertaking research in it, without success. The first number of the new *Journal of Parapsychology* appeared in April 1937, from the Duke University Press, with papers by four members of the emerging community of parapsychologists.

One member of that community who was not happy with the new journal was Gardner Murphy. The project had originally been conceived in a discussion between Murphy and Rhine, and Murphy had then tentatively agreed to serve as its editor.[52] Unlike Rhine, however, Murphy did not feel that the research presently under way was "spectacular" enough in its conditions to gain respect from the hypercritical psychologist, and for the time being he would have preferred simply to see mimeographed reports of ESP research circulated privately among parapsychologists.[53] By the fall of 1936 he had read both Carpenter's and Warner's articles and was convinced that they were entirely unsuitable for publication.[54] Eventually he wrote to Rhine suggesting that he withdraw and leave the journal to Rhine and McDougall. "We have constantly disagreed on problems of publication, and this is likely to continue. . . . Probably 50% of what you and McD[ougall] feel should be published I will have to oppose or ask to have rewritten. I think at times your enthusiasm causes your critical standards to fall perilously low. Is there any merit in the tiresome casting of negative votes; isn't tension fairly sure to arise?"[55] When the journal appeared, it was under the joint editorship of William McDougall and Joseph Banks Rhine, with Charles E. Stuart as assistant editor.

The first number of the *Journal of Parapsychology* began with an "Editorial Introduction" bearing all the marks of McDougall's authorship that surveyed the state of psychical research as of 1937.[56] As McDougall's own work it has particular interest, for nowhere else did he express so carefully his judgment of the course taken by modern psychical research. He began by looking at the history of the field and by praising the work done by the amateur societies; but he repeated the theme of his Clark University paper of a decade before, that psychical research could properly be carried on only in a university. Only there could students have the opportunity "to live and work in an atmosphere of many-sided research, critical and sceptical, yet tolerant, understanding, open-minded to all possibilities." Moreover, university research in a novel field was bound to proliferate, to give rise to similar work at other institutions, a feature of real importance to psychical research. Today, McDougall insisted, these aims were being recognized at Duke, where "it has proved possible and practicable to promote intensive research in the laboratory of the psychological department, enlisting the keen assistance of a number of senior students without unduly interfering with their academic careers or in any way disrupting the work of the department. The growth of the work in scope and personnel is by natural and gradual steps leading to the formation of a sub-department." From this beginning a growing surge of research was to be expected, for which the *Journal of Parapsychology* was designed.

McDougall went further than this, however, and stressed his endorsement of Rhine's particular methods and of his narrow focus of research. The *Journal of Parapsychology*'s title was chosen, he explained, to mark off an area within psychical research as it had historically developed,

> the more strictly experimental part of the whole field implied by psychical research as now pretty generally understood. It is these strictly laboratory studies which most need the atmosphere and conditions to be found only in the universities; and it is these which the universities can most properly promote, leaving the extra-academic groups the still important task of collecting and recording all such reports of phenomena apparently expressive of unusual mental powers as occur spontaneously, obscure warnings and premonitions, veridical phantasms of the living and the dead, and other sporadic manifestations of mysterious origins.

He concluded with the recommendation that future readers of the journal first familiarize themselves with *Extra-Sensory Perception,* since although not all parapsychological work need necessarily involve ESP, it was likely to do so for the near future. In effect, therefore, McDougall was identifying the intellectual paradigm offered by *Extra-Sensory Perception* with the science to be practiced by the new professional community of academic parapsychological investigators and was putting forth the *Journal of Parapsychology* as the vehicle for that community's scientific contributions.

As a forum for parapsychologists, the *Journal of Parapsychology* certainly enjoyed some immediate success; as a means of communicating with and gaining the respect of the outside world, it was less clearly successful. Advance notice of the journal had been given to the public in *Science* and in the New York press in February 1937;[57] subsequently, at the time of actual publication, a number of articles by science writers appeared that generally hailed the scientific importance of the new periodical. As a result, the parapsychology laboratory at Duke received a flood of inquiries about the journal, culminating in some 375 subscriptions[58]—an unusually high number for a new, specialized publication. Exactly *who* the subscribers were is difficult to say; it seems, however, that they included a fairly heavy sprinkling of professional people but few if any academic psychologists. This did not of course mean that it was unavailable to psychologists, for as of 1942 (the first year for which subscription lists survive) it was being ordered by the libraries of thirty-two colleges and universities, and it had been included in *Psychological Abstracts* since its inception. Two questionnaires circulated to psychologists in 1938 suggest that perhaps half of the profession was aware of the *Journal of Parapsychology* and had formed some sort of judgment, favorable or not, about its contents.[59] Yet psychologists would not publish in the journal. During a public controversy in 1937 and 1938 a number of critical or neutral psychologists prepared papers on ESP and the Duke work; only one was offered to the *Journal of Parapsychology*. Their authors evidently preferred to address themselves to their disciplinary colleagues through the "reputable," established journals rather than to begin a dialogue with the members of a new and unlikely field.[60] What communication and respect the journal had achieved was clearly still limited.

Nevertheless, J. B. Rhine remained convinced that the foundation of a new journal was a tactical necessity if the new field was to win recognition and support. In other ways, too, he and Murphy continued to try to broaden the professional basis of the subject. They were particularly eager to see more academic positions created for parapsychology than merely the one or two at Duke, and the most likely locales for this were of course Stanford and Harvard, where money earmarked for psychical research was already known to exist. Consequently, the news that J. E. Coover would retire as Stanford's Fellow in Psychical Research at the end of the 1936/37 year and that the psychology department there was instituting an apparently unprejudiced search for his successor was a reason for hope. In particular, Gardner Murphy was optimistic, insisting to Rhine that "the departure of the incumbent seems to mean an absolutely fresh start," and he suggested two of his graduate students at Columbia as possible candidates.[61] Rhine, who had once hoped that Carpenter or Warner might be placeable at Stanford, was now inclined to recommend Hans Bender for the position, and he proposed it to Bender only to learn that the latter was committed to a program of medical studies in Germany.[62] To be sure, Rhine was far more cynical than Murphy about the supposed open-mindedness of the Stanford department.[63] And in fact none of the parapsychologists' candidates was selected; instead, the new Fellow in Psychical Research was a Stanford undergraduate and Yale Ph.D. in neurophysiology with no previous experience in psychical research, John L. Kennedy, who joined the department in September 1937.

Kennedy's appointment was made under a broadened formulation of T. W.

Stanford's indenture of 1911. This had given money, rather vaguely, "for the investigation and advancement of psychic phenomena." In 1937, however, according to the search announcement, the research of the new fellow could include "mediumistic phenomena, telepathy, clairvoyance, hallucination, dreams, dissociation of personality, subconscious mechanisms, motor automatisms, subliminal perception, and any other phenomena that may be assumed to have a bearing on the psychology of the 'occult.' "[64] This enlarged statement of purpose has been understood by some as a deliberate attempt by Stanford University to evade the terms of the original gift, but that judgment is really too severe. Certainly by the standards of 1911 virtually all the topics in the new statement did pertain to psychical research, and it was only gradually that most were taken over by psychology; in fact, many of the topics listed (such as dream research) are in 1980 still the subject of serious parapsychological attention. What the new formulation did do was to allow Stanford to look beyond the tiny community of parapsychologists, if it chose, and to select a trained psychologist with an appropriate research field to fill the position. Even though the new appointment did not automatically mean another academic position for parapsychologists, therefore, it proved to initiate a period of more serious concern with the subject at Stanford. Kennedy took his duties as fellow earnestly, far more so than Coover had done in his later years. He was certainly out to expose ESP if he could, but he gave the growing parapsychological literature a close study, did experimentation of his own, and raised at least one issue that led parapsychologists to a refinement of their own procedures.[65] When Kennedy left Stanford in 1939, he was succeeded as fellow by D. G. Ellson, who did most of his work on hallucinations but published a brief critique of some of the Duke statistics—and Ellson's successor was C. E. Stuart. Clearly the restatement of the fellowship's purpose was not meant to exclude the possibility that its holder could make positive contributions to psychical research, and Murphy's hope that parapsychologists might be eligible for the position was not unfounded.

As regards Harvard, Rhine and Murphy took a more aggressive role in trying to gain support for parapsychological research, and they had more success there than at Stanford. Their target was of course the Hodgson Fund, about which Rhine had already begun to make renewed inquiries in 1935. It was Gardner Murphy who, taking up Rhine's initiative, finally got the Harvard faculty to agree to an arrangement under which Harvard would provide twelve hundred dollars annually from the Hodgson Fund for the support of a student who would carry out research at Columbia under Murphy's supervision. The first appointee, a Columbia undergraduate named Ernest Taves, who had recently been helping Gaither Pratt there, began his tenure in the fall of 1937.[66] Harvard continued to subsidize parapsychology in this way, though it did not pursue the research itself. That first award started Gardner Murphy on an experimental project of his own for the first time in nearly fifteen years; yet its symbolic importance was even greater, for Harvard's decision to make its carefully administered resources available at last to psychical researchers suggested that parapsychology might not unreasonably look forward to a broader establishment within the academic world.

William McDougall died of visceral cancer on 21 November 1938, after a painful illness. His influence upon psychical research, particularly in America, had been unsurpassed for nearly twenty years. Never an active experimenter, he had instead used his intellectual eminence and academic position to battle for a scientific status for the field, confronting both unsympathetic spiritualists and hostile scientists in the process. The example of his own involvement with psychical research won it the tolerant interest of many observers who were attracted initially by his psychological system. His support of J. B. Rhine at Duke had unquestionably made possible the entrenchment there of an independently funded, university-sanctioned unit for parapsychology. McDougall could have taken considerable satisfaction in the thought that at the time of his death, serious studies of telepathy and clairvoyance were at last being carried on in a number of colleges and universities in the United States; he undoubtedly did take satisfaction in the Duke work that had begun the experimental demonstration of the independence of mind from body. In *The Riddle of Life,* completed in his last year, McDougall alluded to "the new experimental and seemingly irrefutable evidence of telepathy" in support of psychophysical dualism, which in turn for him made "an intelligible possibility" of post-mortem survival.[67] His basic position had changed little in thirty years save to have become very slightly less cautious. During his final illness, in one of his last campus walks with Rhine, McDougall confided that "I suppose if I had to make a decision [about the possibility of post-mortem survival] it would have to be on the positive side—but I'd rather not make a decision."[68]

Although a personal tragedy for those who knew him, McDougall's death was not a serious blow to the science he had fostered so long; in fact, it might even be argued that by now McDougall's open association with parapsychology was doing it some damage among psychologists, for in the 1930s more and more of them were coming to regard his psychological views as outmoded and wrong-headed. The field was of course most secure at Duke, where Rhine had achieved a modus vivendi with his departmental colleagues and had won the firm backing of the president and administration, but there were enough signs of professional acceptance that parapsychologists could look forward to a slow opening up of positions for them—so long as money could be found to support them and their research in the short run.

III

The Duke-centered community of parapsychologists had come into being to the accompaniment of constant and consistently favorable publicity for ESP in

American newspapers and magizines; in fact, Rhine owed many of his initial contacts to that publicity. Academic psychical research had never before received quite this sort of attention. Coover, Troland, and Estabrooks had all written essays and articles on experimental psychical research meant for a general audience, besides publishing a technical account of their investigations, but none of them had achieved anything even distantly approaching national renown. To be sure, to a newsman Rhine's work was of a different order from theirs, since it claimed such strikingly positive results; still, *Extra-Sensory Perception* had been written for a serious readership and had certainly not been presented in a blatantly popular form. It is remarkable how rapidly a new associate professor with few publications and no professional standing achieved this fame. In fact, Rhine was being celebrated in the press even before his work had received a hearing in psychical-research circles.

Publishers had long recognized the enormous public fascination with occult and psychic matters, and during the 1920s in particular newspapers and magazines had given these subjects regular coverage. Perhaps the most important conduit then for channeling news about psychical research to the public was *Scientific American,* which in those years was by no means the relatively uncompromising, semitechnical publication it is today. Its style approached popular journalism much more closely, and it gave descriptive rather than analytical treatment to the great technical marvels of the age—the great dams, the skyscrapers, airplanes. The attention it gave to psychical research was very much in keeping with this popular orientation. In Orson D. Munn it had an editor sympathetic to the field, and Munn appointed a succession of psychical researchers as corresponding editors. In the early 1920s J. Malcolm Bird was on the staff; Bird led *Scientific American* to emphasize reports of mediums and eventually to sponsor the contest that resulted in the discovery of Margery. But following the furor over Margery's performances and the resignation of Bird to join the American SPR staff in 1925, *Scientific American* redirected its coverage of psychical research; it gave up the search for spectacular mediums and began to stress the more scientifically respectable testing of mental abilities, telepathy and clairvoyance. Given this reorientation of the magazine, Walter Franklin Prince was a natural choice as a new corresponding editor. One of Prince's first contributions to *Scientific American* was a lengthy article published in September 1927, "Specimens from the Telepathic Mind," in which he sketched the history of experimental telepathy from the 1870s down to the very recent work of René Warcollier and G. H. Estabrooks.[69]

Just when Rhine was deep into his testing of Hubert Pearce, in March 1933 *Scientific American* began to promote a series of mass tests for telepathy, soliciting subjects from among its readers. These tests, designed by Prince in consultation with the science writer E. E. Free, were based upon the assumption that underlay Rhine's contemporaneous work, that telepathic ability must be widespread and could be better established by a statistical analysis of a large body of responses than by an accumulation of striking personal anecdotes. A first test asked readers to attempt to transmit telepathically to a percipient the number shown on the face of a thrown die; in a second, participants were to make use of more complex mental images as targets, in imitation of the material recently described by Upton Sinclair in *Mental Radio*.[70] In July 1933 an initial report on the first test was published by Prince's committee.[71] In their opinion, the first responses from readers were promising: "The test results show something that cannot be ascribed to pure chance and [indicate] that certain of the findings should be followed up for further and more comprehensive study." Although a tattoo of publicity for psychical research continued throughout the fall of 1933, accompanied by anecdotes from the "hundreds of our readers [who] have taken the time and trouble to set down in writing some of their experiences that appear to be linked with telepathy,"[72] no such follow-up ever appeared, since the responses did not continue to be positive. In January 1934 the editor observed soberly: "The report of our first test (July, 1933) showed a slight tendency toward the operation of something other than chance; results of our second test ['Not Proved'] appear on this page. It now appears that further tests, to be of the greatest value, must be conducted under controlled laboratory conditions."[73]

The readers of *Scientific American* were thus ideally prepared to profit from an account of Rhine's experiments, and Walter Franklin Prince, who was seeing *Extra-Sensory Perception* through the press, undertook as corresponding editor to describe them. His article, which appeared in July 1934, just three months after publication of the monograph by the Boston SPR, presented Rhine's work as the fulfillment of the long search for scientific proof of telepathy and clairvoyance.[74] Prince stressed the fact that the tests at Duke had been carried out in a university psychology department headed by a world-famous scientist; that Rhine's departmental colleagues had collaborated with him; and that the mass testing under careful conditions had yielded not only sensational results but also what appeared to be important correlations between the manifestation of psychical ability and psychological and physiological factors. The "controlled laboratory conditions" prescribed by the editor of *Scientific American* seemed now to have been admirably met. During the next few years the magazine continued to make occasional reference to Rhine's work, always entirely favorable.[75]

But although *Scientific American* had helped to prepare a sympathetic reception for ESP, it was not first to announce the work to the general public, for by the time Prince's article appeared, Rhine's monograph had received broad newspaper coverage. Lydia Allison, of the Boston SPR, gave one of the first copies of *Extra-Sensory Perception* to Waldemar Kaempffert, science writer for the *New York Times,* and on 20 May 1934 Kaempffert devoted the largest portion of his Sunday science column, "The Week in Science," to ESP.[76] His review of the book was wholly lacking in sensationalism and did not even hint that ESP might have startling implications; it treated the Duke work as perfectly normal science carried on under the scrutiny of a psychology department. Kaempffert described the card-guessing procedure briefly and emphasized the strength of the conclusions to which it would lead: "For the first time the psychologist and the mathematician have enough statistical material on which to base conclusions." Much of his report consisted of a straightforward account of the psychological and physiological factors that Rhine had found to affect ESP performance. Kaempffert was recognized as the dean of newspaper science writers, and it was perhaps this article, with its tone of matter-of-fact acceptance, that had most influence in establishing Rhine's fame nationally. Soon thereafter other reviews began to appear, emphasizing the radical consequences that would be entailed by the establishment of ESP. At the end of the month a short editorial titled "In 2034" was published in the *New York Herald Tribune,* written by the organizer of the first *Scientific American* telepathy test the previous year, E. E. Free. Free asked which scientists and scientific and technological achievements would be hailed one hundred years in the future, as Michael Faraday's work of a century before was recognized in 1934. Free singled out Rhine's just-published work as one likely possibility: "Were we ourselves to assay the prophetic broad jump about what totally new thing might be discovered we think we would suggest the understanding and use of human senses now unknown such as the curious mixture of telepathy and clairvoyance of which Professor J. B. Rhine recently has obtained experimental hints and which he has christened extra-sensory perception."[77] By this time letters of inquiry and invitations to write for newspapers and for national magazines were pouring into Durham. Publishers approached Rhine for a popular book; the National Broadcasting Company approached him about a radio test. When Prince's *Scientific American* article appeared in July, it could only amplify what was already an explosion of national publicity for Rhine and ESP.

However well public attention to psychical research had been prepared by earlier publications, therefore, it was unquestionably due to the efforts of science writers that Rhine's work was brought and kept prominently before the public. Free and Kaempffert were soon joined by the regular science writer for the *Herald Tribune,* John J. O'Neill, and these three became the core of a widening network of writers who gave detailed and sympathetic accounts to their readers. But the three New York writers did more: they attempted to justify parapsychology as legitimate science by relating its strange discoveries to recent developments in the physical sciences.

By the mid-1930s the great revolution in physics associated with the theory of relativity and the formulation of quantum mechanics had nearly run its course. If it was not yet clear just what vision of the world would emerge from the new physics—indeed, whether the world was any longer to be visualizable in the ordinary sense—it was abundantly apparent that the nineteenth-century picture of a world neatly composed of definite entities such as matter, space, time, force, and energy could no longer stand without serious modification. The Euclidean space and equably flowing time of Newton and his successors had fused and been transformed into the space-time continuum of relativity theory, and so seemingly firm and obvious a concept as "matter" had been dissolved by the new physics. At the beginning of the century, what had been the fundamental distinction between particulate matter and the continuous waves of radiant energy had been destroyed by Planck and Einstein. With special relativity, mass and energy became interconvertible; with quantum mechanics, the fundamental material unit, the electron, disappeared into a statistical wave function, taking with it the determinism of the materialistic universe.

The architects of the new physics were themselves profoundly aware of the radical changes they had brought about, and a number of them tried to convey to the literate public what had happened and what the implications of the new science were for other areas of human thought and activity. In the late 1920s and the 1930s many of these physicists were insisting that the old deterministic, materialistic universe had been destroyed once and for all and suggesting that while the new world might at first seem forbidding in its strangeness, it would come to encompass more of human experience than its predecessor.[78] In particular, they hinted that there might now be room for spirit, once so systematically eliminated from the materialistic world. In his Terry Lectures at Yale in 1927 Robert Millikan had used the new physics to challenge materialism on just this point:

> Some time ago I was one of the speakers at a forum, and in the course of my address I used the word "spirit" a number of times. When questions were afterward called for, a man arose in the rear of the room and with a somewhat hostile air asked if the speaker would define what he meant by the word "spirit." I replied that if the interrogator would be good enough to define for me the word "matter" I would attempt to define for him the word "spirit." The attempt was not called for. And, in fact, in view of the growth of twentieth century physics and the changes in our conception of matter that it has brought, it is today quite as difficult to find a satisfactory definition of "matter" as of "spirit."[79]

In 1931, in his own Terry Lectures, Arthur Holley Compton deduced the existence of free will from Heisenberg's uncertainty principle, and from this, the independent existence of mind and the possibility of its continued survival after death.[80] The diffusion of the antideterministic and antimaterialistic implications of the new physics into the contemporary consciousness was rapid after 1925; it can be recognized, for example, in the philosophy of Alfred North Whitehead or in the writings of Joseph Wood Krutch and Carl Becker.[81]

Yet while some physicists and intellectuals found in the new physics implications for antimaterialism and for the independence of mind or soul, they did not go further and argue for the scientific validity of psychical research, at least not in the United States. It was rather newspaper science writers, like Free, Kaempffert, and O'Neill, who explicitly cited the new physics in defense of the new parapsychology. Perhaps it was their peculiar role as scientific popularizers that made them open-minded and even sympathetic to parapsychology. They were accustomed to taking a synoptic view of all science, and they had had some scientific training, but they had never been academic scientists. They were certainly not wedded to the particular viewpoint of any one science. And by virtue of their wide acquaintance with science, these writers were in an excellent position to apply recent scientific developments to the defense of parapsychology. The arguments they put forward are well exemplified in a long article written by Waldemar Kaempffert for the *New York Times Magazine* in October 1937, a week after he had reviewed Rhine's second book, *New Frontiers of the Mind*. Here Kaempffert emphasized what he saw as a changed relationship between physics and psychical research, brought about by the discoveries of the past three decades. The modern physicist, he pointed out, had come to recognize his error in trying to construct a mechanistic universe out of space, time, and matter and now taught that "the universe is largely of the mind's creation. From which it follows that as his conception of the universe changes so must our conception of the mind."[82]

There were other reasons why the science popularizers would have found the psychical researcher an attractive subject. As newspaper writers engaged in selling science to the American public, they promoted an image of the scientist that conformed to national ideals: the scientist (usually undifferentiated from the inventor) was a modern pioneer. Like other pioneers, he was marked by individuality and innovation; moreover, he was a subduer of nature, concerned in this case to master its secrets and powers. As John J. O'Neill wrote some years later of the inventor Nikola Tesla, "He made the electric current his slave. At a time when electricity was considered almost an occult force, and was looked upon with terror-stricken awe and respect, Tesla penetrated deeply into its mysteries and performed so many marvelous feats with it that, to the world, he became a master magician with an unlimited repertoire of scientific legerdemain so spectacular that it made the accomplishments of most of the inventors of his day seem like the work of toy-tinkers."[83] Just as the pioneer of old needed his freedom from social constraints, so the pioneer scientist required intellectual latitude so that he might better explore the mysterious border territories of scientific knowledge and ultimately improve and enrich his society.[84] The psychical researcher was thus a naturally heroic figure.

Moreover, all three of the science writers who first took up Rhine's work had been sympathetic to psychical research before they knew of ESP. Free we have seen associated with the *Scientific American* test of telepathy; Kaempffert, it may be remembered, had been on the board of trustees of the American SPR in the early 1920s. O'Neill's interest had been a private one, but even he had characterized himself as "an amateur dabbler in the field" in an unsolicited letter of congratulations to Rhine on *Extra-Sensory Perception,* and he had gone on to tell of various psychical experiences in which he had had a part.[85] O'Neill would subsequently become a member of the board of trustees of the American SPR, like Kaempffert before him, and would contribute articles broadly interpretative of psychical phenomena to its *Journal* throughout the mid- and late 1930s.

With occasional pauses, parapsychology proceeded to enjoy wide and favorable coverage in the press for the next several years. Rhine continued to gain support from science writers for metropolitan and national publications, and he himself began to accept invitations to describe his research in print and in public lectures. During the first year after the appearance of *Extra-Sensory Perception,* for example, in addition to regular newspaper reports, parapsychology was written up in *Time;* in six articles commissioned from Rhine by the *Forum;* in the Sunday supplement *American Weekly* (of which E. E. Free was science editor); in *Liberty;* in *Reader's Digest;* and finally in *Scientific American* again, this time in a report by Rhine of his recent studies of Mrs. Garrett.[86]

The publication in September 1935 of Alexis Carrel's book *Man, the Unknown* came very opportunely for this growth of public interest in parapsychology. Carrel had received the Nobel prize for medicine and physiology in 1912 (the year before Charles Richet) for his research on tissue cultivation in vitro and was now working at the Rockefeller Institute on organ cultivation under the same conditions. His book indicted modern science (and society generally) for its overemphasis upon materialism and urged a new attention to the study of man and his potential, which, he promised, would result under the guidance of an intellectual elite in the physical and spiritual perfection of humanity. One human faculty that Carrel took as demonstrated was the "sixth sense"—telepathy and clairvoyance—of Richet. Carrel had performed his own experiments, he explained, and was convinced of the value of the work of the SPR and the IMI—and, he added, "the Department of Psychology of Duke University has undertaken some valuable metaphysical [*sic*] researches under the direction of Dr. J. B. Rhine."[87] The science writers seized upon this one aspect of Carrel's book as soon as it appeared and publicized it as a tract on psychical research, stressing the endorsement of extra-sensory perception by a Nobel laureate.

Two other endorsements by academics in mass periodicals stand out among the publicity that ESP continued to receive. E. H. Wright, chairman of the English department at Columbia University, rented a reputedly haunted summer house at Cragsmoor, in the Catskills, and the Rhines spent their summers there from 1934 to 1936. Wright developed an educated interest in the Duke work and undertook to write two articles describing it for *Harper's Magazine,* which appeared in November and December 1936. They were well-written synopses of Rhine's experiments that added nothing new to the discussion of ESP; but they were the first to appear in a magazine of *Harper's* stature, and they brought parapsychology to the attention of a wider group of American intellectuals than ever before.[88] Simultaneously, Gardner Murphy prepared an anecdotal article justifying the field for the November 1936 issue of the *American Magazine.*[89]

The Wright and Murphy articles stimulated a new flurry of letters to the Duke laboratory that persisted well into 1937.

In these years, parapsychology in America was a one-man affair in the popular mind. One might occasionally see articles about wonderful psychics such as the California boy Pat Marquis or the Latvian marvel, Ilga, but it was Rhine who was regularly presented as the exemplar of the serious psychical researcher. Nor was this popular perception seriously mistaken, given the state of the field. Rhine had not originally sought out this publicity; rather, it had come to him because of the sympathy of science writers and the interest of magazine editors. Once it came, however, he did not try to avoid it—and what tenuously supported researcher would have? Rhine was merely following in the tradition of the many scientists who have felt it important to communicate accounts of their work and its implications to the public—Millikan, Compton, and Carrel are pertinent examples. Much of William McDougall's writing at Duke had been of the same type. Rhine could scarcely be unaware that publicity given to his research could attract badly needed financial support from outside the university, and at the same time he was conscious that the same publicity could be of advantage to Duke. In January 1936 he went on a lecture tour along the East Coast, speaking, for example, to Washington's Cosmos Club with great success, and his talks stimulated new bursts of publicity in the cities he visited. On his return to Duke, Rhine described his tour and its consequences to President Few, explaining, "I mention all this because I have been told that from the standpoint of public relations the work has a certain interest to the University. Of course, I realize myself fully that it is not upon this publicity angle that the work depends in any way, except perhaps for that of its financial support; but so long as it is a respectable publicity and one that is favorable to the University, I am happy that it has advanced to the degree that it has."[90] Finally, given the moribund state of the amateur societies and the expected resistance of orthodox psychology, sympathetic articles in newspapers and magazines seemed to be the only feasible way of drawing the attention of potential collaborators to the triumphs of experimental parapsychology.

Yet acquiescence in national publicity was not without a danger. The community of American scientists had so far largely ignored parapsychology, but many of them, though silent, were deeply suspicious of the self-proclaimed new science. This latent hostility was always on the verge of passing into open denunciation, and too much public attention might bring out this opposition. For the popularization of parapsychology was different in one crucial respect from the normal popularization of contemporary science: Rhine did not yet have professional acceptance. Courting publicity for parapsychology ran the risk of being considered an attempt to pursue an illegitimate path to acceptance of the field—an attempt to avoid evaluation by professional peers, to go over the heads of the hostile scientific community directly to the general public. This danger was realized in 1937–38.

IV

> Prediction: Within a few months mental radio—telepathy and clairvoyance—will be America's leading indoor sport.[91]

While this assessment of January 1937 proved to be somewhat overenthusiastic, the coming year did see even greater public attention given to parapsychology. Newspaper and magazine coverage continued, and it was accompanied by publicity in other media. This rise in popular interest at last fanned the smoldering hostility of many psychologists into open flame late in the year, and the resultant controversy made parapsychology still more obviously newsworthy.

By early 1937 Rhine had begun to learn how to work with the science writers to mutual advantage in presenting a favorable picture of parapsychology to the public. This was particularly obvious as he mobilized his press contacts in publicizing the new *Journal of Parapsychology,* the first issue of which appeared in April. With the enthusiastic blessing of Duke's director of public relations,[92] Rhine gave a preliminary announcement of the new journal to his science-writer friends in February. In March he sent out advance copy of the first issue to his favored contacts, including Kaempffert, Free, and O'Neill, and early in April he mailed them page proofs, urging Free, for example, to present the journal "to the widest audience you command."[93] As a result of such activity, there was a flood of highly sympathetic editorials and stories just prior to and just after the appearance of the journal. Typical (if a little more portentous than most) was Kaempffert's in the *New York Times.*

> A more propitious moment for launching the Journal of Parapsychology could hardly have been chosen. The physicists have discovered that "objectivity" has not the validity that they thought it had; that their supposed laws of nature are merely statements of statistical averages; that the old machine universe created for us by GALILEO and NEWTON is not only creaking badly but falling apart.... There is a growing realization, fostered by Professor WHITEHEAD and his school, that the artists and seers understand nature better than does the scientist; for in their inadequate way they deal with the whole concrete fact—a landscape, a face, a strain of music—and not with lifeless abstractions. Moreover, Dr. J. B. RHINE'S now classic experimental and statistical study of telepathy and clairvoyance leaves no doubt that there are mental processes which bear no relation to known mechanisms and which are independent of space and possibly of time. We have here the kind of investigation that the Journal of Parapsychology will foster.[94]

By the time the first issue of the *Journal of Parapsychology* appeared, Rhine had moved to secure more direct access to a wide audience: he had written another book. Publishers had begun to encourage him to write a nontechnical account of his work as soon as he had attracted public attention, and in the summer of 1936 he began a sequel to *Extra-Sensory Perception* in which he planned to unveil his research on precognition. Though he abandoned this project, he began a similar one in January 1937. The new manuscript was scarcely a sequel to *Extra-Sensory Perception*, however. The 1934 monograph had been chronologically arranged, but it was fairly technical in presentation and aimed at a restricted audience; the new work, on the other hand, which was eventually entitled *New Frontiers of the Mind*, was a well-written, frankly narrative popularization of the Duke research.

By mid-summer 1937 Rhine had also become involved with the Zenith Radio Corporation in plans for a series of radio-broadcast tests of ESP to be put on during the fall. The president of Zenith, Commander Eugene F. McDonald, was a man who had had earlier connections with psychical research. A self-made man and something of an adventurer, a sailor and explorer, McDonald had gone into radio when it was in its infancy. Perhaps it was his mother's psychic abilities that led him, in his first two years of broadcasting, to sponsor the first American radio tests for telepathy, carried out by Gardner Murphy and another psychologist, Robert H. Gault of Northwestern, in the spring of 1924.[95] It is not clear whether McDonald himself initiated the ESP broadcasts in 1937, but he certainly took up the idea with enthusiasm and was soon brimming over with visions of how Zenith's profits could coincide with the public good. "We have a big job on our hands," he wrote to Rhine, "to entertain the American Public by making them think. If we can do that, it will be the first time it has been done on the radio. . . . However, I feel that we must lead the public very slowly into this subject."[96]

The general format for thirteen half-hour-long programs had been sketched out roughly in May. The first three were to be "audience builders," dramatizations of historical cases of scientific pioneers. The remainder were to focus on parapsychology proper and would include the tests as well as accounts of spontaneous psychic phenomena. Rhine began almost at once to flesh this sketch out with his own ideas, and for this his acquaintance with the history of science now served him well. He worked out themes for the first three broadcasts: "Intolerance," "Rediscovery of Things Previously Known," and "Progress of Mankind." Dramatizations of cases like that of the Viennese physician Semmelweis, struggling with the medical establishment for recognition for his treatment to prevent puerperal fever, were to set the stage for the subsequent presentation of parapsychology. A plan was also proposed to try to associate with the broadcasts not only eminent persons in science and invention, like Carrel and Guglielmo Marconi, but a group of academic psychologists as well.

The Zenith Corporation came to Rhine's aid in one further respect. It had originally been planned to prepare ESP cards for distribution with *New Frontiers of the Mind,* but this plan had fallen through, and a rumor had cropped up that a "Research Guild" was undertaking to market its own version of the cards. At this point, McDonald and his staff took in hand for Rhine the patenting, large-scale manufacture, and distribution of ESP cards; Rhine was of course concerned to protect his own interests and hoped that the sale of such cards might realize important support for the William McDougall Research Fund. The Zenith people found a producer to manufacture ESP cards in four different styles, ranging in price from ten to fifty cents a pack, and arranged to secure the patent rights for Rhine, meaning to integrate closely the sale of the cards and the radio tests in the fall.

These several enterprises began to coalesce into a nationwide surge of publicity on the evening of Sunday, 5 September 1937, at 10:00 P.M., EST, when the Zenith broadcasts began.[97] The tests of telepathy were begun on the program of 26 September. *New Frontiers of the Mind* was published on October 4—it had already been singled out for heavy promotion by its selection as a Book-of-the-Month-Club choice.[98] The ESP cards had been manufactured on schedule and were mentioned on the Zenith programs, offered for sale in every bookstore that sold Rhine's book, and sent out automatically to Book-of-the-Month-Club members who ordered it. Farrar and Rinehart, which had secured *New Frontiers* for its list, had also agreed to print a handbook of experimental test procedures and methods by Stuart and Pratt as an auxiliary volume so as to enable readers to carry out their own parapsychological tests with the Zenith cards in accordance with the methods of the Duke laboratory, and this appeared in December.[99] It would have been difficult to give parapsychology a much greater exposure than it enjoyed during that last quarter of 1937.

The radio tests continued all through this period, until 2 January 1938, while the Zenith hour was carried on for nine months more. It should be emphasized that the tests took up only a small portion of each program, which in the main was given over to accounts of unusual psychic incidents. A variety of test material was used as targets for guesses: five randomly selected black and white spaces on a roulette wheel the first week; vegetables the second week; black and white spaces again the two weeks following; various ESP symbols (in pairs) the next six weeks; and so forth. The listening audience was instructed to send in its guesses for each broadcast to Zenith's Chicago headquarters for grading. The first set of returns, for the test of 26 September, revealed an impressive level of audience involvement, totaling 46,433 responses. For the second test, however, the number of returns dropped sharply to 15,360, and it remained at that level through the eleventh test (5 December). The next week there was another sharp drop, to 6,644, and later returns fell even below this figure. Almost all the weekly tests resulted in a high proportion of successful guesses, and Zenith chose to interpret this as an indication of telepathy at work. Nevertheless, Louis Goodfellow, the Northwestern psychologist who had agreed to evaluate the data for Zenith, argued that the positive results could be explained in perfectly normal ways, by chance cues from the announcers or by patterns of guessing in the subjects that happened accidentally to coincide with the targets that were selected.[100]

Even before the Zenith enterprise had gotten under way Rhine had begun to have second thoughts about it. Initially he had been enthusiastic about the assurance of communicating with a large audience, but it was not long before he became aware of the dangers he and parapsychology would run if the venture were mishandled. If the programs seemed commercially motivated or if they were carelessly run, they could only discredit the Duke laboratory in the mind of the scientific world. President Few, at an early stage, advised against Rhine's involvement with the project, urging him to "go right on as you have been going presenting the evidence; and the more quietly it is presented the more convincing it will be, other things being equal. Do not be swayed from the cause you have set for yourself, even by influences from the 'windy city.'"[101] Subsequently Rhine did his best to keep from being associated with the programs in any seemingly official role, explaining to Commander McDonald that "my appearance on the program in any conspicuous way will be bad for me and my work, because the program will naturally concern itself so much with us at Duke. It will look easily enough like an organized effort at propaganda, which will be resented by some, if not, indeed, by many of the college people."[102] Yet limiting his involvement inevitably meant that he could exercise very little supervision over the actual content of the series, and he became particularly frustrated at his inability to change aspects of the telepathy tests that he felt were uncontrolled and therefore meaningless. When the first batch of test results appeared to be significantly positive, his enthusiasm returned momentarily, but he still had reservations, which were soon reinforced by warnings from friends and supporters concerned that the generally unscientific air of the programs would do damage to the reputation of parapsychology.[103] By mid-November, therefore, Rhine had broken off his official connection with the programs, though he remained on friendly terms with McDonald. Nevertheless, his original association with the series would not easily be forgotten.

It is difficult now to assess the public impact of the Zenith programs and tests. They had been very heavily promoted by the Zenith Corporation, and that they had a huge audience can hardly be doubted; being broadcast on Sunday evenings gave them access to the largest audience of the week. Yet the press gave the series virtually no coverage. No doubt a continuing series was less easily publicized than a single event, particularly a series that never announced sensational results, and consequently the Zenith programs never challenged other contemporary events—for example, the recession, the Japanese invasion of China, or the appointment of Hugo Black to the Supreme Court—for the headlines. Still, from Rhine's correspondence in these months it is obvious that the series was successful in further stirring up public interest in ESP. When the programs were two months old, Kenneth Baker, a psychologist at the University of Minnesota, wrote Rhine that "my interest in the field of Extra Sensory Perception has led to my being literally deluged with questions pertaining to your work and especially, since the Zenith programs, to the results of the tests conducted by radio.... As you are probably aware, everyone is talking and asking about these experiments."[104] This was no doubt gratifying from Rhine's point of view; unfortunately, the response to the tests does not seem to have lived up to McDonald's visions. The falling number of test returns and the unimpressive results they yielded presumably precipitated the corporation's decision to cut short the actual testing in January, although it continued programming into the summer.

Rhine had been quite right in supposing that the Zenith tests would draw harsh criticism from scientists, from psychologists in particular. To many of the latter they were a final affront that could no longer be ignored. To the psychologists in the Chicago area the insult seemed particularly pointed. A number of them, led by Louis Goodfellow, had volunteered for critic conferences that would analyze the Zenith procedures in advance for sources of error and would offer constructive suggestions for test design; most had participated in a spirit of open-minded curiosity. Unfortunately, few if any of their suggestions were adopted, although Zenith continued to proclaim that the tests were being directed by "eminent scientists." Finally, in 1938 the Chicago Psychological Club created a committee, led by Goodfellow and R. H. Seashore, to appraise the claims that had been made; the committee reported angrily in the fall of that year "that the Zenith Foundation used without justification the prestige of science and psychology ... that they misrepresented the position of psychologists on the topic of extra-sensory perception ... that they were dishonest with the radio audience ... that they promulgated a superstition by an appeal based on the 'fal[l]acy of the great name' ... that they embarrassed the few psychologists who attempted to cooperate by announcing ... untrue discoveries."[105] The report was aimed at Zenith and did not mention Rhine, but the innuendo was clear.

Nevertheless, the immediate public and professional reaction to the Zenith programs was far overshadowed by the attention given to *New Frontiers of the Mind,* attention colored by the growing recognition that parapsychology was becoming scientifically controversial. On the day the book was published, *Time* ran a feature story on J. B. Rhine. The magazine noted the appearance of the book but at the same time signaled the existence of a vocal body of critics.

> It was clear last week that psychologists, physicists, mathematicians and kibitzing outsiders are beginning to line up in earnest on one side or the other of a prickly question which has already attracted wide general interest: Is Dr. Joseph Banks Rhine right, or is he wrong, in assuming that the evidence amassed by him at Duke University is sufficient to prove the existence of telepathy, clairvoyance and associated faculties of the human mind grouped under the heading of "Extra-Sensory Perception"?[106]

The publicity had done its work: ESP was at last being judged at the bar of psychology. A professional controversy began that fall with publication of a critical article written by the McGill psychologist C. E. Kellogg in the popular *Scientific Monthly,*[107] and it continued with fervor during most of 1938. Thereafter the coverage by the press of the scientific implications of parapsychology largely reflected the increasingly public debate, though many of Rhine's friends among the science writers remained obviously on his side.

The controversy provoked by ESP did, however, bring out one influential vehicle of scientific popularization in opposition to the new field. In October 1937, immediately on the heels of Kellogg's attack, the several voices of Science Service began to be directed against parapsychology. Science Service had been created in 1920, an idea of E. W. Scripps supported by the National Research Council, the National Academy of Sciences, and the American Association for the Advancement of Science; in the aftermath of World War I, it was intended as an agency that would keep the American people conscious of the value of basic science for material progress and national welfare. Beginning in 1922 the agency issued a weekly bulletin, *Science News Letter,* which published short accounts of contemporary scientific and technological achievements and longer feature stories on issues of particular interest; its coverage extended from engineering and agriculture through the basic sciences to anthropology, archaeology, and psychology. By 1930 it had some ten thousand subscribers, principally schools and libraries—meaning that its potential readership was vastly larger. In addition, Science Service distributed syndicated scientific features to over one hundred newspapers, in this way reaching a still wider audience—seven million, by its own estimation. Finally, in 1924 the agency initiated a series of fifteen-minute radio broadcasts on particular scientific topics on the Columbia network. All in all, Science Service probably reached more Americans than any other medium of science popularization.[108]

The first editor of *Science News Letter* was Watson Davis, an engineering graduate from George Washington University, who in 1933 also became director of Science Service. While Davis directed the editorial policy of the magazine and of syndicated releases, staff writers had responsibility for covering specific scientific fields: it had been Frank Thone, the Rhines' fellow graduate student at Chicago, who had accepted their account of the work with Lady as a news release for biology early in 1928.[109] With this exception, however, Science Service, unlike other sectors of science popularization, for a long time chose to ignore developments in parapsychology.[110] Finally, in 1937 the staff writer for psychology, Marjorie Van de Water, prepared a general survey of the Duke research and showed the result to Rhine, who was a little uneasy to see that it lacked "the straightforward, uninhibited tone which science writers like Kaempffert, Free, O'Neill, [Howard] Blakeslee, and others are using with respect to our work." Van de Water insisted that she felt a keen interest in the subject, but explained that "we [at Science Service] are restricted in our writing, not alone to a noncommittal attitude but also to avoidance of any sort of comment that could be construed as 'editorial' in nature. We can only report what happens; we cannot inject our own opinions or personality, insofar as we can avoid that."[111] Six months later (perhaps inspired by Kellogg's critique) Watson Davis himself issued a news release on ESP that belied everything Marjorie Van de Water had claimed about Science Service's commitment to objective reporting. It was front-page newspaper material on 1 October and was expanded and reprinted in *Science News Letter* the next month.[112] Davis did all he could to prevent "the present telepathy epidemic" from being taken at all seriously, implying that "the critical scientific attitude" was incompatible with a belief in telepathy and clairvoyance and reminding his readers of the current commercial packaging of the subject. He complained that parapsychologists could not explain ESP, moved on to suggest possible alternative explanations for Rhine's success (without actually examining the experiments in detail), and concluded ironically that "another possible but not probable explanation of the results is that telepathy does exist!" In the longer version of his article, Davis went on to sketch out Kellogg's critique of Rhine's statistics and repeated Kellogg's warning that the designs on ESP cards might have been read from the back. For an article in a publication supposedly insistent upon factual reporting, there was curiously little comment upon the experimental basis of Rhine's claims, and this, together with Davis's sarcastic approach to the subject, suggests a deliberate abandonment of the agency's professed policy of editorial neutrality. Davis may have felt that Science Service's links to the scientific establishment gave him a responsibility to attack the heretical, or his engineering training may have convinced him that only the tangible could possibly have reality. Whatever the reason, he saw to it that Science Service became an open adversary of parapsychology during the scientific debate then beginning over its merits.

No less interesting than the reaction that *New Frontiers* received from scientists and science writers is the savage attack that it evoked from many literary reviewers.[113] Lewis Gannett's heavily sarcastic review in the *New York Herald Tribune* can serve to introduce one widely diffused theme in their critiques.

> Mr. Rhine is an earnest young man who started out to be a parson, but lost faith and turned to science. From biology he turned to psychology, and at the nicotine college in North Carolina he came under the spell of the great William MacDougall, who once taught at Harvard. (Mr. Rhine appears to believe that this fact makes Dr. MacDougall an unimpeachable scientist; having spent some years in Cambridge myself, I regret to report that while Harvard has and has had great teachers and scientists not quite all who have ever taught at Harvard are great men.) MacDougall had been a psychic researcher before he ever left old England; he continued such in New England and in the tobacco country. But, since psychic research had lost academic standing, our ESP scientists have invented a new phrase for it, 'parapsychology,' which Dr. Rhine aptly defines as meaning off side or unconventional psychology.[114]

Gannett's review reveals the influence of the scorn and condescension that the northern intellectual establishment had felt for Duke ever since its establishment. A number of articles in the *New Republic* and in the *American Mercury* had lampooned Duke as a parvenu school in a cultural backwater, an instant creation of the Tobacco Trust, with an affluent but stultified student body and an administration ready to crack down on any signs of liberalism or criticism of the Duke family's business activities.[115] Before the explosion of publicity for ESP, parapsychology stood condemned, for liberals, by its association with Duke; once some psychologists had begun publicly to attack ESP, however, it was Duke that could be ridiculed by the literati for its commitment to parapsychology.[116]

Liberals also reacted harshly to *New Frontiers of the Mind* because Rhine seemed to be counseling his readers to turn away from an active search for practical solutions to the world's identifiable social ills; he singled out instead a spiritual malaise as the most urgent problem of humanity and looked for its solution to the attainment of an inner understanding. Norbert Guterman, reviewing the volume in the *New Republic,* challenged the soundness of Rhine's mathematics and of his experimental techniques, but he concluded his review by attacking not Rhine's science but his social program.

> While the achievements of social sciences are underrated, the solution of "the most urgent problems of our disillusioned and floundering society" is expected to come not from any practical human activity, but from the revelations of some ultra-physical source, from what the author, after all the occultists, calls "a profounder kind of self-knowledge." These passive conceptions and hopes reflect a definite social attitude. Modern man, unless he acts and thinks intelligently, feels just as helpless in the face of the blind forces of our society as primitive man was helpless in the face of nature. The modern rationalized and systematized religions are not enough for an anxious mind that seeks a shortcut to supreme happiness; and astrology, crystal-gazing, tea-leaf reading, etc., are flourishing in the midst of our scientific civilization. The "scientific" language used by the latest variety of psychic research must not hide from us the fact that its social thinking is on the same backward level.[117]

The irony was that Rhine—a defender of Sacco and Vanzetti in the twenties and of Upton Sinclair in the thirties—was by no means a social or political conservative.[118] But behind Guterman's complaint was of course the old metaphysical issue. Guterman's ideal of social reform laid stress, not upon an inward search for values, but upon the practical possibilities of the social sciences, sciences designed to deal with a society of "blind forces." It is extremely curious to see the literary critics attacking Rhine by upholding materialistic orthodoxy as the essence of science at the same time that the science writers were hailing parapsychology as the natural concomitant of the new scientific antimaterialism. The daily book reviewer of the *New York Times,* Ralph Thompson, introduced his account of *New Frontiers* with the statement that "J. B. Rhine's long-awaited treatise in mundane magic is one of the most interesting I have read, and I take no stock in it whatsoever. Not that the author's integrity is in question; everything he says is said not only well but temperately. The trouble is that his whole argument, which if true, would bring the world of matter and ideas crashing down about our ears, is simply not proved. It will, of course, be accepted as gospel by hungry thousands everywhere."[119] Twelve days later, in the same paper, Waldemar Kaempffert in effect replied to Thompson by arguing that developments in twentieth-century physical science had already undermined the world of scientific materialism.[120]

In *Science Today and Tomorrow*, published two years later, Kaempffert made it clear that he saw scientific unorthodoxy as a sign of a healthy and progressive society, not (as had Guterman) as a symptom of retrogression or backwardness. His book, a popular exposition of contemporary science and its future, emphasized that the scientific community had always to temper its natural tendency to accept the status quo with willingness at least to listen to evidence that cut against it. This open-mindedness that must characterize science he took to be a concomitant of democratic liberalism; reading the connection back into the history of science, he declared, "it is no accident, therefore, that science, as we know it, should be an offspring of democracy, no accident that the discoveries of Galileo, Newton, Lavoisier, and others were made during revolutions fomented by liberals."[121] Kaempffert found the present-day status quo to be the classical mechanistic world view, against which the new physics was advancing "revolutionary" claims. "Because the old 'naturals' have broken down within the atom, because there is nothing like a machine, our whole conception of the universe has changed. The more revolutionary physicists rejoice. To them cause and effect—the idea of the machine—is a relic of a savage way of thinking. We no longer believe that Boreas puffs out his cheeks to make the wind blow from the north, that angels push the planets around, as Kepler believed. Similarly, according to the revolutionists, it is time that we give up the childish notion that every effect has its cause."[122] If to its opponents parapsychology was a relic of the superstition that science had outgrown when it entered its modern phase, to its supporters, the science writers, it was a symbol of the sort of unconventionality by which modern science had been born and through which its continued advance was assured.

During the first half of 1938, as the discussion of ESP among psychologists grew more outspoken, the coverage given to parapsychology persisted. Rhine's defenders as well as his enemies within the press found plenty of material suitable to their purposes; by now both sides had staked out their positions. Publicity could accomplish little more, for ESP had become part of the common American consciousness independent of the newspapers and was beginning to be taken for granted in a way that spiritualism had never quite managed. Perhaps it was the public familiarity with the subject that gradually made ESP seem less newsworthy to the press. When in September 1938 the American Psychological Association met and (among other activities) debated the merits of the new science, the session received virtually no news coverage. Waldemar Kaempffert expressed his bewilderment at the lack of reaction: "The meagre accounts that were put out by Science Service were not enlightening, and the press for some reason seemed to ignore the meeting, despite its news interest. Probably the events in Europe so overshadow everything else that scientific research must receive short shift these days."[123] No doubt the prelude to Munich did come first in many minds, but if it had not been Munich, it might well have been something else. For the abnormal depends precisely upon its abnormality, its freakishness, for its fascination. By the end of 1938 the news of parapsychology had grown steadily more familiar: continued laboratory testing, no sudden breakthroughs, the same inconclusive professional commentary. Having forced ESP on the attention of the nation, newspaper coverage of the field was almost bound to decline.

CHAPTER SEVEN
The Articulation of Parapsychology

I

It is not surprising that the tiny group of professional and semiprofessional parapsychologists that began to build up around J. B. Rhine in the mid-1930s should have carried out research generally along lines suggested by Rhine's monograph of 1934, *Extra-Sensory Perception*. We have already argued that this book provided experienced psychical researchers with something of a paradigm for further investigation; it could do no less for students coming fresh to the field. For in that book Rhine had formulated a number of concrete research problems concerning the mode of operation of ESP and the conditions under which it was made manifest, while ruling out still others as unanswerable. Would-be experimenters had a wide range of possible psychological and physiological relationships to investigate. How was ESP related to target shape and size? to health, relaxation, and personality of the percipient? to his age, race, and background? Such problems were indeed explored by American parapsychologists in the next several years, so far as the performance of their experimental subjects would allow. Broadly speaking, these efforts were perhaps less important for the novel conclusions they yielded than for the development of experimental and analytical technique that they encouraged, but the result was in any case a first articulation of the paradigm sketched out in 1934.

Rhine himself, of course, had not thought of *Extra-Sensory Perception* as paradigmatic, and rather than pursue a program of narrowly focused research into the problems posed by his monograph, he chose initially to move on to the exploration of other phenomena. The manuscript of *Extra-Sensory Perception* was mailed to Walter Franklin Prince in October 1933, and by January 1934 Rhine was already immersed in two new projects: the study of the operation of ESP in time (which he first termed "previsionary clairvoyance" and later shortened to "precognition") and the study of the interaction of the mind with the physical world (which he termed "psychokinesis," or "PK").[1]

To a degree, the search for precognition at least could be seen as a natural extension of the earlier research on ESP. That original work had demonstrated that scoring was unaffected by the distance separating target and agent; might not the same be true of the *time* separating them? The experiments Rhine now began to carry out were simple: the subject called the cards in an ESP deck, not in the order he thought they were at the time of calling, but as he thought they *would* be after five cuts of the deck had been made. Hubert Pearce and another subject were soon getting highly significant scores in such tests. The psychokinetic work, which was less obviously related to the study of a "perceptive" faculty, had long been associated with physical mediumship. Rhine was put onto this latter topic by a young man who had written to William McDougall claiming to know how to influence the fall of a pair of dice; invited to Duke, he had some slight success in proving his claim, but Hubert Pearce proved equally well able to influence their fall, and eventually so indeed did the Rhines! By the summer of 1934 not only had Rhine involved his immediate students at Duke—Pearce, Pratt, Stuart, and the Zirkles—in this work but he was also encouraging his correspondents to take up related problems. Rhine's psychical-research associates around the country—O. K. Merritt in North Carolina, E. P. Gibson in Michigan, students at Tarkio College—soon became caught up in the new research.

The factors unique to postulated precognitive and psychokinetic abilities naturally suggested new possibilities for experimental design. Because the study of psychokinesis was based on work with dice rather than with cards, a number of different variables were available to the investigators. They tried controlling different numbers of dice, as well as dice of different sizes (and weights), in order to determine whether the psychokinetic "force" followed the laws of mechanics—that is, whether there was an inverse relation between their "effect" as measured by successful prediction and the mass or number of the dice being rolled. In fact no such inverse relation was discovered. Rhine achieved his personal high scores in tests when ninety-six dice were thrown at once—the largest number of dice thrown in that particular experiment. These results seemed to demonstrate the nonmechanistic nature of mind even more strikingly than did telepathic and clairvoyant phenomena.

The study of precognition could employ more traditional experimental strategies. The original technique used in precognition tests at Duke (a technique eventually labeled "PDT") was to have the subject call "down through" a deck of Zener cards in the order they would have after having been shuffled or cut. At first the cutting and shuffling was done by the subject himself. Gradually, however, the experimenter took over the responsibility of controlling the randomizing act, and by the end of 1935 a mechanical shuffler was occasionally being used. The one new variable that could be studied in this situation was time, since it was now possible to shorten or lengthen the time between guess of card and selection of target, and the Duke team attempted to study what effect varying this time had upon the success of the subject.

In any case, Rhine was as interested in establishing a common approach to the study of the various branches of extra-sensory perception as he was in pursuing their unique features, and by 1935 he and his students were trying to bring them into conformity. This was of course easiest to do with precognition, in which the basic experimental situation was close to that used in studying clairvoyance. The "down through" (DT) method, with which Rhine had begun

J. B. Rhine testing Hubert Pearce by the DT ("down through") method. Pearce is calling uninterruptedly down through a pack of twenty-five Zener cards.

his inquiry into precognition, had been widely used by him in the experiments on clairvoyance reported in *Extra-Sensory Perception,* but it was to some degree supplanted in 1935 by a new testing procedure suggested to Rhine by the success of G.N.M. Tyrrell's English work, which came to his attention that spring. Tyrrell had tested for a motor element in ESP by asking his subject actually to point to the one box in a group of five that had (secretly) been selected as a target. Rhine applied much the same approach to the guessing of Zener cards by what he called "matching" techniques. His subjects were asked to match up a deck of Zener cards with the five key cards; the key cards were placed either face upward before the subject, so that he knew what symbol was displayed on each ("open matching," or "OM"), or face downward so that the subject could have no more sensory knowledge of the key, or target, cards than of the deck he was matching to them ("blind matching," or "BM"). The new matching techniques made possible some further variety in ESP testing and helped reinforce the desired atmosphere of play, but they also enforced a motor rather than a cognitive response—especially in blind matching, where the subject could have no normal knowledge of either target or matching cards.[2] With one adaptation, it proved possible to apply essentially these techniques to the study of precognition as well: by asking a subject to match his deck of cards against targets that would be identified with specific symbols only *after* the matching. Both matching and DT tests were used in the Duke precognition work of the mid-1930s. It was less easy to develop experimental comparisons between the "perceptual" phenomena and psychokinesis, although tests were done to see whether caffeine and alcohol affected PK performance as they did performance in telepathy and clairvoyance.

Charles Stuart (right) testing Joseph Woodruff by the BM ("blind matching") procedure. (Courtesy of Burke M. Smith.)

Once the possibility of these psychical abilities was accepted, the question immediately arose of how to distinguish their effects. Rhine had first recognized this problem when beginning to study telepathy and clairvoyance and had soon devised a way of eliminating the possibility of telepathy in a clairvoyance experiment. Now the problem had recurred in a far more acute form. To give an example: it had occurred to one of Rhine's students in 1934/35 that a difficulty was bound to arise in trying to interpret significantly high scoring in the standard PDT test for precognition. Might this not be due to a "psychic shuffle," that is, to an ordering of the cards in the shuffling process that corresponded closely to the subject's guesses (already recorded), rather than to the precognitive abilities of the subject? And, then, how in fact should this "psychic shuffle" be understood? Was it a phenomenon of clairvoyance, in that the shuffler "knew" when he had achieved a close match with the guesses already made and hence "knew" when to stop? or might it be a PK effect, in that the subject was somehow able to influence the mixing of the cards so that they yielded a close match to his guesses? Precognitive, clairvoyant, and psychokinetic explanations all seemed conceivable. Clairvoyance could to some degree be ruled out in tests for precognition by using a mechanical shuffler—it was principally for this reason that it was introduced—but it was more difficult to eliminate the possibility of psychokinesis. One technique eventually used was to cut the shuffled deck at a spot determined by figures in the daily newspaper (butter and egg prices, weather reports, and the like) before matching it to the subject's guesses.

It would ordinarily be unusual for a research scientist to open up a new and fruitful field of investigation, as Rhine did with telepathy and clairvoyance, only to move on to still other untouched fields. But it must be remembered that the goal that Rhine had kept in mind for more than ten years was not the merely scientific one of producing and replicating psychical phenomena. Rhine's aim might better be described as metaphysical: he was committed to studying the human mind and to seeing whether it transcended its material setting. It was this commitment that had made him so unwilling to agree with Saltmarsh that post-mortem survival was unprovable, for if proven, post-mortem survival would provide the fullest possible demonstration of the independence of mind from matter. Rhine was by no means sure that the survival hypothesis was correct, but he was convinced that it deserved investigation; and in moving on from telepathy and clairvoyance to precognition and psychokinesis, where mind freed itself not only from spatial but from temporal and material limitations, he was approaching a confrontation with that ultimate problem of the psychical-research tradition. That this was his reasoning is apparent in a letter to Frances Bolton outlining his plans in the summer of 1934:

> The general plan for the summer has been to extend now the (already excellent) evidence for *prophecy,* or precognition with the extra-sensory perception-thru-*space*. In this we have clearly shown that in some of its processes, at least, the human mind is *not materially limited;* i.e., the cardinal properties of matter, space and time, do not limit mind in this function. This is a clearly and objectively experimental demonstration of the peculiarly non-material (i.e., spiritual, in its commonest usage) character of (at least part of) mind. You see that such a step *has* to come for the scientific world—and we are all going scientific, more and more,—*before* it can accept survival of mind after material dissolution—if that occurs and can be demonstrated (as it seems to many, and as I mean eventually to investigate vigorously and devotedly).... Survival of mind without its nervous system and its muscles, *either* would mean complete removal from all universal intercourse—a detached mind could do nothing to the world of matter and energy (could not communicate thru mediums)—or else there would have to be "psycho-kinesis," mind energy affecting the known energies associated with matter.[3]

In this light, the visit to Duke in April 1934 of the English medium Eileen Garrett is of particular significance: the study of ESP in mediums was to provide the next step beyond precognition and psychokinesis.[4] During her stay in Durham, Mrs. Garrett and her trance "control," Uvani, were both tested for clairvoyant and telepathic abilities. The results seemed to show that Mrs. Garrett and her control exhibited strikingly significant telepathic ability, much less striking clairvoyant ability, and in both cases the pattern of scoring over time was markedly similar.[5] In his report on this work, Rhine pursued the implications of his data for the survival hypothesis, though with proper caution. The study of the psychical abilities of a medium in her waking state as well as in her trance state was a necessary (and hitherto unexplored) precondition to the scientific evaluation of the survival hypothesis. For if the medium normally exhibited telepathic/clairvoyant abilities, this would lend strength to the hypothesis that prescient mediumistic trance utterances were merely the continued exhibition of the medium's own mental psychic abilities rather than the intrusion of a supernaturally gifted spirit into the séance room. Rhine noted, moreover, that the similarity of score patterns between Mrs. Garrett and her supposed trance control, Uvani, lent evidence to the ESP (as opposed to the spiritualistic) explanation. On the other hand, Uvani's personality traits appeared to be distinct from those of Mrs. Garrett, and, in general, Rhine felt it was much too premature to offer any verdict on the spirit hypothesis for this or any other medium.

Demonstrating that a medium was telepathic could of course be taken to *weaken* any case for survival that had been based on her apparent communications with the dead, since she could now just as well be understood as communicating rather with the sitter's unconscious mind. Rhine faced this problem directly in his article "After Death—What?" published in *Forum* early in 1935, trying to explain how demonstrating telepathic power in a medium might actually be turned to the support of the survival hypothesis.

> Many have been baffled by the very inconceivability of discarnate personal existence. Without the sense organs, how could minds intercommunicate? If we find that telepathy is a natural capacity of the human mind, that is one answer to the question. How can the incorporeal personality know what is going on in the objective world, without brain and sensory endings? One answer would be clairvoyance, if clairvoyance is truly a natural property of mind. And so we might go on if we had the facts. The very survey, then, that serves to make it more difficult for the survival theory gives it what may be ultimately much more important to it, a partial rational foundation in conceivability.[6]

As late as this, evidently, Rhine was keeping in mind the scientific investigation of post-mortem survival.

Nevertheless, Rhine was also beginning to look back towards more established ground. He was convinced that he had demonstrated precognition and psychokinesis, although none of the evidence was superficially so striking as Pearce's best work in clairvoyance, but it was continuing to prove difficult to distinguish experimentally between them. More importantly, he foresaw the likelihood of controversy over any extreme antimaterialistic claims, controversy that would endanger the scientific acceptance of parapsychology at this early stage. Pursuit of the survival hypothesis, he decided, would have to be deferred for a time.[7] About psychokinesis and precognition he was torn, as he explained to Tyrrell in March 1936: "If we make a false move in this delicate matter, I think it would be very bad for the whole subject. On the other hand, if we can make sure of the case, even though it takes many years, it will be one of the most phenomenal discoveries, I think, that science will have made. Of course it will be a matter for many laboratories before the thing is done. I think the more of us who can work calmly and surely and not publish too hastily, the better it will be.'"[8] In the end, Rhine asked his colleagues who knew of his work in these fields not to publicize his results, and in fact no public exposition was made until the experimentation on precognition was announced in the spring of 1938 (the PK work was not published until 1943).[9] From 1936 onward, research on extra-sensory perception per se, particularly clairvoyance, was again predominant at Duke, as Rhine and his students tried to refine their understanding of the way it functioned.

II

The initiation of experiments in precognition and psychokinesis had inevitably meant a temporary distraction from concentrated research on telepathy and clairvoyance, but a still more serious blow to any hopes for a continuing research program was the apparent loss of psychical abilities by all Rhine's major subjects by the summer of 1934. In one case, indeed, when experimenters relaxed the test conditions somewhat, the subject seemed to be taking advantage of the relaxation to obtain sensory information about the order of the target cards. Rhine reasoned that since the conditions for this experiment had been far looser than they had been for much of the work published in *Extra-Sensory Perception,* these suspicions could have no bearing upon the validity of the earlier work, and he quietly dropped the matter so as not to humiliate the student (although disconcerting rumors occasionally surfaced thereafter about the presence of cheating at Duke). During this disappointing year Rhine put considerable energy by default into semipopular writing, preparing articles about the survivalist and antimaterialist implications of his work.[10]

Nevertheless, in the fall of 1935 Rhine was able to begin the academic year with renewed enthusiasm. While he himself had been unable to duplicate his first striking successes, other investigators were reporting having done so elsewhere—his correspondent at Bard College, C. R. Carpenter, for example. The most encouraging such reports, however, came from overseas. News of G.N.M. Tyrrell's success in England with the mechanization of ESP had arrived in the spring, and three months later had come the word that in Germany the young Hans Bender had independently found experimental support for clairvoyance.[11] While Bender's report stood out as the most interesting psychologically, all these independent confirmations of the ESP principle were welcome and suggested to Rhine experimental problems on which his own research team might fruitfully be set to work.

Bender's experimental accomplishment was especially encouraging because it hinted that even normally-trained psychologists might be able to evoke ESP so long as they remained open-minded. Certainly Bender's own studies had not arisen out of a devotion to psychical research, even though he had read some of the literature of the field in preparing his doctoral dissertation at Bonn on "automatic spelling" with a planchette—a topic that he had chosen after hearing Janet's lectures on abnormal psychology in Paris in 1929.[12] In the course of research for his dissertation, carried on under the supervision of the philosopher Emil Rothacker, Bender found that some of his subjects claimed to be able to find the target letters without visual perception, and he decided to inquire further into this. Accepting the methodological distinction of earlier psychical research between telepathy and clairvoyance, he decided to test for the latter, since it seemed the simpler. His early investigations (June 1932–June 1933) yielded nothing particularly striking until he introduced one of the subjects, a philosophy candidate, Miss D., to a new experimental situation, in which she was to guess the letters written on cards and hidden from her sight. Miss D. began to develop mental impressions of the letters sought that proved to correspond surprisingly well to the actual targets. Miss D. now became Bender's principal subject in a course of experimentation that continued, intermittently, from May 1933 to November 1934.[13]

Bender carried on his work in graded stages. Using capital letters of the alphabet printed on cards as his targets, he asked Miss D. to identify them first when sealed in envelopes placed behind her back and then when placed beneath an opaque cloth. He allowed her several minutes for reflection and took extreme care to record her introspective judgments as to how the letter became clear to her—for she seemed to "see" an optical image form in her mind. In order to define the process as closely as possible, Bender introduced variables of letter size and of ink and card color to see whether these affected her perception of the target. Concomitantly, he presented identical cards to Miss D.'s sight under an illumination below her sensory limen and then gradually increased the illumination, asking her to describe how these visual images took on clarity in her mind. In all, Bender recorded 134 guesses, describing the experimental situation with great care and giving the subject's words (often with her accompanying drawing) in every case. His results convinced him that Miss D. was indeed receiving information clairvoyantly (so as to rule out telepathy, Bender had made sure that no one, himself included, had any knowledge of the order in which the lettered cards were to be presented). His relatively uncritical method of evaluating correspondences between guessed letters and target letters let him conclude that Miss D. was right about 25–50 percent of the time;[14] but it was not merely this high percentage of success that convinced him, and he explicitly questioned whether statistical evaluation could really be decisive in such experiments, in which letter preferences would help determine choices and the number of trials had to be severely limited.[15] What evidently won him over, as a psychologist, was his recognition of marked parallels between image formation in apparently clairvoyant and in visual perception, especially considered in the light of Gestalt theory.

Shortly before his experiments ended in November 1934, Bender was sent a copy of J. B. Rhine's book by the German psychical researcher Gerda Walther.[16]

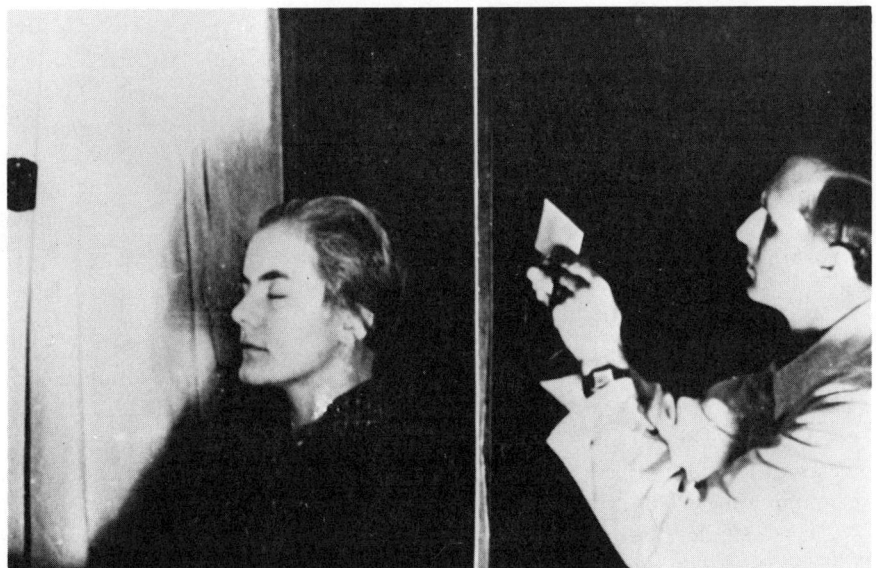

The first phase of Hans Bender's clairvoyance experiments of 1933–34. On the left, his subject; on the right, targets (lettered cards) being exposed behind a curtain. (Courtesy of Hans Bender.)

Extra-Sensory Perception clearly made a strong impression on him, as his very favorable review of the work for the *Zeitschrift für Psychologie* reveals.[17] As Bender saw it, Rhine had approached the problem in the only way that could be scientifically profitable: "The energies of most experimenters who have devoted themselves to this field have been exhausted by the repeated necessity of justifying their research activity through the accumulation of mere facts of 'extrasensory perception.' Thus few have succeeded, as Rhine recently has, in pressing forward to the formulation of problems for a methodological investigation of the biological, physiological, and psychological conditions of these phenomena."[18] In his turn, Rhine took great satisfaction in Bender's experimental report, which he saw as providing corroboration of ESP by an entirely independent psychological laboratory, and he immediately directed the attention of his associates and supporters to it.[19] He was particularly gratified that "it helps to establish the more difficult end of the work, the clairvoyance; that is, most people are much more ready to accept telepathy than clairvoyance."[20] Beyond mere confirmation, however, Rhine was encouraged to find in Bender's report suggestions that a new approach might prove fruitful, one which examined the subject's interpretation of his own perceptual activity in the light of Gestalt psychology. He arranged to get Donald Adams's opinion of the work and at least temporarily turned his mind to the possibility of devoting more Duke energies to introspective experimentation.[21]

Recognition of this sort encouraged a renewed effort by the Duke laboratory between 1935 and 1937 to develop the promise of *Extra-Sensory Perception*. The research group now included George and Sara Zirkle; Charles Stuart; Margaret Pegram, who had come to graduate school from Guilford College in 1934; and Burke Smith, who had graduated from Duke in 1934 and entered the graduate program in 1935. (Gaither Pratt had of course gone to Columbia to work with Gardner Murphy.) Much of their effort was centered upon understanding the psychological context of successful ESP. George Zirkle, for example, had his subjects take ESP tests under both relaxed and stressful conditions and asked them to make first deliberate and then hasty calls in motorized tests like Tyrrell's, in order to see which conditions made for higher success. Margaret Pegram used data she had collected in 1934/35 for an M.A. thesis in 1937, in which she considered the problem of whether ESP was conscious and volitional and whether any psychologically interesting pattern could be identified in the sequence of guesses in an ESP run. Burke Smith, who also completed an M.A. thesis in 1937, returned to the testing of ESP ability under drug influence, examining the effect of benzedrine upon ESP performance.

These studies of the psychological context culminated in 1938 with an experiment carried out jointly by Gaither Pratt and Margaret Price (a new addition to the Duke group) that seemed to indicate that the personal relationship established between experimenter and subject was crucial. That the experimenter had to maintain a positive or encouraging attitude towards the phenomena had long been accepted in psychical research; now, however, it was argued that there were some experimenters who were actually incapable of causing a subject to manifest ESP even though they themselves displayed the right attitude. This "subject-experimenter effect" became one of the explanations parapsychologists could offer of why experimental replication was usually so difficult.[22]

All these research problems taken up at Duke in the years 1934–38 grew logically out of *Extra-Sensory Perception*, and the experimental techniques employed remained generally faithful to those that had been laid down in the monograph. There was, for instance, no interest shown in following Tyrrell in mechanizing the test situation, nor was there much attention given to isolating agent and percipient from one another, since Rhine had already concluded that an easy, relaxed atmosphere was a prerequisite for ESP success. Yet despite their care to follow these supportive methods, the Duke workers were prevented from developing a broad experimental program by the reduced quality of their test results. Neither Zirkle nor Smith was able to win statistically significant results from his subjects; and while Pegram's (obtained before she had come to Duke) were significant, her "high" scorers managed to score only 5.30 correct guesses in every 25—a sad decline from the averages of 6 to 10 achieved by Stuart and Pearce only a few years before.[23] It is certainly true that high critical ratios can result from such averages, given enough runs; nevertheless, these comparatively low scores were weak and insensitive indicators of the roles that specific experimental factors might play in eliciting ESP or causing it to vary.

Yet a few good subjects continued to be reported from Rhine's correspondents carrying on psychic research in other localities. To be sure, the accounts of their performance were never entirely satisfactory, for the experimenters usually lacked a sense of experimental rigor. This made J. G. Pratt's discovery of an extraordinarily promising subject in New York all the more significant in that it suggested that scientists as well as perhaps overenthusiastic amateurs would be able to confirm and extend the Duke research. Pratt had embarked upon a variety of activities upon moving to New York in 1935. He was still busy completing his Ph.D. dissertation, but he had also begun a broad reading program in the psychical-research literature with Gardner Murphy and hoped to conduct some ESP experiments as well. The last project was held up for a time by a lack of good subjects. However, in March 1936 Pratt began to work with a "Mrs. M." (the wife of a Columbia graduate student), who had initiated the contact with Pratt but seemed to have no special commitment to ESP. Pratt described her attitude in these terms in the draft of an enthusiastic report of his work with her: "At the onset, she showed a general interest; an intense satisfaction from getting high scores; a reluctance (which [Experimenter] was glad to find and tried to have her keep) to admit that there was anything more to it for her than just pure luck. This, generally, remained her attitude throughout."[24]

At first Mrs. M.'s scores were highly variable, but early in May she began to get consistently high scores, averaging 7.1 per 25 for 2,000 trials of OM tests on May 3. Pratt then tried her on a new technique of his own devising: Screened Touch Matching ("STM"), in which Mrs. M. pointed successively at one or another of five ESP target cards placed in an aperture at the bottom of a screen interposed between herself and Pratt, trying to call the successive cards of a shuffled ESP deck; Pratt placed each card opposite the target card she indicated. Her scoring in this, too, ran consistently above chance (an average of 7.0 per 25 in 22,875 trials of STM), and Pratt thereafter used this technique almost exclusively with her.

Mrs. M. obviously afforded an opportunity for experimentation unmatched since Hubert Pearce, and Pratt bent all his efforts to testing her with regard to every conceivable parameter of ESP: the effect of witnesses (they depressed scoring, except when they took an active part in the testing), of distance (scores dropped to chance), of conscious distraction by conversation or by engaging her in another task simultaneously with the ESP tests (those distractions that engaged her attention spontaneously actually improved her scores). Perhaps most interestingly of all, when she was tested on a mixed deck of ordinary ESP cards and cards sealed in opaque envelopes, Mrs. M. scored significantly (although *below* chance expectation) on the open cards and only chance on the sealed ones, even though she was presumably unaware of the nature of any given card, since the method of testing was STM. Pratt tried to get her to introspect, but, aside from eliciting from her the information that the ESP symbols "raced through her head" when her scores dropped, this was of little use in illuminating the process of ESP in her mind. Further experiments with Mrs. M. were carried out by the Rhine group that summer during a working vacation at Cragsmoor, New York.

Louisa Rhine tried tests comparing extra-sensory and sensory perception, using packs with ESP symbols of different sizes. Mrs. M.'s scores were consistently higher with the relatively larger symbols, even though she was not informed which symbols were being used in any given test. And she continued to differentiate between open and sealed cards in her scoring without consciously knowing which was which.

Pratt's experimental style throughout this work was very much in accord with Rhine's, just as the types of experimentation he carried out were in line with the "traditional" ESP work being developed at Duke. Primacy was given to establishing the psychological context needed for successful ESP; secondarily, different guessing techniques were studied to find out what best suited Mrs. M. Specific features such as the testing of open versus sealed cards and the summer comparison with sensory perception were, of course, direct outgrowths of the Duke work. It is noteworthy that Pratt limited himself exclusively to the card-guessing tests developed at Duke, since at just this time he was working under the supervision of Gardner Murphy, whose own preferred method of testing had been to attempt the transmission of pictures or other emotion-laden mental images. Murphy had not abandoned this interest; at just this moment, in mid-1936, he was supervising the translation into English of René Warcollier's *La Télépathie*, which was constructed upon experiments attempting the transmission of images. Yet there is no evidence of Murphy's earlier experimental style in Pratt's work with Mrs. M. In general Pratt's findings corroborated Rhine's conclusions about the psychological conditions needed for ESP, although some of his conclusions (such as the disparity in scoring with different-sized symbols or the ability to differentiate between sealed and unsealed cards) were unprecedented.

In the fall of 1936, after six months of high scoring, Mrs. M.'s abilities began to wane. Pratt had recently tightened the experimental conditions of his STM tests by shielding the target symbols from the subject's view during the run, and while Mrs. M. continued to score significantly under these new conditions, her scores were lower than they had been.[25] In November Pratt decided to return to the earlier, somewhat less rigorous situation, conducting the tests in a small cubicle that had been set up in Murphy's office for this purpose, but after a few weeks he decided that it might be possible for a subject to escape from experimental control under these conditions. When Mrs. M. failed to maintain successful scoring, Pratt brought his experimentation with her to an end. In the abbreviated report on this research that Pratt prepared for the first issue of the *Journal of Parapsychology*, in April 1937, he called attention to one possible weakness of his original experimental procedure:

> This original procedure [usual STM—target cards visible] does not, however, completely safeguard against all modes of sensory perception of the order of cards. This is because of the fact that there was an aperture left to allow the pointer to be seen by the experimenter. This aperture might have permitted vision of the cards if the subject brought her head to the level of the table. It is true, the experimenter aimed so to hold

the cards as to render them invisible even through the aperture, and usually the pointing was in advance of the removal of the top card from the pack. Nevertheless, this possibility of vision is enough to warrant rejection of this condition as adequate to exclude sensory perception.[26]

Recognizing that not all the STM research had rigorously excluded the possibility of sensory contact with the cards did not, however, invalidate the experimental work done under other, tighter conditions. This indeed had been the general position adopted by Rhine since the preparation of *Extra-Sensory Perception* in 1933–34.

This same position was stated more explicitly by Rhine as the result of an exchange with John Thomas in the spring of 1938. Thomas had heard rumors concerning experimentation by Duke researchers that suggested that some subjects had tried to deceive experimenters, and he wrote worriedly to Rhine that such behavior might force a revision of all the Duke work. Rhine responded to Thomas in a note inserted in the June issue of the *Journal of Parapsychology,* in which he defined the attitude he believed the parapsychologist should take towards a subject found trying to deceive him.[27] He insisted that the ESP investigator had always to anticipate the possibility of deception and to be prepared against it. "The answer, then, to all those who suspect the honesty of subjects is: If under the conditions deception is humanly possible, the conditions are not adequate to establish the degree of confidence required for so weighty a conclusion as the occurrence of extra-sensory perception. Consequently, if it be discovered (and we have no anticipation of it being so) that one or more of the subjects in the ESP research turns out later to be of a questionable character, it will have no bearing upon conclusions based on conditions in which sensory perception was clearly excluded as a possibility." Some people, he went on, had urged that incidents of cheating be exposed with as much detail as any program of research; he denied, however, that this could serve any purpose. As long as the test conditions were properly secured, he argued, parapsychologists were "under no obligation to concern ourselves either with the ethics of the subjects or with the morbid curiosity of a few individuals."

Not only Rhine and Pratt but William McDougall too had helped formulate this policy. In October 1938, only six weeks before he died, in a long letter to John Thomas, McDougall explained his own views in terms very close to Rhine's.[28] For one thing, he argued, subjects made no routine pledge *not* to cheat, and it would be a mistake to demand one. "Every subject should be allowed to cheat us if he can"; an experiment is inadequate if it is not so devised that attempts to deceive the experimenter will be immediately obvious. When we find that given conditions permit deception, we must discard all work performed under such conditions. "Fortunately," McDougall added, "as I am informed on good authority, no published work has ever been thus involved." Furthermore, he went on, the experimenter must always remember his obligations to the subject. If a subject who had tried to cheat in a test were to be publicly identified and exposed, the ensuing threat to his reputation and career would be a far graver moral offense than the original trickery. "Summing up: our primary scientific concern is to safeguard the experiments against all possibilities of deception, but it is our moral concern likewise to protect as far as we can, the reputation of the occasional subject who may show weaknesses. This being my view, I seem to have nothing further to say on the subject and would only beg you to do what you can to make the higher morality prevail against the vestry Sunday School atmosphere which I seem to sniff." This last may have been a little unfair to Thomas, whose concern sprang less from moral indignation than from the psychical researcher's longstanding interest in eliminating every suspicion of trickery from a field that historically had suffered much from its involvements with fraud.

Psychical researchers had long recognized that their subjects—even apparently successful ones—might try to cheat. Fraudulent mediums were of course the most familiar examples of this, manufacturing false phenomena for their sitters' fees, but it was by no means unusual for participants in psychic "experiments" to be discovered, or to be strongly suspected of, cheating. (The Creery sisters and Eusapia Palladino are cases in point.) Often these subjects were found cheating after a long series of experimental successes in which any similar cheating seemed to have been ruled out; the question was, then, whether a subject's willingness to cheat in one experiment necessarily impugned the validity of all his other work, even if that other work seemed to be fraud-proof? Some individuals believed that it did, arguing that even one episode of cheating cast enough suspicion upon a subject's character to make all his work untrustworthy. Yet many serious psychical researchers were willing to discard only the work that had been proven to be fraudulent, insisting that the subject's other experimental successes that could not have been fraudulently produced in any conceivable way should still be taken at face value. A subject's willingness to cheat could have any of several explanations. It might simply be true that a genuine medium would find it less of a strain to produce psychical phenomena fraudulently than by exercising his psychical abilities and that mediums choose to take the easy way out whenever possible. Or it might be true that as mediums begin to lose their psychic abilities they begin to cheat in order to keep up the claims and the status they have enjoyed for so long. Certainly neither of these explanations seemed psychologically implausible.

Hence Rhine was by no means less demanding than the early psychical researchers when he decided that a subject's attempt to deceive the experimenter should not be taken to invalidate experimental results obtained under tighter conditions, where such deception could not possibly have succeeded. On the other hand, his unwillingness to publicize incidents of suspected cheating, shared by McDougall, was something of a departure from traditional procedures. Richet and Lodge, for example, had not hesitated to expose the tricks of Eusapia Palladino, while insisting that not *all* of her results could be explained away in these terms. Yet the prospective subjects of the Duke research were not professional psychics making ambitious claims for themselves. It is certainly arguable that Rhine's position was ethically sound in safeguarding the test subjects; more problematical was the question of how much information the experimenter needed to report to make possible an evaluation of his results. Nevertheless, the Duke policy remained constant: only those data collected under conditions rigorous enough to satisfy the researcher should be used to support an experimental study and where rigor was in question, the data, however significant, should simply be ignored.

III

The publication of *Extra-Sensory Perception* and the attendant publicity the Duke work had received had led to the establishment of communications between Rhine and a network of individuals throughout the United States who had learned of ESP through the press and were anxious themselves to make contributions to its study. As we have seen, this network comprised both young psychologists and nonacademics from a variety of backgrounds, and in the mid-1930s each group was contributing actively to the discussions of experimental methods and procedures. However, for the first number of the *Journal of Parapsychology*, devoted to outside corroboration, the professional aspect of the field was emphasized; Rhine drew upon the work of the psychologists C. R. Carpenter, Lucien Warner, and J. L. Woodruff (a psychology student of Wilfred George's at Tarkio), together with Gaither Pratt, rather than upon the amateurs, no doubt in the expectation that the former would provide a more professional appearance for the journal. And in fact the articles they published in that first number of the periodical presented parapsychology as a science that was gradually and methodically exploring and refining its subject matter. While Carpenter, Warner, and Woodruff reported much less conspicuous success than Pratt had had with Mrs. M., they could at least claim a measure of success, and they put forward either confirmations or extensions of Rhine's original conclusions in *Extra-Sensory Perception*.

C. R. Carpenter had been the first of this group to launch a research project in the new field, in February 1935. On the basis of Rhine's published data, Carpenter naturally assumed that ESP would be comparatively easy to discover among his students. And in fact his preliminary results were encouraging; within two months he had discovered a particularly promising subject in a pre-law senior, Martin Goldstein.[29] Most of the data published in Carpenter's *Journal of Parapsychology* paper were collected in tests with Goldstein in the first half of 1935.

By the end of the summer, however, Carpenter's initial enthusiasm and ambition had begun to decline slightly. For one thing, he had become aware of the pressure of orthodox scientific opinion and how it might affect his career, and he was beginning to wonder whether it would be safe to publish data supporting ESP. While he was eventually dissuaded from giving up his research, he cannot have been entirely happy with his experimental results, which forced him to curtail his original experimental design somewhat. A preliminary report submitted to Rhine in December included only the survey and testing portions of the original plan, plus the determination of an empirical norm (which Carpenter found to be 5.002/25 in 10,000 trials). He had carried out a sequence of eight tests with a variety of subjects, each test more rigorously controlled than the preceding one, for a total of 74,000 guesses. In the first test, an "introductory, exploratory series" using ESP cards, the subject was permitted to see each card after he had made his guess, while in series two and three the subject was not allowed to learn the actual targets until all his calls had been recorded, though he continued to hold the deck. By the last two series (seven and eight), two observers were being used, one handling the cards to be guessed by the subject (and screened from him by a three-by-four-foot panel), the other recording guesses. Carpenter noted with some disappointment that as test conditions were made more stringent, initially promising scores tended to fall to chance levels. In series seven and eight even Goldstein's scores dropped to a level of nonsignificance. Carpenter urged that the last three series be repeated with Goldstein, however, since they had been done under awkward and possibly unfavorable conditions. He concluded cautiously "that the data collected by these experiments do not permit of broad theoretical conclusions, but they confirm in part some of the findings of Dr. J. B. Rhine. The interpretations of the facts found is quite a different matter which must await more information of a greater variety."[30] Nevertheless, he was clearly not disillusioned by his ambiguous results. Carpenter spent the early part of 1936 on a primatological expedition to Panama, but when he returned to Bard he renewed his ESP research and revised his preliminary report for publication.

That final version is peculiar for several reasons. In substance it remained fairly close to the preliminary draft, except that it included Carpenter's 1936 data and profited from a much more detailed statistical analysis (by H. R. Phalen, a mathematician at Bard). More crucially, it omitted any mention of the inverse relation between strictness of experimental controls and significance of scoring, which had been well brought out in the 1935 draft. Consequently, the tone of the new version shifted from a guarded statement that more work was needed to a firm assertion that the Duke work had been replicated.[31] J. B. Rhine was entirely satisfied with the revision—he was of course eager for outside corroboration and had already referred to Carpenter's research as confirmation of his own work. He had previously recommended to Carpenter that the positive results be more heavily stressed, since the increasing emphasis upon the difficulties posed by the conditions of the later series would be expected to depress a subject's scoring ability.[32] In any event, Carpenter undoubtedly believed that at least some of his results had been confirmatory of ESP and deserved to be published. Yet he was still sensitive to what his professional colleagues might say about parapsychology—as he put it to Rhine, he was more timid about publishing a paper on ESP than about spending eight months in the jungle.[33] It was not long before he gave up his involvement in parapsychology to devote himself entirely to the study of primate behavior.

Unlike Carpenter, Lucien Warner lacked a stable academic base. Nevertheless, by the fall of 1935 he too had undertaken experimental work on ESP, examining a variation on a problem that Rhine was recommending to several of his correspondents: the study of possible parallels between ESP and sensory perception with tests designed to see whether the fundamental law of experimental psychophysics, the Weber-Fechner law, applied to extra-sensory as well as to sensory discrimination between stimuli.[34] Warner concentrated upon weight discrimination: after determining the sensory limen for each of his subjects, he first presented test weights to them for sensory discrimination and then attempted to see whether they could transcend their limen by telepathically gaining information from the experimenter. His first conclusion, submitted to Rhine in March 1936, was that no significant effect had been observed during his research. But in going over the data, Rhine discovered more promise than Warner had seen and urged him to develop his report further. Two months later Rhine had in hand a more positive manuscript, tentatively entitled "Telepathy in the Psychophysical Laboratory." Rhine was particularly struck by the implications that Warner's results had for studies of normal sensory discrimination: if telepathy were indeed at work between experimenter and subject during such tests, it would be virtually impossible to evaluate the results. Anticipating the paper's potential impact, Rhine suggested that Warner verify a quotation attributed to Wundt—"If telepathy be true, then all my work is in vain"—to use as an epigraph.[35]

Warner's paper made its way to a number of psychologists, as it were as referees, and their reactions to a paper supposedly so potentially disruptive are of some interest. Rhine sent it first to Gardner Murphy, expecting an enthusiastic response, and was taken aback when Murphy was unimpressed, complaining of Warner's failure to maintain precautions against even tiny cues and of the limited number of tests he had performed. Murphy suggested that the present paper might constitute an introduction to a more general one pursuing the best subjects under more tightly controlled conditions. Here Murphy was, as he himself admitted, being sensitive to the effect that any but the most scrupulous and thoughtful report might have upon the world of psychology. Ridicule of the paper would not reflect merely upon Warner: "He does this as your follower (and in time many may know with your funds) as one corroborating your work. If he is discredited, it might *perhaps* help to discredit you. We are not dealing with an impartial jury, but with a prejudice-laden world of academic psychologists who will accept every possible excuse for not taking you seriously."[36] Subsequently he reacted to Carpenter's paper with much the same reserve.[37] Murphy had always been uneasy about how psychologists would react to psychical research, and he saw in these papers a serious threat to the professional acceptance of the field.

In fact, the first reaction of psychologists to Warner's work was not at all severe. At Rhine's suggestion, Warner asked two experimental psychologists, E. G. Boring of Harvard and R. S. Woodworth of Columbia, to give him their opinion of his paper. Both responded with a benignity that belied Murphy's fears. Woodworth made one or two minor suggestions and mildly objected to Warner's suggestion that *all* psychophysical experimentation might be influenced by telepathic communication; he did not shower praise upon the paper, but he raised no serious objections to it and certainly showed that he had taken it seriously.[38] Boring was, if anything, more positive than Woodworth.[39] These responses to Warner's article by prominent experimental psychologists must have been heartening to Rhine, coming as they did only a few weeks before its publication in the first number of the *Journal of Parapsychology*.

How should we explain the marked difference between Murphy's response to Warner's work, on the one hand, and Woodworth's and Boring's, on the other? In part the difference was due to real improvements that Warner had made in the organization and descriptive sections of his paper under the impact of Murphy's criticisms. In particular, Warner incorporated a section that attempted to confront the possibility that sensory cues could have affected his results. Beyond this, however, it must be remembered that in early 1937 experimental psychologists like Woodworth and Boring were still less sensitive to the problem of adequate safeguards in work of this sort than was Murphy, who had had twenty years' experience with the difficulties of psychical research. After all, as long as the research could be easily replicated—and at this point Woodworth and Boring had no idea, as Rhine and Murphy did, how hard it was to secure good, reliable subjects—that replication would provide the best test of the experimental conditions.

Warner's conclusions substantiated what Rhine from the outset had expected to be the case, that ESP and SP behaved very differently. If it were to be shown that ESP and SP were parallel, after all, this would imply that they shared a common material neurological basis, that both implied a mechanical transmission from some outside source to a receptor and a pathway from that organ to the brain. Louisa Rhine's results from similar tests with children at about this time confirmed this distinction.[40] Despite an occasional return to the issue, most notably in the Pratt-Woodruff experiments of 1938/39 (which likewise gave no evidence of correlation between the two), the question of parallels between extra-sensory and sensory perception henceforth was not a prominent element in the Duke research program.

The initial research of Joseph Woodruff provides a third clear case of work designed to develop ideas set forth in *Extra-Sensory Perception*. Woodruff was a junior at Tarkio College when he wrote to Rhine in January 1935 for advice about a project for a course in experimental psychology. Rhine had sent Woodruff's teacher, R. Wilfred George, a recent Duke Ph.D., a copy of *Extra-Sensory Perception*,[41] which George in turn had loaned to Woodruff. To Woodruff's eventual inquiry Rhine returned encouragement and a most comprehensive list of possible research topics, and the two continued to correspond as, under George's supervision, Woodruff began to test his friends as possible subjects. Rhine regularly advised Woodruff on the course of his investigations, urging him to tighten his experimental controls gradually and to specify his test situation as precisely as possible.[42]

In October 1935, at the beginning of the new academic year, Rhine offered Woodruff a research grant of two hundred dollars and suggested that he commit himself to pursuing the implications of G.N.M. Tyrrell's recent success in asking his subject to make rapid mechanical selections from among ESP targets, a success that hinted that ESP might be expressed by a motor response more readily than by a cognitive one.[43] Rhine proposed that Woodruff test this implication by comparing a subject's card-guessing scores under the variety of testing techniques that had been developed during the previous spring and summer: with open matching (OM) and blind matching (BM), both screened and unscreened, and with the older BT (before touching) procedure as a further basis for comparison. By the end of November Woodruff had found three satisfactory subjects besides himself and had begun the process of testing them under a variety of procedures. Five months later he was able to report that higher scores were attained with OM and BT (screened or unscreened) than with BM, thus supporting the hypothesis that cognitive tests were better exhibitors of ESP than merely motor ones. Rhine immediately urged him to prepare his material for publication, and it was the resultant article—with considerably more data and Wilfred George's name added to it as coauthor—that helped to inaugurate the *Journal of Parapsychology*.[44]

Carpenter, Warner, and Woodruff were the first people outside the Duke laboratory to report confirmatory evidence of ESP and to attempt to explore its consequences. It is not surprising, certainly, that Rhine urged publication upon them and linked them together with Pratt in the first number of the *Journal*. It should be pointed out as well, however, that this research and publication implied something of a professional as well as an intellectual commitment to a new science, as they all realized. Carpenter, of course, eventually decided against the commitment and left the field; Warner, however, remained in it for several years; and Woodruff too decided to prepare for a career in parapsychology. In the fall of 1936 he entered the Duke graduate school in psychology, intending to study with Rhine, while George continued to supervise some further undergraduates in research at Tarkio. The process of articulating the new science was, in fact, defining the community of parapsychologists.

In the next few years still other individuals were drawn into the nascent discipline by the success of their parapsychological research. This was the case, for example, with Bernard Riess, a psychologist at Hunter College. Riess had held the skeptical attitude expectable from academic psychologists until he heard Rhine lecture at Hunter in December 1936. While remaining skeptical, he found his interest piqued, and when he was told about a psychic friend of one of his students, he decided to test the friend for GESP under well-controlled conditions. At irregular intervals a total of twenty-one hundred trials were made over a distance of about a quarter-mile in two series, one lasting from December 1936 until April 1937, the other lasting from July 1937 until the tests were stopped in late August. In the first series the subject, "Miss S.," achieved astonishing results: 1,349 hits in 1,850 trials. Illness, apparently both organic and psychological, forced the hiatus between the two series, and when Miss S. returned for the second series her scores fell to chance levels. In late August she moved to the Midwest and refused to participate in any further experiments.[45] Riess thus had no chance to set up a program of tests like that J. G. Pratt was carrying on with Mrs. M. to explore the way in which ESP operates; but the two series by themselves are of extraordinary interest. Riess defended his work at the 1938 meeting of the American Psychological Association and remained so far interested in parapsychology as to join Gardner Murphy in 1939 in assuming the editorship of the *Journal of Parapsychology*.

At Earlham College in Indiana research into parapsychology was carried on in 1936/37 by John A. Clark, a professor of philosophy, and Betty M. Humphrey, a sophomore philosophy major. They discovered an outstanding subject who achieved an overall average of 8.26 per 25 guesses in 9,500 trials and, contrary to other experiences, had a higher rate of success on group experiments than on individual ones.[46] Eventually the subject's scores dropped to chance levels, and she disappeared from the research scene; Betty Humphrey, however, continued her own ESP projects during the remainder of her undergraduate career at Earlham and in 1941 went to Duke to do graduate work in parapsychology.

More influential than either of these abbreviated pieces of research was a project begun at the University of Colorado early in 1937, one that not only seemed to confirm the reality of ESP but hinted at still further avenues for research. Interest in ESP had first been aroused at Colorado in 1935 by John Schoolland, an instructor in the psychology department there who was completing a Duke Ph.D.[47] Karl Munzinger, the departmental chairman, was not unsympathetic to Rhine's efforts, and he interested a graduate student of his, Dorothy Martin, in initiating some experimental work. Martin began with DT testing of 39 subjects and a total of 76,525 trials, most (but not all) witnessed. The average score for all subjects was 5.80 per 25 guesses, scoring that she showed to be strongly significant. Martin introduced a new methodological twist by using two concurrent comparisons as controls: the matching of two decks of ESP cards with each other and the matching of the DT guesses of her subjects with the *reverse* order of the cards in the deck for the particular run. The results for both controls came out very close to the chance value, 5. That the matching of successful runs with the reverse order of the test deck should give only chance results seemed particularly striking, and Martin wrote Rhine that "in the opinion of interested scientists here, [this] is perhaps the most cogent argument for the existence of extra-sensory perception."[48]

In the fall of 1937 Martin began intensive testing of one of her earlier successful subjects, a student named Jencks. Jencks continued to get significant scores, this time under well-controlled conditions (screened DT), in two series of 25,000 trials each. In the first (September–November 1937), the DT procedure was normal; in the second (January–May 1938), the procedure was reversed, so that Jencks was asked to guess *up* through the pack from bottom to top—whence the acronym UT. In both series Jencks's scores continued to be amazingly reliable, even when the direction of guess changed. In the first, DT series he averaged 6.89 per 25 guesses; in the UT series, after an initial warm-up, his scores averaged an even more impressive 7.39 per 25 guesses.

Martin's treatment of her data was elaborate. Critical ratios were computed with regard not only to the theoretical standard deviation but also to deviations empirically derived from reverse and matching control series, carried out as in her first experiments. Daily scores were reported, and fluctuations were noted; symbol frequency was tabulated, and frequency of correct guesses was correlated with card position in the deck. This last correlation proved particularly interesting. In both Jencks's series it was found that the frequency of hits was greater for the top half of the deck than for the bottom, to a statistically significant degree; no such correlation was found for the matching series. This result was in contrast to Rhine's original study, which had identified a U-shaped curve for DT work, with best results at the top and bottom of the deck and a sag in frequency of hits in the middle. As Martin realized, her results suggested that "the dominant factor here involved may rest upon a physical basis rather than upon a psychological one."[49] The relative sophistication of Martin's analysis was recognized at once, and it was subsequently held up in answer to critics who complained that parapsychologists were inattentive to such matters.

It remains to consider one individual who did *not* pursue experimental research on parapsychology and yet even so was an extraordinarily influential member of the growing parapsychological community: Gardner Murphy. Still only an assistant professor of psychology at Columbia,[50] Murphy was heavily burdened with departmental and professional responsibilities, and he complained frequently to Rhine about how little time he had left for parapsychology. But Murphy had in any case always felt that he lacked the gift of eliciting psychical phenomena, so necessary in an active experimenter. What time he did have therefore went largely into supervising and criticizing the work of others, particularly of course Gaither Pratt's in 1935–37, and guiding into print the English translation of René Warcollier's collected articles from the *Revue métapsychique;* the collection finally appeared in 1938 as a publication of the Boston SPR. For much as Murphy appreciated the statistical judgments made possible by the Zener cards, he remained fascinated by the richer psychological conclusions that the different material used by Warcollier (and others—Bender, for example) seemed to make possible. In May 1935—a year after the publication of *Extra-Sensory Perception*—Murphy had conducted some group experiments in telepathy and clairvoyance through the Boston SPR; true to his basic interests, while he used some of the Duke techniques (such as DT), the objects guessed were playing cards rather than ESP cards and, in two experiments, passages of poetry and picture postcards.[51]

At the beginning of 1937, however, Murphy drew up a sketch for a parapsychological research program of his own, a copy of which he sent to Rhine for comments.[52] It was again typical of his orientation that his suggested research should have been far removed from the series of experiments done by others more directly under the influence of *Extra-Sensory Perception*. Murphy was less interested in testing directly for ESP, in finding a supply of good subjects, or even in ascertaining the most favorable conditions for eliciting ESP than he was in analyzing some general psychological factors and correlating ESP test results with them. The factors he chose were "perseveration" (the "hangover effect" of one activity on a successive one), "ideomotor suggestibility" (the carry-over of an attitude appropriate to one activity to its successor), and unconscious automatisms, which he called "general dissociability." In his proposal, each subject would be tested for these three factors and then given a battery of ESP tests. Even in these ESP tests, Murphy proposed much more diversity in target material than was normal with the Duke group, suggesting, for example, that playing cards and pictorial material be analyzed in Warcollier's manner. Rhine's response to the proposal was decidedly unenthusiastic, since it seemed to him far more elaborate and ambitious than could be justified until positive results were being regularly produced in a simple test situation.[53] The study of the psychological (and physiological) concomitants of ESP was dependent on reliable subjects, who were no longer to be found. To Rhine, it was far more important to discover and define the conditions under which psychic ability could manifest itself. Murphy pressed ahead with his own ideas, however, and something of their influence is to be seen in the design of the experiments that Ernest Taves began at Columbia that fall.[54]

IV

The development of a science does not occur merely as a consequence of its own internal logic; there are always social and professional pressures at work that help to shape its approach and its aims. In the case of parapsychology, some such pressures came from mainstream science. In order to present parapsychological research as persuasively as possible, the Duke group, for its part, attempted to show that parapsychology was an organized, coherent discipline whose results made consistent sense and whose methodology could be used by any careful researcher. The Stuart-Pratt *Handbook* of 1937 [55]—admittedly aimed at a larger public audience than the scientific community but keeping the latter in view—put the card-guessing and matching techniques of parapsychology into orderly form and summarized the psychological conditions that Rhine was convinced were necessary for manifesting ESP. To be sure, the *Handbook* was not a successful venture. It encountered a traditional difficulty of psychical research, that of deciding upon an appropriate audience for its appeal, and did not really succeed with either scientists or the general public. The *Journal of Parapsychology*, founded earlier the same year, was more consistently aimed at scientists and was more concerned to display the development of parapsychological research as an orderly, cumulative, entirely "normal" attack upon experimental problems, wholly comparable to other scientific disciplines.[56]

But parapsychology conceived as part of the larger scientific enterprise had to be responsive to whatever serious criticisms emerged from science. These first became acute in late 1937, and they compelled parapsychologists to come to grips with two specific issues: (1) the design (and reporting) of an adequate ESP experiment and (2) the proper statistical evaluation of the results. Neither of these problems had previously been ignored by the Duke researchers, but they had not been fully prepared for the extreme demands that might be made. In meeting these demands, the nature of parapsychological investigation had inevitably to change somewhat.

Parapsychologists had not been unaware of the importance of sound experimental design. In *Extra-Sensory Perception* Rhine had called particular attention to those of his experiments in which he felt conditions had been tightest as providing the best evidence for psychical abilities. He had placed some stress, for example, upon the PT work done by George Zirkle and Sara Ownbey, in which the two participants were separated by a wall, signaling was done by a telegraph key, and an electric fan was kept going to safeguard against unconscious whispering. In the case of his best subject, Hubert Pearce, he had gone so far as to describe and even tabulate the results Pearce had obtained under successively more stringent conditions: from tests in which he had held the target cards in his own hands, Pearce had moved on with undiminished success to screened work and DT experiments.[57] Still, these relatively careful descriptions were the exception, not the rule. Having presented the details of a few such ironclad experiments, Rhine saw no need to provide the same for the rest, which, though they might have been conducted more loosely, were still part of a mass of consistently confirmatory material. He prefaced his summary of the work of the five major subjects besides Linzmayer, Pearce, and Stuart with the explanation that "in presenting the data in this chapter, I shall not give consideration to the question of fraud or deception, since that has been perhaps overdone in the earlier chapters. Not that all these results were obtained without any precautions—some of them had the best of conditions. But because it involves too much unnecessary duplication to describe conditions repeatedly, and because we are beyond the question of proof and are after the explanation and conditions.'"[58]

This inconsistency bothered the first reviewers of *Extra-Sensory Perception*, psychical researchers and psychologists alike. R. H. Thouless, who was both, had reviewed the book for the Society for Psychical Research; he commented severely upon Rhine's sketchy account of how tests were conducted, offering his own suggestions as to conditions to which particular attention should be paid.

> If Dr Rhine is to carry general conviction of the truth of his finding as to the commonness of extra-sensory perception, it is absolutely necessary that he should state clearly how many of his 18 subjects showing extra-sensory capacity were tested

under critical experimental conditions. The minimum requirement for a critical experiment would seem to be: (A) that an experiment in which the card is visible to the subject should never be carried out by means of a pack of which the subject has previously seen the back of each card and been informed as to what was on its face; (B) the subject should not be informed as to what cards have been drawn until the whole pack is completed; and (C) the back of the cards should not be visible to the subject at all unless it is absolutely certain that the figure on the face has made no perceptible modification of the surface of the back. It is quite impossible to discover from Dr Rhine's book how much of his evidence is derived from experiments of this kind and it is entirely possible that even though one or two of his subjects had genuine extra-sensory power, the others were getting successes through inadequate control of the experimental conditions.[59]

The psychologist R. R. Willoughby criticized the same feature of Rhine's presentation of his work, attributing it to the manner in which the monograph had been organized.[60]

Rhine, of course, believed that since ESP was a fact established unquestionably by, say, the best of Pearce's work, and since clairvoyance and telepathy were so widely and regularly demonstrable, it was unnecessary to demand (and to describe) all possible niceties of control in every experiment. Very few scientists, indeed, ever do so. The experimental work he had outlined in the 1934 monograph was not meant merely to demonstrate the fact of ESP but to go beyond this to draw some tentative conclusions about the way in which it operated. As Rhine saw it, the very consistency of his results in all possible experimental situations, loose as well as tight, was evidence that the same underlying process was at work throughout.[61] Charles Richet had defended this against Pierre Janet, some thirty years before, as normal scientific reasoning. Rhine had also come to agree with the conclusions of earlier psychical researchers in stressing that ESP only functioned well in conditions of relaxation or play. This meant that rigorous conditions and a fixed experimental plan could not well be imposed at the outset of work with a subject if positive results were hoped for; it was necessary to establish a mood of ease and confidence and to stiffen the conditions gradually, perhaps unknown to the subject. The very notion of defining a standardized experimental situation was in a way inconsistent with parapsychological investigation.

In these first years, therefore, Rhine was of mixed minds about how strict one should be in collecting and interpreting parapsychological data. He certainly warned his earliest associates to take care about their test conditions. But he was also prepared to give them (and their data) the benefit of the doubt once they had what seemed to be positive results in hand, as his encouragement of C. R. Carpenter shows.[62] Still, the general effect of the early criticisms of his methods was to lead the Duke laboratory by 1936 to urge the regular use of relatively rigorous conditions: preventing a subject from learning of his successful guesses until the end of a run; use of a screen; checking of all results by the experimenter. Rhine remained convinced that such conditions had to be introduced delicately, and he continued to oppose independent witnessing because this seemed to disturb most subjects' scoring abilities.[63]

Despite this early attention given them, methodological issues were not really an overriding concern for the Duke parapsychologists until late 1937, when for a variety of reasons their techniques underwent the sharp criticism of a number of American psychologists. The most pointed of these criticisms followed the claim by B. F. Skinner and others that both Rhine's original homemade cards and the commercial ESP cards put out under the aegis of the Zenith Corporation were imperfect in that their symbols could be read from the back under certain conditions. Might not Rhine's amazing results have been due to sensory cues' reaching the subjects? Other questions followed quickly: Could errors in recording the scores have occurred, considering that the subjects themselves sometimes helped record them? Had *all* the data been reported, and, if so, had any of them been improperly discounted?[64] How sensitive the Duke group became under these circumstances is graphically illustrated in Rhine's correspondence with John A. Clark about the research the latter was carrying on at Earlham in 1937. In June Rhine played down the need for uniformly high standards, arguing, as was his custom, that consistency of results proved adequacy of conditions. Five months later, in the midst of the storm, Rhine wrote much more insistently about the need to provide adequate descriptive detail. "Were any of the tests done with screens? Was the subject able to see the back of each card in every trial? Can you give more information about the conditions in the group experiments so as to enable the reader to judge whether sensory cues from the cards were fully inapplicable there? . . . There is an incredible amount of skepticism, of course, among certain of the psychologists, and we want to do all we can to meet this skepticism in the article itself."[65] From this point on, parapsychologists would become intent upon designing and reporting research that was technically irreproachable.

The Duke group responded to the methodological criticisms one by one. The most important of these was the complaint that sensory cues might have been possible in the apparently successful tests for ESP, and this issue was taken up immediately. The December 1937 number of the *Journal of Parapsychology* included several articles describing successful experiments in which the exclusion of sensory cues appeared to have been complete. Rhine showed that the best of the early Duke experiments (those in which the targets had been totally concealed or separated by a considerable distance from the subject) were still strongly positive, while Bernard Riess described his successful distance work with Miss S.[66] Lucien Warner offered a series of clairvoyance tests of a single subject designed to refute all conceivable criticism. The subject and Warner were on different floors of the same building, with the subject locked in her room and able to communicate with the experimenter only by a (one-way) electric signal. Instead of using a single deck of test cards, Warner made each target selection by cutting a different (and newly shuffled) deck of cards. The cut card was not exposed and recorded until after the subject signaled that she had made her guess;

all guesses were published. With all these precautions, the subject still scored a strongly significant 93 hits in 250 trials.[67] At the same time, however, Rhine acknowledged the general soundness of the psychologists' position, admitting that the commercial cards could indeed have defects and warning would-be researchers to use only screened tests in serious experimentation—"the screen should be so uniform a condition that card imperfections are not a matter for experimental concern."[68]

The other experimental criticisms of the psychologists were promptly met as well. The June 1938 issue of the *Journal of Parapsychology* gave detailed instructions designed to ensure some degree of uniformity and high standards of accuracy in recording data:

   1. Wherever possible, make independent records of calls and cards.
   2. The checking process should be witnessed by at least two persons when feasible.
   3. When the subject is present at the checking process he may act as a witness but should never handle the cards or actually do any of the recording or checking.
   4. The call-series column should be covered when the card-series is being recorded in a column next to it. The correspondences between the two should be checked afterwards. In checking down the column it is found advantageous to point with the pen to each pair of symbols to be checked.
   5. In checking correspondences between call- and card-series on sheets where there are [sic] more than one run, lay a ruler or paper over the adjoining column on one side at least, to help in focusing.
   6. Records should, if it is convenient, all be made in ink.[69]

The question raised about the data themselves—that is, whether all the experimental data in each set of tests had been reported and evaluated—was not so simple to answer, for it touched on issues that were partly statistical, issues of random sampling and optional stopping. Most of the criticism had focused upon the data summarized in *Extra-Sensory Perception;* in response, all the parapsychologists could do was to assert that all data that had been collected had, in fact, been tabulated and evaluated. But in at least some research published in 1938 experimenters went beyond merely tabulating their data to publish the raw data *in extenso,* as some critics had urged. Indeed, Lucien Warner had already done this in 1937 for his "test case"; in 1938 the same method of reporting was used by Martin and Stribic in their second report and by Margaret Price in tests of blind and sighted subjects.

By 1938, therefore, parapsychologists had markedly refined and clarified their experimental strategy. While they still felt it essential to carry out tests in an informal, relaxed, even enthusiastic atmosphere, they were now more self-conscious than ever about the necessity for standard conditions to exclude any sensory contact between subject and target cards, and they made changes in their methods of recording data too. Although some of these refinements certainly were developed in response to external criticism, they were instituted quickly and ungrudgingly, for their use was fundamentally consistent with longstanding policies of the psychical-research tradition and with the aims of experimental parapsychology.

Even more directly than the refinement in experimental procedure, the increasing sophistication of statistical techniques used in parapsychology during 1934–1938 came in response to criticism and controversy. To answer their critics, parapsychologists and their mathematical allies developed a broader and stronger mathematical basis for ESP. In writing *Extra-Sensory Perception* in late 1933, Rhine had borrowed probabilistic algorithms from the psychical-research tradition and elsewhere to evaluate his subjects' success in guessing the five targets in his standard test deck. As he saw the situation, the use of a 5 × 5 deck (five cards of each of five different symbols) made the necessary computations vastly simpler than the use of a deck of playing cards (in wide use up to that time) would have done. He had reasoned, straightforwardly, that "pure chance" guessing ought to result in one correct guess of an ESP symbol out of five tries; this probability of a correct guess (1/5) multiplied by the total number of guesses made in an ESP test would give what he termed the "normal chance expectation" for that number. For a run of twenty-five guesses, for example, the mean chance expectation would be five correct.

In order to evaluate the statistical significance of a collection of trials whose successful guesses deviated sharply from the mean chance expectation, either positively or negatively, Rhine had taken over a standard algorithm: the deviation divided by the probable error (PE)—the latter defined by Rhine as "that deviation from the mean (chance) expectation at which the odds are even (1 : 1) as to whether pure chance is operating"[70]—produced a number, which he called first the "anti-chance index" or simply "X" and subsequently the "critical ratio" (CR) and which could be translated into probability values by appealing to published tables. An X of 2 yielded odds of 4.6 to 1 against chance; an X of 4, odds of 142 to 1; an X of 5, odds of 1,300 to 1. An X of 4 he accepted as the threshold of significance.[71] Finally, faced with the need of summing anti-chance indices for different groups of data (that is, the various series of guesses of a single subject), Rhine could find no algorithm in the literature and so devised an empirical one of his own, which, as he elatedly pointed out, indicated so high an anti-chance index for his best subjects' guesses as to establish a "relative certainty . . . for the Extra-Sensory Perception principle."[72]

As this should indicate, Rhine had had to pick up what facility he could in the use of techniques appropriate to his particular problems very largely on his own. This was almost inevitable in the early 1930s, when statistics was by no means the familiar scientific tool that it is now. Consequently, his description of what he had done was not complete. He gave no information, for example, about the distribution of guesses in the runs he had observed, nor about specific runs of cards and calls. More seriously, he failed to explain the mathematical derivation of the PE for a series of guesses, and he likewise failed to point out (or perhaps had not yet seen) that his choice of probability tables entailed the assumption that random deviations from mean chance expectation (MCE) would follow the *normal* probability integral.

The reception given *Extra-Sensory Perception* by friendly as well as hostile critics soon made it clear to Rhine that the mathematical foundations of his argument needed support. From England, Whately Carington had written immediately to praise Rhine's experimental achievement, adding some reservations about his statistical approach.[73] More influential, however, in leading the Duke group to think through its use of statistics were the criticisms of R. R. Willoughby, in articles published during 1935 and 1936. Reading Rhine's monograph in 1934, Willoughby had come to feel that Rhine's report of tremendous odds against chance was ipso facto suspicious.

> Don't your "astronomical" odds ever give you the suspicion that something may be wrong? Must we trust alleged "scientific method" even when it tells us things we know can't be so? Consider the figure $X = 10^{-50}$, which is by no means the largest of your figures; have you ever tried imagining a proposition to which odds of that size could reasonably be ascribed? That the sun will rise tomorrow morning?—not at all; if we had all the records back to the nebular precursor of the solar system they wouldn't give us warrant for half these odds. . . . And so on—no proposition is conceivable which could have any such odds; and yet clairvoyance, doubted by nine-tenths civilized mankind, is a fact that is trillions of times more certain than that, e.g., I am making black marks on yellow paper.[74]

Willoughby himself was not really at home with statistics, but he did his best in the next few years to investigate Rhine's monograph for statistical flaws, and his correspondence with Rhine and C. E. Stuart, continuing into 1936, did much to force them to begin to examine the assumptions on which any such use of statistics would have to be grounded.

Like many psychologists, Willoughby was from the beginning highly skeptical of the use of a theoretical chance distribution (and associated tables of probability values) to compute anti-chance values for test results, as Rhine had done. He argued that no theoretical distribution could be justified as well as an empirical distribution determined at the same time and under the same conditions as the experimental series. Gradually he became convinced that a new experimental format was called for, one that would control such features as stiffness of cards or peculiarities of shuffling that might correspond with guessing tendencies on the part of the subject. "I want to control the thing directly and experimentally," he explained, "not indirectly and logically."[75]

Aside from his objections to assuming *any* a priori theoretical distribution, Willoughby became increasingly sure that Rhine's use of the normal distribution in particular (implicit in *Extra-Sensory Perception* but first stated outright only in his article on Mrs. Garrett) was illegitimate. The exploratory card-guessing series that Willoughby and Hunter had carried on had naturally yielded numbers of successes skewed strongly towards numbers higher than five rather than lower, which Willoughby interpreted as inevitable consequences of the inequality of the values of p and q put into the binomial theorem (.2 and .8). He asserted that Rhine's supposedly "normal" curve could clearly be no such thing. To resolve this matter, Rhine turned to Charles Stuart, who was able to show that any skewness of the ESP guessing curve rapidly became smaller as the number of trials increased and insignificant by twenty-five trials. By September 1936 Willoughby was compelled to concede that his statistical objections "won't hold water."[76]

By this time a second and even more influential critic of the Duke statistics had appeared, the McGill psychologist Chester E. Kellogg. The most important element of Kellogg's analysis was his assertion that the theoretical distribution of random hits in ESP guessing could not possibly follow either a normal or, for that matter, a binomial curve. The binomial distribution might govern random sampling from an infinite pool of ESP cards, but the ESP decks were not infinite; rather, they were limited to the number of permutations that five symbols each occurring five times could undergo. Moreover, the subject (knowing the general nature of the deck before him) was likely to try to guess a roughly equal number of each type of card. Since individual guesses were thus interdependent rather than independent, the assumption of random sampling was also invalidated. As he came to recognize during correspondence with Rhine and Stuart, a valid ESP statistics required a distribution that would describe the frequency of successes to be expected when two ESP decks were matched against one another.[77]

To respond to these criticisms, Rhine began to consult a number of outside mathematicians, more or less at random. In preparing the American and English editions of *Extra-Sensory Perception,* he had written to Irving G. Gavett and R. A. Fisher for advice on the problem of combining CR's and reassurance that the method on which he finally settled was sound. This same problem led him in the next few years to ask for help also from P. J. Rulon at Harvard (whom he met while giving a seminar there in 1936) and Churchill Eisenhart, a young statistics student at University College, London, whose father had interested him in the Duke work. Within the field of psychical research, both Whately Carington and S. G. Soal offered advice. Rhine was at first forced to depend on these scattered consultants because of the shallowness of statistics at Duke, although he did have some contact there with E. R. C. Miles, an assistant professor of mathematics, who taught the school's one course in probability. Charles Stuart, of course, whose undergraduate work had been in mathematics, had first taken on the task of answering Willoughby in 1934, as well as of supervising the statistical evaluation of the laboratory's continuing projects. But by 1937 Rhine had managed to secure the services of Joseph Greenwood, a young geometer in the Duke mathematics department who gradually made himself over into a statistician at Rhine's instigation. It was Greenwood and Stuart who came to assume the responsibilities of studying the mathematical criticisms being made of the Duke work and of deciding what changes or developments in theory and technique ought to be undertaken.

The extent to which the Duke statistical program had grown more knowledgeable by early 1937 can easily be recognized in an article by Greenwood and Stuart that appeared in the first volume of the *Journal of Parapsychology*[78]—the very first attempt by the parapsychologists to explain at all systematically the basis for their mathematical reasoning—and owed its existence to the need to respond to the published arguments of Willoughby and Kellogg. Although neither of these critics wrote with great clarity, their objections may now be seen to have been aimed at two general if initially ill-defined problems: (1) the nature of the theoretical distribution appropriate to purely random guessing of a series of 5 × 5 decks of ESP cards and (2) the best way of evaluating actual sets of subjects' guesses. The problem of the theoretical distribution had been raised indirectly by Willoughby when he argued that since a random distribution of scores was bound to be skewed towards scores above the presumed mean of five, "no computation of odds based upon the normal curve is in the least applicable to the data."[79] Stuart had been able to show that the skewness involved was insignificant, but Kellogg had then raised the same issue in a much more direct fashion. He raised doubts as to whether the MCE was truly 1/5 and argued still more strongly that the obvious interdependence of targets would affect what he called the "dispersion," or "spread," about the mean.[80] While Stuart demonstrated that even under this accurate analysis of the ESP experimental situation (soon christened the "matching hypothesis") the MCE remained 1/5, he and Greenwood did acknowledge that the "spread," or variance, under the matching hypothesis was greater than under a strictly binomial distribution; Greenwood eventually proved that the maximum variance (under the matching hypothesis) was 4.167 rather than the 4.000 under the binomial distribution and the standard deviation was hence 2.041 rather than 2.000.[81] This scant difference suggested that the exact distribution under the matching hypothesis would closely approximate the binomial, Kellogg's warnings notwithstanding, but the details of the new distribution still remained to be found.

The second statistical issue that Rhine's critics had initially raised was that of the evaluation of the data: What criteria should be employed in determining the significance of a given result? In *Extra-Sensory Perception,* Rhine had fixed his attention on the extent to which a subject could guess above chance in a run, or runs, of guesses, measuring significance in terms of a CR that could be converted into a probability value. In the next year or so, the Duke team modernized their technique to the extent of basing their calculations upon the standard deviation instead of the probable error (a movement generally true of statistics in the 1930s), but they continued to concentrate upon measuring the improbability of a ratio of successful guesses to total guesses.[82] However, first Whately Carington (in correspondence) and then Willoughby and Kellogg (in print) argued strongly that a more rigorous and comprehensive way to evaluate card-guessing success would be to study the *distribution* of successes per run of twenty-five cards over a series of such runs, using the chi-square method. Because in his original research Rhine had not kept records of the number of successes in every run (much less of the specific calls and cards in every run), this technique could not be applied generally to existing data. Fortunately, complete records had been kept of perhaps the most successful of Rhine's early experiments, the Pearce-Pratt long-distance trials of 1933, comprising seventy-four runs of twenty-five cards each. To satisfy Willoughby and Kellogg, Greenwood and Stuart performed a chi-square goodness-of-fit test to show that the scores of the Pearce-Pratt runs were not explicable merely as chance variations from the binomial distribution.[83] By 1937 chi-square methods had become an accepted part of the Duke statistical armamentarium.

In the fall of 1937, Kellogg, now growing rather unpleasant in tone, returned to the question of the theoretical form of the matching hypothesis in two new articles (one of them semipopular). With the assistance of a student, Kellogg had performed the tedious calculations necessary to work out a set of approximate probability values for the range of scores obtainable under the conditions of the matching hypothesis, and he now published the resulting table of values in juxtaposition with a table of values for the normal and the binomial distributions. Unlike Willoughby earlier, Kellogg recognized that the binomial approached the normal curve with larger values of the distribution sample, but he claimed that this approach did not hold for scores located at the upper extreme of the distribution. The probability value of the normal curve for these scores was much lower, he claimed, than those computed from the binomial or, more particularly, from the matching hypothesis, and to make his point clear he published comparative p-value tables for the three distributions.[84]

By the end of 1937, however, an increasing number of relatively prominent mathematicians had begun to take an intelligent interest in the statistical problems raised by parapsychology, and these men (who in the main found Kellogg's attitude offensive) now took up the problem of the distribution with some care. E. V. Huntington at Harvard, Rhine's most enthusiastic advocate among the mathematicians, worked out the distribution for the matching hypothesis applied to 3×3 and 4×4 decks, and on this basis T. E. Sterne of the Harvard observatory proceeded to derive the first four moments of the 5×5 case—the first of which agreed with Greenwood's earlier calculation of 4.167 for the variance. Huntington and Sterne published their work in *Science* in November 1937.[85] Here Sterne gave the exact frequency values for 21, 22, 23, 24, and 25 successful guesses in a deck of 25 cards; gave the first four moments; and by fitting a Pearson Type I curve to the mean and moments, produced close approximations to the exact frequencies of scores of 0 to 20. As for the relation between the matching and binomial distributions, Sterne concluded that "to distinguish observationally between the true values and the binomial values, in the 5×5 case, would require some 30,000 trials." Within the next month both T.N.E. Greville at Michigan and Bancroft H. Brown of Dartmouth had computed the exact values for the frequency of all scores under the matching hypothesis,[86] correcting the approximations Kellogg had offered previously. Finally, using Sterne's work, Huntington constructed a "rating table" for the 5×5 case, which he presented to the December meetings of the American Mathematical

Society in Indianapolis.[87] With this table, the theoretical proportion of cases with a better score than any given score in a specific number of runs could be computed exactly, and hence the probability values of any given score for any given number of runs could be computed as well. At the same time, Huntington's friend and former student Burton Camp presented Sterne's work on the matching distribution.

The new contributions of the mathematicians made possible a review article by Stuart and Greenwood that December focusing upon a comparison of the three theoretical distributions that had come under debate.[88] With the main statistics of the matching distribution known, this comparison could now be carried out in detail. The probability values of the three were compared with respect to four different CR's and three different run lengths, and they showed that in one respect at least Kellogg had been justified: there was a nontrivial difference between the normal distribution and the other two for low CR's and small numbers of runs. The probability values of the normal distribution were lower than the others, and thus scores computed for significance using the normal probability integral would be assigned slightly greater significance than was proper. Consequently, Greenwood and Stuart suggested that the threshold for "significance" for scores of numbers of runs of less than twenty-five be raised from a CR of 2.5 to one of 2.8, and for scores of numbers of runs greater than twenty-five, from 2.5 to 2.65. A few months later Greenwood published a method of computing the variance of Zener card-matching applicable to *any* distribution of calls.[89] It had been known for some time that in ESP experiments the possible variance of the distributions of runs varied from 0, when only one ESP symbol was matched against a 5×5 array of symbols, to 4.167, for the matching hypothesis (when $n = 25$). Greenwood now generalized the method by which he had worked out the variance of the matching hypothesis in order to compute the variance for any distribution of ESP symbols in any test matching situation. There could no longer be any reasonable doubt as to the nature of the probability distribution of ESP test scores: the mean was known, the variance could be precisely determined, and the other moments could be closely approximated by comparison with one or another of the three main distributions.

At this point—early 1938—it seems apparent that the four years since the publication of *Extra-Sensory Perception* had refined and enhanced the role of statistics in parapsychology. The mathematical techniques appropriate to the field had now been worked out, and henceforth there could be a sense of genuine security about the algorithm in use. To be sure, Kellogg and a few others continued to argue against the probability mathematics, but by 1940 even Kellogg had largely given up, concluding that "the battle of the probabilities need hardly be fought again."[90] What is of more importance than this feature of technical development and refinement, however, is the more intangible reorientation of parapsychology as a science. Rhine and his associates had now come to appreciate something of the subtlety and difficulty of the chain of arguments necessary to support their conclusions. This in turn had a variety of implications for future parapsychologists: they would need a rigorous training in statistical reasoning or at least the close support of competent statisticians, and they would have to understand the mathematical consequences of experimental design.

V

Within a few years of the publication of *Extra-Sensory Perception,* then, American parapsychology had come to possess many of those attributes by which we ordinarily recognize a science. It had established a small community of researchers who, despite differences in professional training and careers, were engaged on common problems; it had acquired an institutional niche at Duke and had the potential for immediate support at other schools. This community of researchers shared a common set of assumptions, elaborated out of the psychical-research tradition, and an experimental technique that from their point of view yielded at least intermittently replicable results, permitting them to investigate relationships among their data. What was strikingly absent from this science was any overall model or theory that would serve to unite method, metaphysics, and results—that would *explain,* satisfyingly, just what ESP was and how it operated. This is a deficiency of which critics of parapsychology have always made much, claiming that nothing can pretend to be a science that shows no signs of being able to provide an explanatory matrix for the events it claims to study. Psychical researchers had long been sensitive on this point. In presenting his results, Rhine agreed that as yet ESP had no theoretical basis, that it apparently could not be explained in the light of existing science. He did not feel that this made parapsychology somehow unscientific, however, but remained true to his essential positivism and argued that a true science had inescapably to be founded on data alone. Broad explanations of those data might follow, but he felt that they could not be a first step in the development of the scientific approach to a new problem.

This did not mean that Rhine had no feelings about the form that a general theory of parapsychology would have to take. At the beginning of his Duke career he had been inclined to treat all mental processes as "electromagnetic group-wave" phenomena; the new quantum physics seemed to hint that physical, mental, and psychical phenomena might all prove to originate, not in static, material structures, but in just such immaterial waves.[91] Subsequently, in writing *Extra-Sensory Perception* in 1933, he became convinced that telepathy and clairvoyance could be explained by *no* known physical model, certainly not by the transfer of information by radiant energy. What had led him to this conclusion was his success in conducting ESP distance experiments, which seemed to reveal that communication did not decline with increasing distance, as would be necessary if electromagnetic radiation were involved. The fact that no direct measurement of a transmitted force or energy was possible in these distance tests did not deter him from concluding that ESP operated by some unknown form of energy whose behavior was radically different from any known. Success at DT experiments also seemed inconsistent with a wave theory, since the radiation from all twenty-five cards would present itself simultaneously to the percipient.[92] Then, during 1934, Rhine's work on psychokinesis and, more particularly, precognition encouraged him to think that none of the current laws of physics could apply to parapsychology, since the psychic transmission of information seemed now to be independent of temporal as well as spatial limitations. While believing that ESP must have a "natural" rather than a supernatural explanation, he felt equally strongly that these phenomena were "clearly independent of the so-called mechanistic limitation in some degree."[93]

Rhine expanded on these ideas in *New Frontiers of the Mind,* which, though a popularization, came as close as any parapsychological work to providing an overview of the field and its place among the sciences, based upon the research as it stood by 1937. Again he argued that "there appears to be no known physical condition or process to which [ESP] can be related," especially since neither short- nor long-wave radiation was consistent with successful distance work.[94] Yet, he went on, since physical targets do appear capable of influencing a subject's thoughts, some sort of energetic causation must be involved—and he hinted that some day scientists might need to accept "the concept of a non-mechanical energy or a nonmechanical physics which might include the phenomena of mind."[95] With this conclusion, of course, he felt he had answered his original question, whether psychical research could provide proof of the inadequacy of a purely materialistic philosophy. But—as we have already remarked—parapsychologists now were by no means alone in their perception that materialism might have to be abandoned: the successes of relativity and quantum mechanics were forcing many physicists to a reappraisal of the foundations of their own field.

That psychology had nothing to say about the nature of ESP Rhine was less ready to believe. He had already acknowledged in his 1934 monograph that ESP functioned in conjunction with other mental activities: it was dependent upon volition, attention, and confident enthusiasm and was adversely affected by fatigue and certain drugs. Three years later he was still convinced of this affinity with "the higher, more complex mental processes." On the other hand, he did not try to pursue this further into psychological explanation, in whatever direction. Elaborate psychodynamical models in the Myers or Warcollier traditions (revived to some extent by Gardner Murphy in the 1940s)[96] were not really congenial to him. He was more concerned to decide whether ESP might be understood by analogy with ordinary SP. As he had explained in *Extra-Sensory Perception,* his original expectations in this regard had changed quite early.

> I began using the term "Extra-Sensory Perception" (E.S.P.) at first with the more tentative meaning, "perception without the function of the recognized senses." But as our studies progressed it gradually became more and more evident that E.S.P. was

fundamentally different from the sensory processes, lacking a sense organ, apparently independent of recognized energy forms, non-radiative but projectory, cognitive but unanalyzable into sensory components—all quite nonsensory characteristics.... Hence the present interpretation is rather that E.S.P. is, frankly, "perception in a mode that is just *not* sensory," omitting all question of "unrecognized." I think we have progressed this far with reasonable certainty.

In *New Frontiers of the Mind* he continued to reject a connection between the two, and for the same reasons.[97]

Yet Rhine was not unmindful of the possible relations between SP and ESP. In fact, during 1935 and 1936 he encouraged some of his associates—Pegram, Carpenter, and Warner—in research designed to test for such relations. In November 1936 a visit to Harvard led to a lengthy correspondence with E. G. Boring that revived the issue. Among the points discussed by Boring was the apparent resemblance between ESP and difficult sensory perception, perception near the limen of discrimination: stress, fatigue, and distraction caused both to lose in precision.[98] In the next few years students at both Harvard and Duke were at work exploring the psychological conditions of subliminal or near-liminal perception, and in both places there were overtones of a concern with the implications of this research for ESP.[99] Two of Rhine's coworkers eventually completed doctoral dissertations on the conditions affecting subliminal perception. Neither one directly pursued the comparison of ESP with SP, but both investigated subliminal SP according to the parameters that Rhine had shown to bear upon ESP. J. L. Woodruff, working with Donald Adams, studied the effect of motivation, knowledge of success, attentiveness, and distraction upon it; Burke Smith, extending his M.A. research under Karl Zener, studied the effects of dissociative drugs such as sodium amytal. Smith, in fact, openly acknowledged that one of his reasons for undertaking the research was "the desirability of permitting comparisons to be made between some of the aspects of sensory perception and extrasensory perception,"[100] and Woodruff's immediate postdoctoral career suggests that he shared these concerns. A connection between ESP and SP remained a tantalizing possibility that, unlikely as it seemed to parapsychologists, could not be rejected categorically.[101]

Still, in the late 1930s the lack of a general theoretical foundation in physics or psychology could hardly have seemed a serious defect of this aspiring science. After all, in the few years since the appearance of *Extra-Sensory Perception* there had built up a community of dedicated research students who were beginning to develop limited hypotheses and procedures from their experimental tests. They might be frustrated by their inability to bring the production of psychical phenomena under regular control, but the early successes of the Duke laboratory were not far in the past, and there were signs that the psychological circumstances favorable to psychical abilities were becoming better understood. The parapsychologists' lack of a broad unifying theory did not prevent them from finding a common disciplinary identity in the shared language, issues, and techniques that they were coming to possess.

CHAPTER EIGHT
The Psychical-Research Societies in the Mid-1930s

The Duke program of research into extra-sensory perception received such disproportionate publicity in the mid-1930s that it was easy, particularly in the United States, to forget that there were many clusters of investigators in America and in Europe that had a longer and equally intense tradition of involvement in experimental psychical research: the psychical-research societies. While to many of these groups Rhine's monograph of 1934 certainly seemed to suggest a possibly fruitful technique of investigation, they had their own lines of approach, with which ESP research had to compete for attention. Moreover, the societies—particularly the larger and wealthier ones—had always had the potential to effect the introduction of psychical research into the university, as Rhine had done. Yet this did not happen—or if it did, the societies' decision to involve themselves with the academic world was consequent upon the impact of Rhine's own success. Something like Rhine's single-minded drive and concentration of focus was required to force psychical researchers and scientists alike to consider seriously whether psychical research ought to be recognized and pursued as a science. Lacking that incentive, the amateur societies remained as divided as ever by personalities and issues over their proper methods and aims, and their own contributions to research were necessarily disorganized and limited.

I

The Boston Society for Psychic Research had been virtually destroyed as a research organization by Walter Franklin Prince's death in 1934. As Research Officer of the society Prince had borne the responsibility for carrying on a program of investigation and publication on behalf of its members. Indeed, it might be accurate to say that Prince *was* the society insofar as it was a research organization, for although the membership at large was willing to support research, it was unable or incompetent to pursue it. In the first nine years of the society's existence Prince had written or edited four books and twenty-two bulletins under its auspices, many of them maintaining the skeptical tone that had set the BSPR apart from the American Society for Psychical Research and had earned Prince the international reputation of a "vicious" and "hard-bitten" critic of all mediumistic performances.[1] With Prince's death, however, all this came to an end, for he left no natural successor.

The consequences of Prince's death highlight the absolute necessity of forceful leadership for coordinating the activities of an amateur society. The Boston society had established itself in that city because it was the home of its two most important financial supporters, Mrs. Augustus Hemenway and Mrs. William M. Wood, but many of its members lived in New York City and elsewhere: when the society took stock of its situation in October 1934, after Prince's death, Elwood Worcester of Boston remained its president; John F. Thomas of Detroit was elected vice-president; and Lydia Allison of New York was chosen the society's executive secretary.[2] There remained the problem of a successor to Prince. The membership still cared little for research; the Boston society was essentially a passive audience rather than a collection of eager amateurs sharing a tradition of active investigation, like the SPR. In the absence of a journal that could speak to and for the Boston group and give it an intellectual identity, the appointment of a dynamic research officer was critical. The selection of J. B. Rhine, Prince's closest and most successful protégé, was certainly an understandable one, but Rhine, in North Carolina, was far removed from the other nuclei of the society; besides, he had neither the time nor the inclination to keep up a string of critical reviews of current psychical research like those Prince had regularly edited. Rhine felt that the Boston SPR's affairs might make too many demands on him: it was with some reluctance that he agreed to Lydia Allison's request that he contribute to a Walter Franklin Prince memorial volume, and when he learned of his election he insisted upon being called *honorary* research officer.[3]

To add to these difficulties, Rhine's growing enthusiasm for an independently funded institute for parapsychology at Duke forced him gradually to see that the psychical-research societies would inevitably be his competitors in the search for financial support. By 1936 he had decided regretfully to cut free from the societies, at whatever cost to them. When Gaither Pratt's *Towards a Method of Evaluating Mediumistic Material* appeared as a bulletin of the Boston society in March of that year—the first since Prince's death—Rhine added a postscript in which he explained why mediums could be studied effectively in the psychology laboratory and hinted broadly at the need for such laboratory studies to be permanently endowed.[4] He wrote in similar terms thereafter to Mrs. Wood and Mrs. Hemenway, the society's chief benefactors.[5] He was well aware how his actions might strike his fellow members, and he acknowledged apologetically to John Thomas that "the Boston people... would be very much hurt by what would seem to them a flagrant disloyalty of one of their sons."[6] Thomas urged Rhine not to commit himself to the academic world at the expense of the societies, arguing that there was a role for both, but Rhine insisted that his financial backers had been definite about wanting their funds to be used in the university and implied that interest in the Boston society was not really very deep.[7]

From his point of view, Rhine had every reason to see the society not merely as competing for funds but as wasteful of what money it received. Its diffuse membership remained more interested in hearing about mediumship, survival, and the "supernatural" than in learning of recent advances in the laboratory study of extra-sensory perception. One of Rhine's Boston advocates, Francis Strickland of the Boston University School of Theology, after having given a program to the BSPR in January 1937 on "The Significance of Extra-Sensory Perception for Psychic Research," wrote Rhine bitterly that

> there were only thirty present four of whom were my own students. Not a member of the Council was there nor were some of the more prominent members who have been out to previous meetings. This confirms what I have thought heretofore, that very few in the society are interested in research. They like talk about borderland phenomena but seem but little interested in the scientific study and investigation of these phenomena.... The members are however fond of programs of ghost stories. The November meeting was devoted to a talk by a visiting speaker on "Laying a Ghost." It dealt with haunted house stuff! There were about sixty present and they seemed to enjoy it. Last year we had a lecture on "Banshees" which was well attended.[8]

Such a report could only confirm Rhine's mistrust of the amateur psychical researcher. Curiously, Gardner Murphy at Columbia remained much more hopeful than Rhine that the psychical-research societies could be a useful forum for parapsychological investigation, probably because Murphy had developed his interests within the framework of the societies and had in any case no expectation of support from Columbia for his work. In default of help from Duke, therefore, it was really Lydia Allison, John Thomas, and Murphy who struggled continuously to keep the Boston society alive: Thomas published a condensed version of his dissertation research under its auspices in 1937 (Rhine wrote gloomily to him that it was "probably the last worth-while publication they will have"),[9] and Murphy did likewise with a translation of Warcollier's collected articles on telepathy in 1938.[10] But the BSPR membership remained tiny and disorganized, and the bill-collector was always at the door of its office at 719 Boylston Street.

The New York–based American Society for Psychical Research was in 1934 likewise in no condition to serve as the basis for a concerted research program, for its leadership had no interest in anything but mediumistic investigations. From 1932 until 1942 the same men served as its officers; indeed, some had been officers and trustees of the society since the twenties, when they had become firm friends and partisans of the medium "Margery," Mina Crandon. All were professional men, by now in their late fifties or sixties: William H. Button, a New York corporation lawyer, was president; Mark W. Richardson, a Boston physician, first vice-president; Daniel Day Walton, second vice-president and counsel; Thomas H. Pierson, a civil engineer involved in the construction of a number of major New York buildings, secretary; Lawson Purdy, a distinguished New York tax lawyer, treasurer. Understandably, these men did not themselves have the time needed to undertake extensive psychical research; but in any case their commitment to Margery and her ability to communicate with her dead brother Walter was so intense that they felt little interest in or need for a wider exploration of psychical phenomena. J. Malcolm Bird had resigned from the ASPR in 1930 after suddenly confessing that he had seen elements of fraud in the Margery mediumship from the outset, and, significantly, the ASPR had had no research officer and had subsidized no investigations in its own name since that time.[11]

In these circumstances the individual in the best position to arouse and guide the attention of the society's membership was the editor of its monthly *Journal*. By his selections for publication from among submitted papers, by his choice of materials for republication from other sources, by his own contributions in the form of "Editorial Notes," the editor had the opportunity to direct the thoughts of the society favorably or unfavorably towards whatever subject he chose. In 1930 Frederick Bligh Bond replaced J. Malcolm Bird in this position. An English spiritualist, Bond had directed the excavations of Glastonbury Abbey from 1908 to 1921 by following information purported to have come through automatic writing from the monks who had lived there in the sixteenth century. He had come to America in 1926 and had almost immediately become another of Margery's champions, calling her in 1927 "about the finest medium God ever gave."[12] The new editor's orientation in psychical research thus conformed very well to that of the officers of the society—fortunately, since he was expected to clear the contents of each *Journal* with them. In 1934/35, the first year after the appearance of *Extra-Sensory Perception,* the ASPR *Journal* contained very much the same sort of material that it had from the beginning of Bond's editorship: articles on poltergeist phenomena; on the study of phantoms; on the past achievements of the society; on Richet's experiences with Pascal Forthuny; and, over and over again, on Margery and other mediums.

At the beginning of 1935, however, the society's advocacy of the Margery mediumship was suddenly tested once more when Bligh Bond followed Bird in questioning the nature of some of her phenomena. He learned in January of that year that the British Society for Psychical Research was about to publish an authoritative report confirming that certain fingerprints mysteriously produced in wax while Margery was in trance—purporting to have come from her dead brother "Walter"—were in fact identical with those of her dentist, Dr. Frederick Caldwell ("Dr. Kerwin"). Bond at once implored the society's president, William H. Button, to allow him to defuse the situation by publishing an independent account of the matter, and when Button refused, Bond defiantly published a summary of the SPR report anyway. Bond was by no means a skeptic in psychical matters, and he remained by his own account "as firm a believer in Margery's mediumship as any one";[13] but the trustees of the society could not tolerate the probability that his editorial would invite "irresponsible" attacks on the authenticity of the fingerprints, and they straightway discharged him and repudiated his acceptance of the British report.[14]

In the light of the American society's blind devotion to Margery, it is not surprising to find that it showed no interest in encouraging a program of experimentation upon possible psychical abilities in normal individuals. Inevitably, it felt itself allied in spirit not to the right wing of international psychical research but to the left. When in January 1936 the society's trustees decided to add to the *Journal* a "Letter from England" as a regular monthly feature, they consigned it to Nandor Fodor, research officer of the newly founded International Institute for Psychical Research (IIPR), which had been organized to serve the interests of those who were committed to post-mortem survival (though not necessarily to the spiritualist thesis).[15] Fodor's accounts reported variously on the work of Harry Price, of local spiritualist societies, and of the IIPR itself, as well as on psychical matters dealt with by the English press; but they avoided any mention of the research of the SPR. The readers of the American society's *Journal* were thus able to remain undisturbed by any knowledge of the attitudes and policies of the most critical and exacting—and most nearly "scientific"—of the amateur societies.

Nevertheless, the intense emotional commitment to the search for life after death, which had been a natural response to the tragedies of World War I, evidently no longer had the same appeal in the 1930s. The membership of the ASPR declined regularly, year after year—from 1,128 in 1932 to 872 in 1934 to 505 in 1938. At the same time, the growing publicity being given to J. B. Rhine's work was turning attention to the apparently demonstrated possibility of experimentally studying supernormal faculties present in man. It was inevitable, therefore, that occasional favorable notices of research upon telepathy and clairvoyance should have begun to creep into the *Journal* alongside its transcriptions of séances. Still, it was several years before the *Journal* commented directly upon Rhine's work. *Extra-Sensory Perception,* which of course had been a publication of the Boston SPR and had been sponsored by Walter Franklin Prince, was never reviewed or even acknowledged in the *Journal,* whose first brief references to the Duke experiments appeared only in 1936, often in articles reprinted from other sources.[16] Then, at the end of that year, Jocelyn Pierson, daughter of the secretary of the ASPR, called direct attention to Rhine's research. She had been hired to assist in the production of the *Journal* and had gradually taken over much of the responsibility for supplying its contents. Recognizing with dissatisfaction the narrowness of the *Journal,* she began to incorporate in its pages a wider range of material, including, in the December 1936 issue, her own quite favorable summary of the popular articles by E. H. Wright and Gardner Murphy that had outlined Rhine's experiments, and from this time on, the *Journal* included regular commentary upon extra-sensory perception.[17] Generally speaking, the *Journal* applauded Rhine for having forced the world to admit to the existence of "definite and important problems in the psychic realm," though insisting that "Dr. Rhine's method is only practical for conditioned experimental clairvoyance"[18]—the problem of properly evaluating the "qualitative and subjective" phenomena of mediumistic clairvoyance remained a vexing one—and reminding its readers that his work was not the first in its field (citing *Phantasms of the Living* as "the best, the most scholarly, and the most scientific work done with Telepathy and Clairvoyance").[19] Whatever their limitations, however, William Button was sufficiently intrigued with the Rhine experiments to conduct during the winter of 1937/38 a series of tests of Margery's ability to call playing cards clairvoyantly; he reported with complacence that with "Walter's" aid, Margery had regularly been able to identify twenty-three of twenty-five cards.[20] And when Rhine's second book (*New Frontiers of the Mind*) appeared, it *was* reviewed, and relatively favorably. "Whatever one may feel regarding the value of the Duke experiments," the reviewer wrote, "they certainly offer a distinct improvement on the former mechanistic view of mind."[21] The leaders of the ASPR may not have felt entirely content with Rhine's scrupulous avoidance of the survival question, but they could at least take comfort at his having given one branch of psychical research a renewed respectability and popularity.

During the academic year 1937/38, when Gardner Murphy and Ernest Taves were conducting their first year of research into extra-sensory perception under Harvard's Hodgson Fund, Adele Wellman, the executive secretary of the ASPR, came uptown to visit their laboratory at Columbia and participated in some of their experiments; other members of the society soon joined her. Through her agency, Murphy and Taves were invited to propose a plan of research to a meeting of the society in September 1938. The resultant sketch of a program stressed Murphy's conviction that it was important to employ a wide variety of test materials as targets—colors, pictures, and dice, as well as standard ESP cards—so as to ensure active interest and prevent boredom. Murphy proposed studying a group of volunteers from the society in telepathic and clairvoyant experiments to learn whether "extra-chance results are obtainable from a group of subjects distinguishable from the general public *only in terms of their interest, not in terms of previous demonstrations of high scoring ability*," which no previous researchers—not even those at Duke, he argued—had done. Fifteen members volunteered, and the experiments began immediately at the ASPR offices; sessions were held once or twice a week and were carried on for more than a year.[22]

By the end of 1938, therefore, even the "leftwing" ASPR had come to accept the importance and value of a narrowly experimental study of psychical phenomena in ordinary people. To be sure, the society's leaders still felt some regret that an experimental approach forced researchers to ignore the bulk of the reported psychical phenomena, together with many of the questions that had animated earlier students of the subject; but they justified the procedure, as Rhine had already done, by explaining that such work was a necessary preliminary to the study of post-mortem survival. In introducing the first report of the Murphy-Taves investigations, William H. Button and John O'Neill wrote:

> This approach to the problem may make it appear that only a purely physical interpretation of the phenomena is being considered and that all other interpretations are ruled out, or in other words that participation in the phenomena by extra-physical entities,

or "spirits" is not being considered. Such is not the case. No agency of transmission is ruled out. Science works from the known to the unknown. We know a great deal about physical phenomena. When we have attained a sufficient mastery over the purely physical phenomena involved in Extra-Sensory Perception, we have provided ourselves with an important route for penetrating into the realms that lie beyond.[23]

But clearly Rhine's acceptance by the ASPR had depended upon the scientific and public recognition he had previously received. The society would not have moved towards ESP experimentation to the limited extent that it did had Rhine not already established himself independently.

II

The same sorts of problems of resources and of organization beset most European psychical-research societies in the mid-1930s. They were beginning to lose public attention as well as some of their own self-assurance. A measure of their decline is the relative disappointment of the international congress held in Oslo in 1935: McDougall had been named "President of Honor," but he did not attend, nor did Rhine; the Nazi government prevented German investigators from going;[24] and British psychical researchers were rather thinly represented. As the personal tragedies of World War I moved further into the past, and as the Depression continued, the individual European societies found their membership and financial support steadily dwindling. The Institut Métapsychique International in particular was hard-pressed for money: in 1932 its director, Eugène Osty, had had to beg for special contributions to allow for the continuation of studies on Rudi Schneider,[25] and early in 1934 a special Society of the Friends of the IMI was established, at Charles Richet's urging, in the hope of securing regular donations on a larger scale.[26] Neither of these measures worked entirely satisfactorily. The sudden death of Osty in August 1938 weakened the leadership of the institute still further, and the outbreak of World War II a year later was the final blow; the *Revue métapsychique* ceased publication in June 1940.

The English community of psychical research continued to subsidize the field relatively well; what bedeviled and weakened it was the continuing sharp disagreement among right, left, and center as to what the proper aims and methods of psychical research were, a disagreement that had openly split the SPR in 1931. The English academic world took psychical research seriously enough to be willing now to support responsible and decorous scientific testing of the phenomena; spiritualists, strident and still numerous, wanted only research that was a priori sympathetic to extreme hypotheses; and the Sidgwick leadership of the SPR occupied something of a middle ground, failing in the process to satisfy many from the other two parties. These tensions led in 1934 to the formation of two other associations meant as complementary or alternative to the SPR.

One of these—the more nearly ephemeral—was the International Institute for

Psychical Research, established in February 1934 at the initiative of a survivalist organization. Insisting that it meant to study psychical phenomena from a strictly scientific standpoint, it soon had the support of a number of eminent men; its officers included scientists of some distinction—the biologist Grafton Elliot Smith as president, the physiologist D. F. Fraser-Harris as research officer, Julian Huxley as one of the vice-presidents—as well as the spirtualist Shaw Desmond.[27] It did not follow, of course, that the scientists had much understanding of the internal difficulties of psychical research: Harry Price mentioned sardonically that when Fraser-Harris had "entered this Laboratory for the first time in 1931, [he] candidly admitted that he did not know the meaning of the letters S.P.R.!"[28] By mid-year it had been announced that the scientists had resigned and that the society had been reconstituted by fusion with the British College of Psychic Science. Thereafter it was led by spiritualists, Nandor Fodor serving as its research officer until 1938, and was of relatively little importance to experimental psychical research.[29] But the circumstances of its original establishment make it clear that by the 1930s psychical research still had potentially a much broader base of support in academic orthodoxy in Britain than in America.

Harry Price's research organization was far more important on the English scene, precisely because he was able to enlist this academic support on a continuing basis. The National Laboratory for Psychical Research, which Price had founded in the 1920s, had been funded almost exclusively from his own pocket and had carried on comparatively little research. Nevertheless, it remained a perennial thorn in the flesh of the SPR because Price noisily publicized it as a society that would offer a progressive alternative to the older institution, one that would actively and aggressively investigate all phenomena, including the physical and the overtly spiritualistic. It thus attracted a peculiar mixture of supporters, ranging from those spiritualists who were outraged at what they considered the excessive conservatism and skepticism of the SPR to critics like S. G. Soal who were scornful of what *they* viewed as its unscientific and overcredulous position on mental phenomena. Because he tried to present his laboratory simultaneously in these two very different lights, Price's own convictions remain a puzzle; but his obvious ambition to dominate psychical research and his knack for self-advertisement were by themselves enough to earn him the distrust and suspicion of the SPR. Price's offer of 1930 to merge his laboratory with the SPR, which had been rejected out of hand by the Sidgwick leadership but had won the approval of many of the society's members, had shown the council just what potential strength Price's circle possessed.

Price's failure to consolidate his National Laboratory with a psychical-research society—he had approached the IMI as well as the SPR in 1930—decided him to turn to the academic world for a possible home for his institution. Through the philosopher C.E.M. Joad, Price proposed to the University of London in 1933 that it assume the responsibility for continuing his work in some fashion, in which case he would engage to give the university his library, equipment, and five hundred pounds per year for research. The university referred the proposal to several boards of studies for recommendation, and it was endorsed without comment by the boards of medicine, philosophy, physics, and theology. The psychology board went beyond mere endorsement and defended it at length, arguing that the psychical researchers' investigation of hypnosis, multiple personalities, and automatic writing had already demonstrated the possible value of so-called psychic phenomena to abnormal psychology. Only the physiology board objected to the plan, feeling that the subject lent itself to misrepresentation and fraud and that it might discredit the university. Eventually the board of the Faculty of Science voted for the plan, 11 to 4, and in January 1934 first the academic council and then the senate agreed "that Psychical Research is a fit subject of University study and research."[30]

Yet within a few weeks Price's proposal was turned down by the university.[31] The council's recommendation had been contingent upon locating the laboratory within a college in which there were departments of psychology and perhaps physics or philosophy. When Price's supporters within the university began to look for a specific college willing to adopt psychical research and give space to his materials, they had no success. Cyril Burt found "unexpected difficulties" in getting his own University College to take it on, and Joad at Birkbeck was no more successful.[32] A move to extend the limits of scientific investigation could be easily endorsed as an abstraction; when it became an immediate threat to academic resources and perhaps reputation it was a very different matter. The astronomer Sir Richard Gregory, then editor of *Nature* and a man not unsympathetic to the plan, explained regretfully to Price that "I know a fair number of people in the [University] Senate were in favour of providing accommodation for the laboratory and library, but I expect the majority were fearful of taking up psychical research. . . . "[33]

Price still had the sympathetic good will of a number of psychologists at the University of London, however, and when the plans for a new department collapsed, he found them willing to try to associate psychical research with the university in another, less formal way. In May 1934 a University of London Council for Psychical Investigation (ULCPI) was formed to supervise Price's National Laboratory and to arrange with the university's boards of studies to direct students in psychical research for an advanced degree. The psychologists Cyril Burt, J. C. Flugel, C.E.M. Joad, C. A. Mace, and F.A.P. Aveling, together with representatives from several other boards—Guy B. Brown for physics, E. S. Waterhouse for religion, E. D. Macnamara for medicine—all agreed to join the council.[34] Price and S. G. Soal represented the amateur tradition on the council. Price saw to it that the new group was publicized to widely different movements, ranging from the spiritualist journal *The Two Worlds* to *Nature,* and the apparent fusion of psychical research with respectable English academe evidently engendered high expectations everywhere—except within the

SPR, which inevitably announced the formation of the new group with a certain mild disdain.[35]

However, the ULCPI never in fact lived up to its promise. Its own attempts to carry out research programs in the manner of the old National Laboratory soon led to the familiar tensions between the psychical researcher and the scientist. By the end of 1934 it had indeed launched two projects of some interest: an attempt by Soal to replicate J. B. Rhine's card-guessing experiments and an inquiry by Guy Brown into "water-divining and dowsing generally."[36] But Soal saw in his project an opportunity to reiterate his contempt for the SPR and its approach to psychical research. Already in August 1934 Joad was expressing dismay to Price at Soal's tendency to disparage the SPR wantonly and to suggest "that so eminent and reputable a person as Gilbert Murray didn't play fair."[37] Soal somewhat altered the offending passages in a critique of the experimental evidence for telepathy,[38] but not enough to suit Guy Brown, who complained in February 1935 that "in serious scientific work one never brings in *personalities*—they must go!'"[39] Soal's paper never did appear, a fact for which he came to blame Harry Price.

Price's own irrepressible showmanship, too, had begun to fret some academic members of the ULCPI. In joining the council, Brown had warned Price that his personal situation at University College made it essential that their public statements be restrained. When Price took it upon himself to sponsor a demonstration of clairvoyance by Kuda Bux in the name of the council in July 1935 and to propose a filmed fire walk for later in the year,[40] Brown protested immediately: he objected to Price's tendency to run council affairs by fiat and to his continual search for publicity for the organization, which he feared might damage the tolerance shown by the university. "We only have the title 'University of London' on sufferance; we are not officially connected & any time the Senate could make us climb down—to the great delight no doubt, of the SPR etc., etc.'"[41] Nevertheless, Price's enthusiasms remained very difficult to control.

The supervision by the council of degree work in psychical research also encountered some difficulties, but it was not without some results. Whatever mistrust the London dons may have felt of Price's behavior did not necessarily affect their readiness to provide support for serious psychical research. F.A.P. Aveling had joined the council with the warning that he might be compelled to act as an *advocatus diaboli*,[42] but he soon agreed to direct John Hettinger in a psychical-research topic for a Ph.D. at the University of London. Between May 1934 and September 1937 Hettinger carried out a series of experiments in psychometry, in which a subject was given a material item and asked to give any impressions she received of individuals associated with the item. The correctness of the subject's impressions was evaluated by a variant of the techniques long in use for the evaluation of mediumistic material. Hettinger, who had been trained as an electrical engineer, seems never to have had close ties with any of the psychical-research societies, and he received his Ph.D. without fanfare in 1939—the second man (John Thomas having been first) to obtain a doctorate for work in the field.[43]

Harry Price did his best to encourage the university to change its mind and make a permanent place for psychical research; and when the lease on his Laboratory's quarters expired, he committed himself in 1936 to give his library to the University of London on "permanent loan," arranging through Cyril Burt to have his laboratory apparatus stored in the psychology department. There was even some sign that occasional students might care to follow Hettinger's lead.[44] But apparently the university authorities remained resistant even to this limited incursion of the field into the academic world. A letter from Burt to Price in March 1937 makes it clear that psychical research still had little general acceptance there:

> At the moment . . . it would be rather difficult to link up such a [psychical research] department with a teaching laboratory; and, as you know, we had some difficulty in getting subjects in this direction recognized as proper subjects for University research. The Provost saw me again about the matter yesterday, and said that he thought it would be unwise to arrange anything like séances in a laboratory which was frequented by younger students—not, I imagine, because he himself thinks that they would be in any way disturbed, but unfortunate rumours might get about which could easily upset the working of the Department. We have had similar difficulties in the past when we have attempted experiments on hypnotism or desired to study feebleminded or mentally disordered patients. In fact, we have found it wiser to carry out research of this kind at the clinics rather than at the laboratory.[45]

Burt's support made it possible for some conservative parapsychological research to be pursued in the university's psychology laboratory—Soal carried out the final stages of his experimental replication of Rhine's card-guessing work there—but any hopes that the ULCPI might receive a stimulus from a closer association with the academic community were disappointed.

Eventually the strains imposed by attempting to combine work in normal science with psychical research led to the dissolution of the council. Aveling withdrew in October 1936, apologizing for his consistent inability to find time for its meetings.[46] Guy Brown resigned angrily in September 1938 over a particularly extravagant letter by Price in the *Times*, complaining, "I am afraid [it] will do me a lot of harm" at University College, and commenting on the impossibility of doing two research jobs at the same time: "Psychical research needs people who can give all their time to it."[47] Price was by now becoming doubtful of the benefits that might come from integrating his laboratory with the academic world, and his decision to close the council down was made when the June 1938 *Harper's Magazine* carried an article by Joad (then chairman of the ULCPI) that fell far short of the standards Price had boasted his organization would maintain. Joad had been both careless and inconsistent in describing the personal experiences he claimed to have had with physical mediums, and in one case (as W. H. Salter announced in a 1939 review for the SPR membership) had actually claimed for himself an experience that R. J. Tillyard had reported ten years before.[48] Price announced his intention of terminating the council to E. J. Dingwall, whom he blamed for Salter's attack, and Dingwall replied that *he*, at least, had no expectations that university affiliation would ever solve the problems of psychical research. "If no reply is forthcoming [to the review], I cannot but think that it may cause some slight criticism in University circles, and among those who imagine that by transferring P.R. to a University milieu, we thereupon rid ourselves of some of the difficulties that we have laboured under for so many years outside. I never had any illusions myself, but then my attitude is always characterized as the limit of cynicism.''[49] In the end, neither academics nor psychical researchers in England had believed strongly enough in the importance of establishing the field within the university to ensure this sort of footing. The resolutely amateur SPR was left to itself as the one organization there that could claim to be continuing a program of serious research.

Psychical research on the Continent had been no more successful in rooting itself in the academic world. Only a few European parapsychologists had been successful in establishing even a tenuous relationship with an academic institution. Perhaps the first of these was P. A. Dietz, who was made privatdozent in parapsychology at Leiden in 1932; W.H.C. Tenhaeff was given a similar position at Utrecht a year later. As privatdozenten, however, they were merely tutors to such students as cared to come privately for instruction in the field; their university employment and professional future were still dependent upon their teaching within the normal curriculum of the Dutch universities. The journal that they edited, the *Tijdschrift voor Parapsychologie,* was largely sustained by their own contributions: Tenhaeff's interests tended more to spontaneous phenomena, Dietz's to experimental work, but neither had the opportunity to carry on significant programs of research.

Psychical research had not even this limited entree into the German academic scene. Perhaps Hans Driesch might have been able to formalize its place in the German university system—he certainly believed that it *should* have such a place—had he not been forcibly retired by the Nazi government in 1933. His retirement made the young Hans Bender the man best situated to give psychical research academic respectability in Germany. He had at Bonn what Rhine had at Duke—a junior appointment at a university—and doctoral training in psychology as well. His own careful experimental work had disclosed the presence of psychical phenomena, against his own expectations, and this made his advocacy the more convincing.

What Bender did *not* enjoy was solid institutional backing. Emil Rothacker, his mentor at Bonn's Psychological Institute, was really a philosopher rather than an experimental psychologist, and while Rothacker amusedly accepted Bender's results, he did not feel their fundamental importance as strongly as McDougall did Rhine's. Moreover, Rothacker did not command McDougall's authority within a university or within psychology, and Bender's position was by that token the weaker. Consequently, Bender's first prosecution of psychical research had to be carried on without assurance of solid financial support. In his days as a graduate student he had had to use his own money to pay for apparatus and experimentation, due to the limits of the university budget, and had gone to the point of approaching the SPR (unsuccessfully) for a subsidy.[50] Acquisition of the D.Phil. still did not alter the fact that he was "working under very restricted financial conditons" and had to count every penny in planning an experiment.[51]

What might drastically have altered the German future of parapsychology was a sudden impulse by Harry Price in 1936. Price had been growing increasingly pessimistic about the alliance with the University of London; having read and been impressed by Bender's monograph on *aussersinnlichen Wahrnehmung*, he abruptly decided to see, through Bender, whether the University of Bonn would take over his materials under the terms he had originally proposed to London. Price suggested that he would be willing to transfer his library and make a grant of money if Bonn would establish a department of psychical research. He valued his offer at twenty-five thousand pounds and insisted that there were no strings attached (although he hinted delicately that "certain officials at the University [of London] were proposing that I should receive a doctorate, *honoris causa*, in recognition of the fact that I had established the first Chair of Psychical Research in Great Britain").[52] Bender followed up the suggestion quickly. The rector of the university was encouraging, and officials of the German ministry of propaganda "concerned with 'Social Hygiene' express[ed] their keenest interest in the matter of establishing a 'Parapsychological Department.'" The matter nevertheless dragged on into the fall—Bender suggested that the German government was hesitating to make certain "that in accepting such a generous offer they are not counteracting [*sic*] to possible interests of a foreign nation"[53]—and eventually Price decided again to get something going at London; he had already transferred his library and apparatus when he heard belatedly in March 1937 that the German government had agreed to authorize a "Department of Abnormal Psychology and Parapsychology" at Bonn, to take advantage of his gift.[54] While he wrote Bender that "there still may be hope that something may be done," the collection in fact remained in London, and with Price's offer no longer possible, the future of the "Department" was not so bright. If the offer had gone through, it might very well have had an influence upon the shaping of the new discipline in the prewar years.

Instead of devoting himself entirely, or even principally, to parapsychology, therefore, Bender now gradually turned his career in another direction. He had already begun training for a medical degree in 1935, with the intention of going into psychiatry. His earlier interests had by no means disappeared, to be sure, and he simultaneously involved himself in a range of parapsychological projects that is simply incredible, given the constraint that medicine imposed upon his time—negotiating with Harry Price, translating a book of William McDougall's into German, studying the young Latvian telepath Ilga K. He also planned ambitiously to continue his dissertation research with "investigations in abnormal psychology, especially experimental investigations of subconscious processes, i.e. of artificial or pathological dissociation of personality.... As I feel that parapsychical phenomena are closely connected with dissociation... I will attend to their eventual manifestation in the Crystal Visions."[55] This study was presented as a *Habilitationschrift* at Bonn in 1941. Nevertheless, Bender's parapsychological work was being increasingly constrained by other studies. When Rhine reported to him in 1937 that William McDougall would be willing to recommend him for the fellowship in psychical research about to be vacated at Stanford by Coover, Bender was obviously pleased, but he was firm in his intention to complete his medical education first.[56] By the outbreak of the war in 1939 Bender had become a psychiatrist with clinical responsibilities.

It is not difficult now to understand why Rhine alone had such marked success in giving parapsychology a start as an academic discipline. It was only partially due to his actual experimental results; several fortuitous circumstances contributed as well. Most important, no doubt, was the position he found himself in at Duke. There was surely no other university in the United States where he could have been given a position in the psychology department and been allowed to devote himself to psychical research. The institutional fluidity inevitable in the establishment of a new school may have helped to make this possible, but what was enormously more significant was the unchecked support of President Few and William McDougall, which ensured Rhine's continuation in the department. McDougall's evident approval was somewhat less influential in the psychological community at large, a number of whose members came to dismiss Rhine's work as "just what one would expect from a disciple of McDougall,"[57] yet it certainly provided Rhine with an initial entree into the world of professional meetings and journals. By the time McDougall died in 1938, Rhine had become a public figure whose claims had to be judged on their own merits.

Contributing only slightly less to Rhine's establishment as the single dominant figure in American psychical research was the fact that there were simply no serious competitors in the United States. Unlike England, where two lively organizations controlled (and divided) the field, America no longer had any psychical-research society that could be taken seriously: the American SPR had been destroyed by the Margery case; the Boston SPR, by Walter Franklin Prince's death. The only man who could have challenged Rhine's emergence as the leader of the new parapsychology was Gardner Murphy, and Murphy was simply not temperamentally capable of doing so, even had he wanted to. He was tremendously busy, deeply involved in a wide variety of other psychological activities, and sensitive to the consequences that aggressive promotion of psychical research might entail. Hence he was quite willing to support the Duke experimental program, with all its promise, and thus to allow Rhine to be seen as the creator of the new discipline in the United States.

One last factor that needs to be appreciated is the role of Rhine's personality in his success. S. G. Soal and Harry Price, each in his own way, have shown how an individual's personality could weaken the force of his contribution to psychical research; Rhine shows us the opposite. He was quite evidently an enormously attractive and appealing individual, capable of inspiring strong affection in his elders (like P. D. Strausbaugh) and fierce loyalty from his juniors (like Gaither Pratt and Margaret Pegram). The close ties that developed between Rhine and McDougall illustrate this perfectly. The young man from the University of Chicago had almost nothing in common with the cultured, aloof Briton when they first came together in Durham in 1927, and Rhine's earliest letters prove how great a gap he perceived between them.[58] Ten years later Rhine had become "Banks" to McDougall, something of another son and perhaps his closest friend and confidant at Duke. This gift for winning the liking and confidence of others—Frances Bolton is another example—made all the difference in his dealings with those who were in a position to help establish parapsychology.

III

The inability of European parapsychologists to establish institutional affiliations in the academic world did not mean that it was impossible for the amateur societies to pursue serious investigations that might conceivably compel the attention of scientists. The British SPR, in particular, prided itself upon its traditional commitment to careful observation and critical interpretation, and in the mid-1930s the leaders of the society—Mrs. Sidgwick, the Salters, and their associates—continued to support some ambitious studies of psychical phenomena. These studies yielded no unequivocally positive results, and this no doubt helps explain why English psychical research made far less impression upon contemporary science than did American parapsychology; but there is more to it than that. The English psychical researchers remained unable to agree among themselves on some of the fundamental issues of the previous decade: specifically, on what sorts of investigations were most appropriate to psychical research and on what sort of data should be collected. In the absence of any internally accepted program for what scientific psychical research should look for and how that research should be carried on the English could scarcely expect the wider scientific communities to take much notice of their claims.

It is apparent, however, that by the mid-1930s the SPR leaders were coming to believe it likely that restricted experimental studies could be of particular value to psychical research and deserved more encouragement than they had previously received. The society had for some time supported a research officer at three hundred pounds per year, but it had not taken an active role in sponsoring original work of any sort; it remained largely a clearing house for the collection and discussion of the independent researches of its members. But in 1934 the gift by an anonymous donor of a thousand pounds to be spent on research forced the society to consider what sorts of investigations were most deserving of support. The gift was first bespoke in May 1934: Theodore Besterman, who had recently been appointed investigation officer of the SPR, had already arranged to make a trip to Brazil to investigate the medium Mirabelli at the expense of Brazilian members of the society, and the council accepted his proposal that some of the new grant be used to allow him to go as well to North America to investigate a medium in Winnipeg and to visit Duke University "to obtain more details as to Dr. Rhine's methods of research"—for of course *Extra-Sensory Perception* had appeared only the month before.[59] As it happened, however, the North American investigations never came off. Besterman decided for personal reasons to cut his trip short, and after his sittings with Mirabelli he went directly to New York and thence to London, pleading illness to the SPR. When the details of the matter became clear, the council asked for Besterman's resignation,[60] which took effect in January 1935.

This sequence of events made it all the more urgent for the society to decide how it would henceforth orient and supervise its own efforts at research. An investigation committee had been set up early the previous year to advise Besterman and to deal with the use of the thousand pounds, and after Besterman's return it was reorganized into a research committee "to advise the Council as to the future organization of research."[61] Upon Besterman's dismissal, this committee urged the appointment of a research officer, but no satisfactory candidate was found,[62] and it was eventually agreed that a research student should be appointed who might be trained in the society's methods with a view to eventual appointment as research officer. Their selection was of very different background from Besterman. C.V.C. Herbert, whose appointment as research student was announced in July 1935, came from an aristocratic family, had read natural science and law at Cambridge, and had subsequently worked in photoelectric photometry at the University of London Observatory. He was named research officer the following year, and he remained in that position until 1941.[63] But during his apprenticeship it fell wholly to the council and to the research committee to decide whether and how the research funds should be put to use.

The research committee incorporated a cross-section of the SPR but included perhaps the society's most vigorous exponent of the experimental approach, W. Whately Carington, who was clearly a man of enormous promise. He had a good scientific background, including graduate training in experimental psychology with C. S. Myers at Cambridge, where he and R. H. Thouless had been contemporaries in the early twenties; his monograph on *The Measurement of Emotion* had been well-received when it appeared in 1922.[64] He enjoyed a broad knowledge of psychical phenomena as well as a philosophical breadth of interest that had been common early in the history of the SPR but was now becoming increasingly rare. Finally, he had a creative intelligence associated with a drive and eagerness that caught him up completely in whatever he attacked. In the early 1930s Carington had become convinced that the quantitative, statistical side of psychical research would provide the key to all subsequent study of psychical phenomena, and he had thrown himself into the study of statistics to the point where, after S. G. Soal, he was the acknowledged exponent of their use in the field; he impressed the council with his expertise to the extent that in 1934 they spent several hundred pounds on a calculating machine for Carington and any other interested member of the Society to use in his computations.[65] With an advocate like Carington on the research committee, it is perhaps not surprising that during 1935–37 it should have recommended that the society use a portion of its research funds to support two large-scale, narrowly experimental projects.

One of these projects was proposed by G.N.M. Tyrrell, a retired engineer who had joined the SPR in 1908, when he was twenty-nine. Of all the active members of the SPR, Tyrrell was perhaps the most enthusiastic supporter of J. B. Rhine's recent work—with the possible exception of Tyrrell's close friend H. F. Saltmarsh. But whereas Saltmarsh was reflective by nature, Tyrrell was drawn to active research. In 1921 he had published in the *Journal* a brief account of some successful experiments in telepathy and clairvoyance carried on with a young woman, Gertrude Johnson, who had displayed marked psychic gifts in everyday life,[66] and it was evidently the appearance of *Extra-Sensory Perception* that now caused Tyrrell to undertake a new series of experiments with Miss Johnson. Rather than have her guess cards, however, he designed a new experiment that would correspond to her own particular abilities. He had noticed that her powers were especially marked when she was given the task of finding something, and so he devised an experimental situation in which she was to discern into which of five boxes he, screened from her, had introduced a pointer.[67] Between October 1934 and February 1935 Miss Johnson scored hits at a rate of 31.21 percent (chance expectation was of course 20 percent), in sessions witnessed by a wide range of the leaders of the SPR—Saltmarsh, C. D. Broad, the Salters, Carington, and many others.

Tyrrell had in mind, however, a number of desirable modifications in this pointer apparatus, and he and Saltmarsh therefore worked out a design for a more fully automatic apparatus that would select the order of targets mechanically while simultaneously recording the results. Saltmarsh presented this design to the research committee, which found it highly satisfactory, and Tyrrell was given a grant of thirty pounds (later fifty) to carry out the construction.[68] The completed "electrical apparatus" was in operation by the beginning of May 1935: now a subject (usually Miss Johnson) was asked to guess which of five boxes contained an illuminated lamp (determined by the random action of a commutator or selector) by lifting the lid of the box in question, and her action caused both the trial and any success to be automatically recorded. At first, Miss Johnson was able to obtain only chance results in the new situation, but by returning to the pointer apparatus to develop her confidence, Tyrrell was finally able to get her to accept the electrical apparatus, and by mid-June she was scoring well above chance level. Hyperaesthesia and collusion between experimenter and percipient, against which the machine had been devised, seemed to be even more definitely ruled out by the high rate of speed at which the two worked; and a delay-action relay, which prevented *any* lamp from being lit until a box lid was opened, rendered normal explanations still more unlikely.

Then in mid-October a discovery by SPR member G. W. Fisk called some of Miss Johnson's work into question. Fisk was of an inquiring, inventive mind, and he enjoyed challenging others' assumptions or deflating their certainty. He had already once played devil's advocate in a debate with Tyrrell, arguing that failure to obey the inverse square law need not necessarily rule out a mental phenomenon like telepathy from being a radiation process.[69] Now Fisk discovered a simple method of guessing in the Tyrrell experiment that let him get strongly extra-chance results when Tyrrell was selecting the targets: he had simply to guess one box repeatedly until he was correct, then do the same with another, and so on. Eventually he and Tyrrell recognized that this had happened because Tyrrell was in a sense more random than a machine in the short run, since unlike a machine, he tended to avoid repeating himself. There seemed to be no real reason to think Miss Johnson had attained her scores by using this method, which in any case would not have worked when the targets were mechanically determined, but unfortunately the discovery that her success could have been achieved by using a "trick" method appeared to have an adverse psychological effect upon her power of scoring, for after late October 1935 she found herself unable to score above chance.[70]

Miss Johnson's sudden decline left the SPR much less enthusiastic about Tyrrell's work than it once had been, the more so because he had associated himself so closely with Rhine's experimental results. No one in England had yet been able to have any success with subjects guessing Zener cards, although many had tried; the council and research committee had asked Tyrrell himself to conduct experiments with Zener cards in the spring of 1935, and these (involving 21 participants making a total of 5,600 guesses) had given only chance expectation.[71] Tyrrell had repeatedly defended Rhine's claims on the grounds that English researchers must not yet have replicated the necessary psychological conditions, a proposition that men like Dingwall and Thouless found somewhat suspicious.[72] And now Tyrrell, carrying on experiments in some respects inspired by the Duke research, was being forced to defend his own lack of results in the same way: extra-sensory perception is a fact and takes place even when telepathy is excluded, but it depends upon the correct psychological setting, including good health, a positive attitude, and rapport with the experimenter.[73] No accounts of further researches with the electrical apparatus appeared in the publications of the SPR, although Tyrrell did publish one more article describing his apparatus in the second volume of the Duke *Journal of Parapsychology*.[74]

The other piece of work subsidized by the SPR was proposed by Whately Carington. Carington was not very well off in the 1930s, living on a small private income, and it was with this in mind that in May 1934, at the suggestion of the research committee, the council decided to allot to Carington one hundred of the thousand pounds earmarked for research, to enable him to carry on. In January 1936 an additional one hundred fifty pounds was approved.[75] The work that Carington pursued was of quite another order from Tyrrell's: it involved the possibility of distinguishing between the personalities of a medium in trance by means of the tests used by psychologists to characterize and study normal and abnormal personalities. Carington had proposed this idea as early as 1920,[76] but the first person actually to try such an experiment was Hereward Carrington, an American, who in 1932 and 1933 carried out a series of tests with the medium Mrs. Eileen Garrett, applying a word-association test both to her normal self and to her trance control, "Uvani." He measured the reaction times and psychogalvanic reflexes of the two personalities in this situation, and having determined that a comparison of these measurements showed much less resemblance than similar measurements taken from Mrs. Garrett in a normal state at a number of different times, he felt warranted to say that Mrs. Garrett and "Uvani" were indeed different personalities—whatever that might mean.[77] His paper, not well thought out and amateurishly written, evoked several critical responses from other psychical researchers. Whately Carington himself praised Carrington faintly for having been first to take up the subject experimentally and then complained of his inadequate psychological, physical, and statistical understanding of the subject—particularly the last.[78] In North Carolina, J. B. Rhine set his student Gaither Pratt to work applying his graduate training to an evaluation of Carrington's study, once again attempting to use psychology to illuminate psychical research, and was delighted with the result, which he considered having published as part of a Boston SPR bulletin.[79] This idea was given up, however, when Whately Carington, who had been sent a copy of Pratt's paper for his opinion, replied rather condescendingly that it was quite all right but that Carrington was too easy a target to take that much trouble with—besides, Carington added, all this became "vieux jeu" now that his own studies were coming into print.

For Carington had now begun to explore experimentally the idea he had put forth ten years before. Between 1934 and 1936 he published three papers on "The Quantitative Study of Trance Personalities"—studying Mrs. Garrett (and "Uvani"); Rudi Schneider (and his control, "Olga"); and Mrs. Osborne Leonard (together with her control, "Feda," and two other communicators purporting to be the late Reverend John Wesley Thomas and his daughter Etta).[80] As Hereward Carrington had done, he used reaction times and variations in the psychogalvanic reflex during word-association tests to try to determine the relationships between a medium's personalities. These ambitious papers were variously criticized by their readers for experimental and mathematical defects alike, and Carington was repeatedly forced to withdraw the conclusions of his previous papers, but at the end he still expressed his conviction that he had presented evidence of some influence extraneous to the medium—"presumably something in the nature of what John and Etta claimed to be." For the interest of such tests was of course in part that they bore on the question of post-mortem survival, still the most important problem for many of the society's members.

Throughout this series of papers Carington stressed the importance of statistical techniques to psychical research, and the reaction to this aspect of his work was not entirely favorable. Even granting the possible value of narrowly experimental studies, many of his readers felt that those culminating in conclusions about mathematical relations or probabilities did not bear on the main themes of psychical research—the nature of the human mind and the possibility of post-mortem survival. One of Carington's most articulate critics was J. Cecil Maby, a member of the society who had set up his own "biophysical laboratory" in his home near Oxford, who directed attention to a point that was no doubt on the minds of many: Can a strictly quantitative behaviorist study of personality (as Carington himself had proudly characterized it) really tell us about anything save the physiological mechanism through which the personality manifests itself? Surely numerical analysis can only provide an artificial approach to the structure of mind and personality. Maby wanted simply to express the feeling that Carington emphasized the quantitative approach too heavily—that "qualities are quite as important as, and sometimes more important than, quantities"; he accepted the intrinsic interest and potential importance of Carington's research without question.[81] Others were less sympathetic. The botanist-naturalist Henry Ridley, who had joined the society in the first year of its existence, complained a few months later: "What we require are facts and observations throwing light on a future life, and the connection of those who have passed away with those at present alive. Instead of this we get long letters of vague theory and futile mathematical calculations about the proportions of guess-work. These may possibly be suitable for mathematical or statistical societies, but have no bearing on the objects we founded the Society for half a century ago."[82] And at this point only two of Carington's three very technical studies had appeared!

Perhaps, then, we may be justified in interpreting W. H. Salter's paper "Statistical and Other Technicalities in Psychical Research," delivered to the society on 30 September 1936, as something of a defensive measure, aimed at justifying the society's support of so technical a subject.[83] Here Salter ingenuously began by saying that his own personal bias "was not at all in favour of statistics," and he reminded his audience of Piddington's statement in 1924 that the scientific expert would not necessarily be the best judge of psychic phenomena, yet in those realms of psychic research where it was essential to eliminate chance as the explanation of a phenomenon, it was (and had been since the 1880s) inevitable that statistics would be brought into play. He concluded by referring specifically to Carington's work and the statistical analysis therein as promising precision in the classification of mental states, having already borne results in the study of controls and communicators. These remarks did not stop protests against the unintelligibility of Carington's papers, protests that increased when the third paper appeared at the end of the year. Because of the outcry, the council considered asking a sympathetic psychologist such as R. H. Thouless or C. A. Mace of the University of London "to write a critical and elucidative article for the Proceedings on this investigation."[84] Carington proposed that he himself write a nontechnical summary of the whole investigation but added that he would have to have further financial support. The council, however, was unwilling to subsidize Carington further from the society's funds and began to inquire whether his work might not now be supported by private donors. Some funds had already been collected by the philosopher C. D. Broad, then president of the SPR, when suddenly Carington announced yet another possible statistical flaw in his research, this time affecting his recently published third study of trance personalities.[85] Consequently, it was decided in April 1937 that before the monies were turned over to Carington, Thouless (who had agreed to report to the council on the investigations) should make it clear that Carington "was proceeding on the right lines."[86]

Two months later Thouless read his report to a private meeting of the society, a report devastating in its reduction to scraps of most of Carington's claims.[87] While he praised the attempt to evolve a statistical technique for studying trance personalities, he chided the author for having been in such a hurry to publish the consecutive imperfect stages of its evolution and then proceeded to explain carefully not only the errors that Carington had already recognized but several others as well. Thouless identified in particular a number of mistaken applications and interpretations of statistical methods, concluding by showing that the data in Carington's third paper seemed actually to support a conclusion opposite to the one that he had maintained: "The results of the experiment are not merely that it fails to show that the communicators are autonomous personalities, but that it conforms remarkably well to the hypothesis that they are not."[88] All this must have been intensely galling to Carington, who had so often defended the importance of mastering statistics. Nevertheless, he took the critique with good grace, and in a postscript to his abridged publication in the October 1937 *Proceedings* he admitted the validity of almost everything that Thouless had said, with only the wry remark that "although I have yet to hear that fruit-trees enjoy the process of pruning, all that matters is that their fruit should ultimately be sound."[89] Then he set to work afresh on a new paper in the series.

One other topic received experimental investigation by English psychical researchers in the mid-1930s: precognition. The conviction that the future can sometimes be known to men is of course a very old one, and the early SPR did not neglect to collect accounts of apparently precognitive experiences. The furor attending the popular reception of relativity theory in the 1920s led many people to give renewed consideration to the topic, in hopes that the newly stated dimensionality of time might make knowledge of the future philosophically more plausible. No one seriously entertained the possibility of subjecting precognition to experimental verification, however, until the publication in 1927 of J. W. Dunne's stimulating book, *An Experiment With Time*.[90] Dunne claimed to have found from his own experience that dreams were composed of images of past experiences blended with others of experiences still to come and that anyone could confirm this by taking immediate detailed notes upon awaking from a dream and then looking for elements of that dream in his life over the next few days or weeks. In order to explain his discovery, he elaborated a theory drawn from C. H. Hinton's *The Fourth Dimension* involving the infinite dimensionality of time and the imperishability of the mental observer of temporal events.

Dunne's book was by no means merely trivial or superficial, and it excited a considerable reaction. The philosophers and the mathematicians gave his theory very serious consideration, although they were eventually unanimous in regretfully rejecting the infinite dimensionality of time (its "serial nature," as Dunne came to call it) as a logical impossibility.[91] Rejection of the theory did not invalidate Dunne's experimental claims, however, and attempts by individuals to master Dunne's technique were quite frequent over the next decade. Interest was strong enough in the SPR for Theodore Besterman to organize a careful attempt to duplicate Dunne's experiments in 1932.[92] None of these attempts met with any marked success, but this did not disprove Dunne's own claims to have had precognitive dreams, and his book continued to excite readers well into the 1930s.[93]

The sudden arousal of interest in precognition also led H. F. Saltmarsh to try in 1933 to make a systematic classification of all recorded cases of precognition, in much the same way as had already been done for instances of telepathy (most recently by Mrs. Sidgwick),[94] and this immediately suggested to Whately Carington a test for precognitive ability that could be given tighter controls and more precise evaluation than Dunne's.[95] He asked his subjects (forty in all) to roll a single die, having first written down the number they thought would come up, and to do this over and over again. Even though his subjects scored exactly at chance expectation, Carington argued that certain patterns in their guessing were statistically significant and that "*some* paranormal factor of precognitive character [was] established beyond any reasonable doubt."[96] Characteristically, he had announced his certainty while flaws still remained in his techniques; the October 1935 *Journal* carried a note from Carington confessing that R. A. Fisher had called his attention to a mistaken assumption that invalidated much of his argument. Equally characteristic is Carington's unchastened paper of January 1936, in which he attempted (not entirely convincingly) to preserve something of his claim for precognitive abilities.[97]

Carington was not alone in his attempt to put precognitive ability to the experimental test; G.N.M. Tyrrell was attempting inquiries of the same sort.[98] It may well be that their experiments owed some share of inspiration to J. B. Rhine at Duke as well as to Dunne's book, for in 1934 Carington and Saltmarsh had learned from correspondence with Rhine of *his* precognitive experiments and apparent success. In turn, Rhine's eventual decision to publish his research in 1938 stemmed from the English work. He had originally withheld word of precognition and PK for fear of the reaction of scientists to such particularly radical claims and had warned his English correspondents against publishing too hastily; but once Carington and Tyrrell had appeared in print on the subject, Rhine released his own results, though he still felt they had acted precipitately.[99]

Here as elsewhere Rhine's reports of experimental success far transcended any claims that Tyrrell and Carington were able to make, and the disappointing results of their research programs were not such as to encourage English belief in the fruitfulness of specifically experimental psychical research. They could not have helped Harry Price in his efforts during these years to establish psychical research among the sciences at the University of London. Nevertheless, they left unaffected a certain interest in the wider field in one particular sector of the academic community. Because the British had always tended to emphasize philosophy over experimental science as a guide to the study of the mind, it is perhaps not surprising that a few academic philosophers of the 1930s should have tried to reconcile psychical research with contemporary thought.

Preeminent among these was C. D. Broad. Broad had entered Trinity College, Cambridge, in 1906 and within four years had taken firsts in natural science and moral science; he was elected to a prize fellowship there in 1911, and in 1922 he became Trinity College Lecturer in Moral Science. He became Knightsbridge Professor of Moral Philosophy (Sidgwick's chair) in 1933, and he remained in Cambridge for a quarter of a century.[100] He shortly established himself as one of the leading interpreters of the philosophical implications of the new physics—his *Scientific Thought* (1923) can still be read with profit—and in this he is to be set off sharply from the other academic philosophers who played such a great role in the development of the SPR; Henry Sidgwick, after all, had been primarily a political and moral philosopher, while F.C.S. Schiller had been preeminently a logician, concerned with his pragmatic theory of truth. Broad's scientific training and his desire to clarify man's understanding of how he perceives the natural world equipped him excellently to help amateur psychical research make the transition to a serious scientific subject.

Broad had joined the SPR in 1920, not because of the field's association with his college (Sidgwick, Myers, and Gurney had all been fellows of Trinity) nor because of any hopeful inclination towards the survival hypothesis, so strong in the 1920s, but apparently because to a degree he felt the same distaste for a totally mechanistic world that the Rhines were suffering on the other side of the Atlantic.

I think that what lies behind my interest in the subject may possibly be this. I feel in my bones that the orthodox scientific account of man as an undesigned calculating-machine, and of non-human nature as a wider mechanism which turns out such machines among its other products, is fantastic nonsense, which no one in his senses could believe unless he kept it in a water-tight compartment away from all his other experiences and activities and beliefs. I should be sorry if anything so absurd and (as it seems to me) so dull and boring were to be true, and if those who take it for Gospel should happen to be right. . . . I should therefore welcome the irrefutable establishment of alleged facts which, if genuine, would be so palpably inconsistent with this view as to leave it without a leg to stand upon.[101]

However, Broad had no very great hopes that scientific psychical research might some day validate the claims of religion. Indeed, in his generally pessimistic view of man's moral future Broad was much closer to William McDougall than to the Rhines. In particular he felt Christianity to be doomed by the progress of the sciences, and he regretted the consequences. "Though I am not a Christian, and never have been one since I began to think for myself, I take no pleasure in this prospect. . . . Ordinary human nature abhors a vacuum, and it will not for long rest content without some system of emotionally toned and unverifiable apocalyptic beliefs for which it can live and die and persecute and endure."[102] These words were published in 1939, in the context of emergent fascism and communism, but Broad's despairing conviction that a triumphant materialist science might have tragic results for civilization was apparently not then new. He does not always strike his reader as a cheerful or a "positive" personality.

In 1930 Broad was elected to the council of the SPR, and in 1935 and 1936 he was chosen its president. These honors were in some sense a recognition that he had become a most influential spokesman for psychical research. In his *Mind and Its Place in Nature* (1925), in which he presented an exhaustive analysis of the mind-body problem, he had accepted telepathy as an undoubted fact of which he had had personal experience (in sittings with Mrs. Leonard); and while he did not credit simple human survival, he did argue there that the evidence was at least consistent with the existence of a "psychic factor" that could unite with a human organism to produce "mind" and yet could persist independently after the death of the organism.[103] Yet Broad's value was not merely as a public figure, for his philosophical gifts were invaluable to the society. Bertrand Russell's review of Broad's first book captures the approach that continued to inform Broad's philosophy, as well as his writings on psychical research: "This book does not advance any fundamental novelties of its own, but it appraises, with extraordinary justice and impartiality and discrimination, the arguments which have been advanced by others on the topics with which it deals."[104] Broad's occasional papers on psychical research mark him out, for his analytical powers and his dispassionate judgment, as the heir to the Cambridge tradition of Henry and Eleanor Sidgwick.

The first work of Broad's to appear in the publications of the SPR, his presidential address of 1935, is a case in point. Entitled "Normal Cognition, Clairvoyance and Telepathy," it was very much a propos in the atmosphere engendered by the discussion of the Rhine-Tyrrell claims. The paper began by considering the tactics that might induce scientists to look into the evidence for mental phenomena and concluded that they might be readier to do so if the parallels and discordances between normal and supernormal cognition were set out plainly.[105] If it were possible to suggest analogies between the two, man's tendency to rule out the supernormal as antecedently very unlikely might be overcome; and it was this task that Broad set himself. It should be noted what a far cry this policy was from the American tendency to insist on the uniqueness of extra-sensory perception; it should also be noted that Broad had judged well—for his own circumstances, at least—in that psychical research gave signs of winning a small place in contemporary English philosophy. From this point on, his writings on the implications of psychical research appeared regularly in *Philosophy*. In 1937, moreover, he was invited by the Aristotelian Society and the Mind Association to "introduce a philosophical discussion on a subject connected with psychical research" at their July meetings; his paper (and his ensuing dialogue with H. H. Price of Oxford) dealt with precognition much as his presidential address tried to deal with telepathy and clairvoyance.[106]

Broad had developed his paper of 1935 by first examining clairvoyance, which under his analysis proved to be totally unlike normal sensory perception. Acknowledging a subject's ability to identify correctly the pips on the sixth card down from the top of an undisturbed pack as being square would require physical, physiological, and indeed psychological assumptions entirely at odds with present knowledge. Nor did it seem any more likely, Broad argued, that clairvoyance should be direct, nonsensory prehension of objects. These were old conclusions, certainly. But whereas for fervent antimaterialists like Rhine and Carington they were cause for exultation, for Broad they were reason enough to continue to look rather skeptically at the "evidence" for clairvoyance, as his summation of the topic implies: "Neither of [these two views] is in the least attractive or plausible, but I know of no other alternative that is even intelligible. I hope that some of those who think that there is adequate evidence of clairvoyance will be inspired to suggest some other view of it which will be equally intelligible and much more plausible."[107] Telepathy, however, was another matter; there were analogies in current science that made it plausible to view telepathy as the direct action of N's mind on M's brain (or perhaps vice versa)—was it not well known that in cases of multiple personality two minds could affect the same brain? For a number of reasons, Broad felt it improbable that N's mind should be able actively to pick through the contents of M's brain, selecting what he wanted; the objections of the survivalists still retained something of their force. Broad's own view, however, was far from survivalistic.

> We must therefore consider seriously the possibility that each person's experiences initiate more or less permanent modifications of structure or process in something

which is neither his mind nor his brain.... The modifications which are produced in this common Substratum by M's experiences *normally* affect only the subsequent experiences of M; those which are produced in it by N's experiences *normally* affect only the subsequent experiences of N. But in certain cases this normal causal "self-confinement," as we might call it, breaks down. Modifications which have been produced in the Substratum by certain of M's past experiences are activated by N's present experiences or interests, and they become cause-factors in producing or modifying N's later experiences.[108]

It is worth quoting Broad in full here because he so seldom went beyond analysis of possibilities to develop conclusions of his own. Yet to the SPR, any generalizations, even highly speculative ones like these, seemed rich with potential. They had always felt the weakness of their field to be that it was all evidence and no explanatory theory, and to some members, at least, Broad gave signs of finally being on the track of such a theory.[109]

IV

One other English psychical researcher was devoting his attention to quantitative experimental work in the 1930s, but largely independently of the SPR. At the end of 1934 S. G. Soal began an ambitious and careful attempt to replicate the experiments of J. B. Rhine on extra-sensory perception, under the auspices of Harry Price's National Laboratory for Psychical Research (later the University of London Council for Psychical Investigation). Soal gave what time he could spare from his position as lecturer in mathematics at the University of London to collecting data, often as much as seven or eight hours a week, first at the National Laboratory's offices at 13D Roland Gardens and then in the psychology laboratory at University College, London. By 1938 his had become the most extensive ESP experiment attempted outside North Carolina, involving over 100 subjects and some 120,000 guesses.[110]

That Soal was the one to take up Rhine's work most intensively is perfectly understandable, but it was not without its effect upon the development of the English attitude to experimental-statistical research. As a mathematician Soal was jealous of his distinction as one of the few working scientists active in psychical research, proud of his professional skills, and anxious to win recognition as the exponent of a fully "scientific," objective approach to this marginally scientific field. Perhaps this last was intensified by a certain self-consciousness about his lack of an advanced degree. Some such mixture of feelings was evident in his meticulous and unsuccessful attempt to replicate Ina Jephson's clairvoyance experiments under rigorous conditions; it was plain again in his decision to ally himself with Harry Price against the SPR. A replication of the Duke research was therefore exactly to his taste: it was a well-defined piece of experimental science that would allow him to demonstrate his ability to construct a watertight scientific experiment (and to point out the flaws in the work of others) while giving him full play for his training in mathematics. Eventually Soal came to see the project as the means of establishing himself as the unchallenged master of experimental method and statistical techniques in psychical research. Throughout the later 1930s Soal maintained an aggressively proprietary attitude towards card-guessing experimentation, doing his best to discourage others from poaching on his territory and to compel them to defer to his superior mathematical expertise.

In broad outline, Soal's work was modeled upon J. B. Rhine's as described in *Extra-Sensory Perception* and in Rhine's paper on Mrs. Garrett. Soal tested his subjects both for clairvoyance by STM and for "telepathy" (actually for undifferentiated ESP). The specific procedures that he instituted, however, were designed to meet the rigorous criteria that the right wing of British psychical research had come to insist upon: agent and percipient were carefully screened from one another, all series were witnessed and recorded in duplicate, and the experimental protocol was described in precise detail to permit exact duplication if necessary. Within these limits, Soal did his best to follow Rhine's increasingly anxious instructions about maintaining a productive psychological mood in the experiment, even though he found it difficult to believe that this actually affected results. The motivation of his percipients—a few of them remarkably confident and forceful personalities—sometimes raised his hopes, but eventually the laws of chance seemed always to prevail.[111] Soal went so far as to offer money prizes for high scores: fifty pounds for fifteen or more correct guesses in a series of twenty-five; five pounds for thirteen or fourteen; three pounds for twelve.[112]

Only in the application of mathematical techniques to card-guessing did Soal make a significant innovation. He had recognized almost from the beginning of his research that Rhine's use of 5 × 5 decks meant that the ordering of each deck could not be strictly random and that consequently Rhine's comparison of his results with the expectations derived from a binomial distribution was improper. To avoid this problem, Soal produced a randomized set of some one thousand cards by listing the final digits of a series of logarithms and assigning a particular Zener symbol to each one. The packs of twenty-five into which he separated these cards were more nearly random than Rhine's in that there was now no assurance that any particular deck would contain equal numbers of the five symbols, and by using these from the outset of his investigation Soal was able to ensure that the binomial distribution would be strictly applicable to his results.[113]

By mid-1936 Soal had accumulated over forty thousand guesses following these methods. He had found no undoubted Pearces or Linzmayers among his fifty-two subjects, although a Mrs. Stewart had briefly seemed to show promise at telepathic guessing.[114] His evidence, however, was by no means unpromising. He reported to Rhine with some enthusiasm two inexplicable features in his results, features upon which he elaborated to the SPR at the beginning of 1937, after having recorded another thirty thousand guesses and forty-one subjects: "a certain tendency for a number of individuals to score below chance expectation" and "a more or less general tendency to score runs of five or more successes in unbroken sequence considerably in excess of what chance would predict." Soal unquestionably felt that this evidence demanded further investigation and actually apologized to Rhine for his earlier expressions of doubt of Rhine's results; in confidence, he expressed his belief that his work might be accepted as a thesis for the Ph.D. in psychology at the University of London—for at this moment there still existed some ties between university faculty and Harry Price's circle.[115]

During the following year, however, Soal's enthusiasm cooled rapidly. For one thing, the features of interest that he believed he had identified received no further confirmation in the course of fifty thousand more guesses. What he himself chose to stress as disillusioning was the particular series of unsuccessful experiments that he was able to carry out in May 1937 with the American medium Eileen Garrett.[116] Rhine had tested Mrs. Garrett in April 1934, reporting remarkable success at both telepathy and clairvoyance—13.4 hits out of 25 in one series—in both the normal and the trance states. It may well have been that having already had one demonstration of the supernormal through a medium, Soal had particular hopes that work with Mrs. Garrett could provide further evidence of extra-sensory perception. As Rhine had done, Soal tested her for both telepathy and clairvoyance in both normal and trance conditions; disappointingly, however, he found no trace of supernormal ability in 12,425 guesses. In carrying out the experiment, he had prepared methodically for Rhine's now customary defense against reports of failure—that favorable psychological circumstances were necessary for success—by using a variety of agents and maintaining a relaxed atmosphere, and Mrs. Garrett actually provided him with a testimonial to the effect that she had preferred the English experiments, where she had not felt the "emotional tension, urging, or strain to produce results, such as is noticeable at Duke." Even so, they did not produce high scores. The statement with which Soal concluded his report on these series with Mrs. Garrett—"I do not wish to convey the impression that Dr. Rhine did not actually succeed in obtaining highly significant results in his work with Mrs. Garrett"—gives every sign of having been written ironically, for over the next few years Soal's reaction to Rhine's claims was always colored by at least a tinge of scorn.[117]

Soal's reports of experimental failure could only have enhanced the tendency to discount card-guessing work that had grown steadily stronger within the SPR since the first flurry of excitement over *Extra-Sensory Perception*. There had been a few attempts at repetition immediately after the book appeared, but with the exception of Soal's they had been terminated within a year. In the absence of any clear successes, those few who were sympathetic to Rhine (like Carington and Tyrrell) were content to leave card-guessing to him and to Soal and to pursue lines of research that seemed personally more promising, while those unsympathetic simply expressed skepticism that such work *could* yield positive results. In any event, both friends and critics agreed that Rhine's book seemed to leave open very real questions about the adequacy of his precautions against the subject's sensory contact with the cards to be identified. This issue of experimental rigor, put repeatedly to Rhine by his English correspondents in the years following *Extra-Sensory Perception,* seemed to him to show that they were "not familiar with the research attitude of the psychological laboratory." Researchers concerned to verify the claims of a medium might need to insist on absolute rigor before admitting conviction, he replied to these correspondents, but experimental psychologists want both less and more than this. Initially, at least, they will be happy with positive results—obtained under any conditions—that can be tested under further constraints in order to demonstrate not merely the reality of an effect but the circumstances under which it can be observed.[118]

Rhine clearly did not appreciate that English psychical researchers, after fifty years of disillusionment in their search for replicable mental phenomena, would naturally demand a rigorous demonstration of experimental telepathy—rigorous by their *own* standards—before committing their attention to it. Hidden behind the legitimate insistence upon experimental precautions was a certain chauvinism, which C. D. Broad expressed with remarkable candor:

> (i) I cannot help wondering why the proportion of persons who seem *prima facie* to have marked powers of extra-sensory perception should be so much greater among the students of Duke University, North Carolina, than among otherwise similar persons experimented upon in England. (ii) The S.P.R. have recently been supplied with samples of the kind of cards used in these experiments. They are disgracefully badly constructed, and are so defective that in certain sorts of experiment a person familiar with them could sometimes guess the nature of the card from merely seeing the edge of it. (iii) It is quite true that in many of the experiments described by Dr. Rhine this defect would not have helped the percipient in the least. It is also true that the way in which successes tailed off after a time with some of his best percipients does not fit in at all with a normal explanation on these lines. But the facts about the defective cards do produce in my mind an impression of general "sloppiness" which makes me doubt Dr. Rhine's competence to devise properly and describe accurately *any* kind of experiment. . . . (iv) . . . A most admirably careful series of experiments, on the same lines as Dr. Rhine's, has been carried out by Mr. Soal, partly on students at Queen Mary College, and partly on the medium Mrs. Garrett who scored an extremely high proportion of successes with Dr. Rhine. In none of Mr. Soal's experiments is there the least trace of any result which cannot reasonably be ascribed to chance. To speak quite frankly, I *know* that Mr. Soal is a highly competent investigator; I have some *prima facie* reason to doubt whether Dr. Rhine is so; and I do not think that any scientist would be prepared to accept as proven in *any* subject a startling claim coming from a youthful American university unless and until it was confirmed by experiments done in older and perhaps more self-critical seats of learning.[119]

The last, rather smug phrase suggests that Rhine's experimental reports would have encountered English suspicion no matter how airtight they might have been methodologically. The English researchers were almost to a man reluctant to believe in something that they themselves had been unable to reproduce—for obvious scientific principle, no doubt, in part, but a sense of affronted tradition and national pride also came into play. As Lydia Allison wrote back to Duke during a visit to London in the summer of 1937, "Mr. Herbert and I believe some of the others think there must be something racial that would explain the results obtained at Duke and in America. Although Soal tried different nationalities here. But the English will never be satisfied until they are able to repeat, and Mr Piddington once wrote something to the effect that he would really only have confidence in an Englishman."[120] The prospects for widespread acceptance of ESP in Britain depended in the long run not so much on reports of striking extra-chance results under the most strict experimental conditions *at Duke* as on some sort of replication of the work by a trustworthy member of the SPR—a Soal, say, or a Carington. It must be remembered, too, that the majority of the society still found experimental card-guessing profoundly uninteresting and even trivial, peripheral to the traditional aims of psychical research.

In the winter of 1937, however, the topic was drawn back to the center of the society's attention in the wake of the growing public controversy in America over extra-sensory perception. E. J. Dingwall, returning from a trip to the United States, took up the cause of the critics and described the American work in a sarcastic letter printed in the December *Journal,* commenting scornfully that the experimenters there had originally used cards of variable size in the same deck and were still using cards on which the impress of the design could be read from the back. In his ineffable manner, he concluded: "It is, I think, to be desired that our colleagues in America should try to pay a little less attention to the statistical analysis of their results and should try to take the trouble, however arduous it may be, to train themselves properly to conduct and report the experiments on which they base these analyses."[121] Dingwall's letter elicited replies from Rhine's few English defenders—Saltmarsh, Maby, and Besterman— complaining that Dingwall's criticism was in large measure simply *ad hominem*. This series of events may have been the catalyst that decided the SPR that a conscientious and systematic study of extra-sensory perception was now appropriate; at any rate, in January 1938 the council voted to ask R. H. Thouless "to review the ESP experiments in general"—no doubt with no great expectations of a positive report. Soal, who had been a member of council for a little over a year, objected strongly to this upon reflection. He pointed out that "Professor Thouless would not be in possession of all the available material as [my] . . . report of [my] own inclusive investigation would not have been published" (he was expecting publication by Harry Price), and the council agreed to abandon this plan. Instead, they decided to conduct their own ESP tests, this time under the supervision of their newly appointed research officer, C.V.C. Herbert. De la Rue, Ltd., made up the special Zener cards that spring[122]—Herbert having already concluded that the American-made decks were "unsuitable for serious experimental work."[123] If any such tests were ever carried out, however, their results were not reported, and Herbert may not even have begun the task.

Indeed, the SPR leadership was by this time becoming irritated by Soal and his fixation upon the experimental-statistical approach. Relations between them had already been somewhat strained by Soal's proclamation of alliance with Harry Price, but there had at least been no previous question of his entirely serious if severely skeptical approach to the study of psychical phenomena. Now, however, his objectives were growing rapidly incomprehensible. In particular, he manifested an obvious ambivalence—one might almost say a two-facedness—on the subject of ESP. In a speech on "'Snags' in Extra-Sensory Perception," made to the Ghost Club (a social "psychic" club that Harry Price had just founded) in March 1938, Soal presented himself as the most obdurate cynic of all with respect to ESP. He argued, for example, that Rhine's clairvoyance experiments had often been vitiated by his failure to conceal the Zener cards from his subjects; that his telepathy experiments had been vitiated by his failure to eliminate any possibility of conscious or unconscious signaling between agent and percipient or to ensure a truly random sequence of targets; that the long-distance work, such as the Pearce-Pratt experiment, was rendered almost valueless scientifically because it had not been overseen at every stage by disinterested observers. He ridiculed the appeal to poor psychological conditions—an excuse he attributed to Tyrrell, though Rhine had certainly employed it—as a reason for discounting chance data.[124]

The Ghost Club address was fundamentally an attempt to discredit Rhine. Soal labeled the Duke claims "extravagant," and insisted that they should not be accepted as genuine until they were confirmed in English laboratories. He endorsed C. E. Kellogg's hope, expressed five months before in the *Scientific Monthly,* that the "craze" for guessing Zener cards would die down before it led many more young psychologists into futile research projects. Yet in a lengthy review of Kellogg's article, published in the June 1938 SPR *Proceedings,* Soal put on an utterly different face for his different audience.[125] He did indeed briefly enumerate the failings in Rhine's experimental technique, but he was much more interested in discrediting Kellogg as a pretender to statistical competence. Soal began by referring to Kellogg's remarks about the "craze" for parapsychological research diverting students from problems of real importance as an indication of Kellogg's prejudice against the subject and then went on to examine the statistical basis for Rhine's work—of which (unlike Kellogg) he on the whole approved. In this context and for the SPR membership he took a much more positive attitude toward the Pearce-Pratt experiment: it was unfortunate, certainly, that more careful supervision had not been arranged, but given the honesty of the investigators (which Soal did not dispute), it appeared that this series established the existence of clairvoyance in man.[126] It must have been very difficult for Soal's exasperated English colleagues to understand just what conclusions his intensive experimentation had led him to.

Rhine, at least, found it impossible to comprehend Soal's position. It was hard to understand a correspondent who in June could write sympathetically and tolerantly of Rhine's work, encouraging him to fight the prejudice of university psychologists in England by pressing ahead with careful experimentation[127]—and then four weeks later could insist that *he* would never be convinced personally of ESP until it had been produced by just such English psychologists.[128] Privately Rhine asked his friend H. F. Saltmarsh for his personal evaluation of Soal's attitude, and Saltmarsh wrote cautiously that he felt "considerable bewilderment" on this point: "I cannot reconcile the attitude which he appears to take in some of his letters to the press with the opinions he has expressed, privately to me, & in his articles published in the Proceedings and Journal."[129] In England as in North Carolina Soal was becoming perhaps something of an embarrassment.

The SPR was all the more suspicious of Soal in mid-1938 because he had apparently betrayed the trust of the society in order to maintain himself in command of English card-guessing experimentation. In May 1938 the council (of which Soal was still a member) agreed to G. W. Fisk's suggestion that the society should sponsor a BBC test of extra-sensory perception reminiscent of its broadcast experiments of a decade before; a committee consisting of Carington, Herbert, and Soal was appointed to look into the details of arranging it.[130] Apparently Soal privately viewed this as an intrusion by incompetents into his own statistical domain; at any rate, he shortly wrote secretly to Harry Price about the plan, urging him to get in first and to arrange for the BBC to do such an experiment with the ULCPI, to be carried out under Soal's own supervision without interference from Carington.[131] When in June the council learned that they had been forestalled, they were outraged. Soal adopted a bold front by claiming (falsely) "that the ULCPI had considered approaching the BBC before the SPR committee were appointed";[132] this did not soothe council members. Bitterness led to a gradual ostracism of Soal in council, and this in turn naturally exacerbated Soal's own feelings of grievance as a sober, scientific worker set upon by fuzzy-minded survivalists. In the fall of 1938 an increasingly offensive and unyielding series of letters from Soal to the SPR leadership brought them to the point of looking for ways to ease him from the council.

The difficulties with Soal were to some degree the fault of his own personality—he was extremely shy, often abrupt and aggressive in manner, defensive and protective of his own work.[133] Given such a personality, he was bound to take offence at anyone not fully sympathetic to his single-mindedness, which was the greater pity since this made it next to impossible for psychical researchers to resolve the issue that he was raising—the perennial question of how properly to make psychical research scientific. To be "scientific" was Soal's ideal for the SPR, and by that he meant depending upon statistical-experimental work; by this token, as he saw it, only he himself, Carington, and Herbert among all the SPR members could really be said to have done anything approaching scientific research.[134] He wrote to Salter in October to precisely this effect,[135] and Salter's curt reply revealed that the long-standing difference of opinion had not yet been settled: "It is not my intention to argue in detail the points you raise. If by 'scientific' you mean that psychical research is a branch of any of the generally recognised sciences, or that a special training in any of them is indispensable to the psychical researcher, I should say that the Society is not scientific in that narrow and limited sense of the word. If to 'scientific' you attach the more general meaning of systematic and impartial, then the Society might claim the right to style itself so.''[136] In more detail Salter wrote to Lord Rayleigh, then president of the society, outlining the disagreeable history of Soal's conflict with the society and summarizing the issue in these terms:

> Both in and outside the Society he is constantly crabbing all the work done by the Society from the Founders downwards which has not been done in conformity with the only methods he approves, that is to say, everything which is not the outcome of experiments, the results of which are capable of statistical analysis. Everything else he chooses to brand as unscientific, although it was not so considered by Myers, Gurney, the Sidgwicks and the rest to whom psychical research owes its existence. He has in this attitude the support of one or two members of our Society, whose work I value a great deal higher than his own, in particular Whately Carington who until he got h[is] craze for "Quantitative Analysis" had a very good understanding of the subject.[137]

Salter proposed to Rayleigh that a meeting of certain council members be held "at which Soal and Carington could put their point of view," but apparently no such meeting took place. Certainly there was no satisfactory understanding reached between the two sides.

V

In the midst of this heated discussion of what it meant to do "scientific" psychical research, it occurred to Whately Carington that it would be valuable once more to assess the pre-Rhinean quantitative work on mental phenomena, this time in a search for evidence whose value might originally not have been fully appreciated. In a talk delivered to the SPR in June 1938 he reviewed the work of Coover, Troland, Estabrooks, and Brugmans and the little-known attempts of Usher and Burt (1910) to transmit cards and drawings over long distances; he concluded hopefully, saying that the material "at least established an incontrovertible case for further and more meticulous investigation."[138] In November 1938 Carington moved to Cambridge, and early in the new year he began to try to carry out such a program of investigation himself. He was guided in his work by the advice of other members of the Cambridge community—Thouless (who had come there in 1938 from Glasgow), Broad, the biologist Oliver Gatty, and the statistician J. O. Irwin—who used to gather regularly in Broad's rooms at Trinity College to discuss Carington's progress with him.

Carington's original thought had been to use Zener cards once again as targets, but upon reflection he decided instead to use drawings—like Warcollier and Sinclair or Usher and Burt.[139] He recognized that difficulties would arise in evaluating successful guesses, but he hoped this would be compensated for by the livelier nature of the material used. In essence, his procedure was to post a different randomly conceived drawing in his study on ten successive evenings and then to ask a group of unselected percipients to attempt to "guess"—or reproduce—the series of drawings evening by evening. The collection of percipients' drawings was then evaluated against the ten originals, but without Carington knowing which particular sketch was meant to correspond to which original; only after each sketch had been judged against all ten was Carington given information enabling him to determine the number of direct hits. As he came slowly to recognize in the course of several such series, while *direct* hits did not occur significantly often, the mass of drawings made by a group of percipients seemed *as a whole* to match the whole set of ten targets at which they had been aimed (regardless of chronological order) markedly better than they matched the ten targets of any other series.

In order to verify this, Carington devised a method by which an independent assessor could evaluate hits and partial hits among these guesses in such a way as to permit a statistical test of his hypothesis; and he was able to show that the set of drawings made for each particular set of originals did indeed correspond more closely to them than to any other originals. He took this as provisionally demonstrating a "paranormal" cognitive process, and he was able to infer from the way in which successes were obtained that the "process involved is not of an all or none character"—that is, that partial transmissions seemed to occur frequently. The hits so evaluated, however, were not made significantly often upon the target exposed *simultaneously* with the making of the drawing. Rather, they tended to fall upon targets exposed just after (or, slightly less markedly, just before) the drawings had been made. Carington presented this as evidence that the process involved incorporated both a precognitive and a less obvious retrocognitive effect, and he suggested that these effects might be responsible for the confusing results so far obtained in experiments using Zener cards.

The scoring of these sets of experiments took place in August 1939—in circumstances "intrinsically somewhat difficult," as Carington laconically put it. With the outbreak of war, Queen Mary College was evacuated in September, and S. G. Soal too moved to Cambridge, though he was separated from the Carington group by a certain amount of personal rivalry as well as by a difference of scientific opinion—for of course, unlike Carington, Soal had become convinced of the futility of experimental searches for evidence of telepathy and clairvoyance. It was Robert Thouless who during that fall of 1939 asked Soal whether he had observed a displacement effect (precognitive or retrocognitive) in his own work;[140] Soal replied that there had been none. Then Carington insisted that Soal inspect all his past data for evidence of such an effect—and Soal, yielding, discovered with a shock that the guesses of two of the 160 persons he had previously tested showed precisely the kind of effect that Carington had urged him to look for. Both Mrs. Stewart, who had briefly seemed so promising in 1936, and a London photographer, Basil Shackleton, whom Soal had tested fruitlessly in the same year, gave strong evidence of a $\pm$ 1 displacement in guessing Zener cards.[141]

Carington and Soal published a short note announcing their discovery in *Nature* in March 1940[142] and a full account of their research, results, and conclusions in the June number of the SPR *Proceedings*. It was an extraordinarily dramatic denouement of a sort rare in the history of science, to find confirmatory evidence of a totally new effect in a mass of supposedly exhausted data—but then it was precisely what Carington had just been trying to extract from the work of Coover and Troland. No doubt most members of the society were as much amazed as delighted to find Soal also numbered among the prophets. Some sense of the satisfaction felt by the traditional "right" of the SPR can be gathered from the essay with which C. D. Broad introduced the Carington and Soal reports in *Proceedings,* stressing their technical rigor, their statistical excellence, and their strongly successful results.[143] Yet Broad still took care to defend the authors to readers who might object to "another mass of boring statistical stuff!", insisting that the study of paranormal cognition had reached the point where a dependence upon exact quantitative measure was necessary, despite the apparently trivial nature of the results obtained; only such conclusive research could win over experimental psychologists. Broad alluded only briefly to the earlier Duke work that had served Soal as a model, but he could scarcely have forgotten how three years before he had said he would remain skeptical until Rhine's conclusions were confirmed in an English institution. It was nearly impossible, in fact, not to see the Carington and Soal reports as broadly confirmatory of extra-sensory perception, and there is at least tacit acknowledgment of that relationship in Carington's paper and perhaps even in Soal's.

By 1940, therefore, English psychical research seemed on the way to consolidation around the experimental technique that had already come to dominance in the United States—that is, a technique dependent upon statistical analysis of hits and misses in guessing objective material. The traditional English preoccupation with anecdotal, mediumistic, and qualitative material was still vigorously alive, but for a time the excitement of the apparently replicable Carington-Soal effect was bound to keep the experimental approach uppermost whenever a moment could be spared from the war for research. There was still disagreement on minor issues—Carington's results had been obtained from a mass of unselected subjects, while Soal had found evidence of "extra-sensory cognition" in only two of 160 and felt strongly that only "sensitives" could profitably be tested—but the open quarrels and personality conflicts that had marked the 1930s were at least temporarily softened.

Broad's hope for an academic entree for the field in England also seemed not unrealistic. In 1940 Trinity College established a studentship in psychical research out of a bequest from Frank Duerdin Perrott made in Myers's memory;[144]

W. Whately Carington ca. 1941.

Broad himself was charged with reviewing applications. Before the end of the year, Whately Carington was named Trinity's first Perrott student. If Carington could continue to present significant results from within the academic world, it would be difficult indeed for contemporary psychologists to ignore the problems of psychical research. This had always appeared as one possible route by which the subject might achieve recognition, particularly in the United States, and, though Broad could not have been fully aware of trans-Atlantic developments, something of this sort seemed ready to happen there as well. In 1939–40, as psychical research was beginning to enter the Cambridge community, it had already attracted widespread consideration from psychologists in America and was beginning to command professional toleration—if not endorsement.

CHAPTER NINE
Parapsychology and Professional Psychology, 1934-38

I

American psychology in the 1930s was a much less tidy science than it had appeared to be in previous years. In the twenties it had been primarily an experimental discipline, dominated by the claims to authority of a few competing schools of thought, from which behaviorism had by 1930 emerged victorious over the early introspective psychology. Yet in subsequent years this apparent simplicity was gradually replaced by an astonishing diversity of outlooks and methods. If the orientation of American psychologists remained chiefly behavioristic, it was a behaviorism much less narrow, much more eclectic in its expression than the Watsonian form of the previous decade—in the influential work of E. C. Tolman, for example, introducing elements drawn from Gestalt psychology. At the same time, new problems somewhat removed from those typically associated with the study of behavior assumed more prominence within psychology, personality being the most important. The preeminence of experimental psychology per se was also weakening, as in particular both clinical and applied psychology grew in importance, perhaps in response to a growing feeling in these Depression years that psychology needed to be employed more aggressively in the service of man. By 1940 only half the members of the American Psychological Association were in academic positions, and a number of specialized professional organizations had come into existence.

The almost uniform refusal of American psychologists early in the twentieth century to make a place for psychical research was not an encouraging precedent for the reception of *Extra-Sensory Perception* by these diverse successors thirty years later. Psychical researchers had come to accept as a matter of course that the academic world would refuse to take their announcements seriously, since they failed to conform to a tidily materialistic view of nature; it might well have been expected that psychologists would react to Rhine's first claims with the same reflexive scorn. Yet as a matter of fact the psychological community of the 1930s proved more open-minded than it had been in the days of Hall and Münsterberg. Why this was so is difficult to say. Perhaps by now psychologists felt more secure as a profession, able to entertain unorthodox opinions; perhaps the new diversity of that profession made unorthodoxy less clear-cut; or perhaps the scientific revolution in physics had weakened the power of strictly materialistic explanations. In any case, once Rhine's claims and their experimental basis became known, a sizable number of psychologists began to try to replicate his work. In spite of several failures to replicate and sharp criticism from a few individuals, and in spite of Rhine's own initial tendency to keep his distance from the psychological community, many psychologists seem to have come to feel a certain tolerance for parapsychology.

At the outset, Rhine himself shared something of the suspicion widely felt by psychical researchers of psychologists and their narrow view of man. Rhine had had no intention of a career in academic psychology when he arrived at Duke in 1927, and his eventual position in the Duke psychology department had not come about in fulfilment of any preconceived plan of his or of William McDougall's. Although Rhine sat in on all McDougall's courses and eventually taught psychology courses himself, he was still (as he expressed it) "shy" of the field. While he could admire the devotion to the experimental method manifested in the new behaviorism, he remained unhappy with the limited and fragmentary model of the human mind that seemed to have been forced upon psychology by its division into aggressive schools of thought, each with its own leader. If one part of Rhine wanted to bring parapsychology into the world of contemporary science, another part was skeptical of the motives of orthodox scientists and still showed signs of the populist orientation that had figured prominently in the makeup of many earlier psychical researchers. Inevitably, he felt, psychical research would eventually be a part of a fully integrated and developed science of psychology, but he believed that the larger field had not yet become broad and flexible enough to accept it.[1]

In these respects, Rhine's attitude was more hesitant than Walter Franklin Prince's: Prince was convinced that psychical research needed to maintain professional standards and to maintain close connections with "science," with psychology in particular. Prince's *Enchanted Boundary*[2] might have adopted a tone of mockery, but it was nonetheless inspired by the sincere conviction that in the end psychical researchers would have to win acceptance from professional psychologists if they were ever to gain widespread respectability. It was unfortunate that the perilous situation of the Boston SPR regularly forced Prince to claim for the society's auspices works like Estabrooks's article of 1927 or Rhine's monograph of 1934, which could certainly have found publication in a wider psychological forum. Nevertheless, Prince did his best to ensure that *Extra-Sensory Perception* reached the psychological community. Ellen Wood provided him with a special fund to permit sending complimentary copies to 400 eminent men of science, and in January 1934 he wrote Rhine asking for a list of 175 American psychologists, physicists, and mathematicians to whom the monograph might be sent.[3]

The list Rhine eventually returned to Prince was a peculiarly revealing one, for it included quite as many acquaintances and chance correspondents as it did scientists. The names included seventy or so psychologists but less than half a dozen physical scientists; the bulk of the list was made up of ministers, theologians, philosophers, and popular-science writers. The psychologists singled out were themselves a rather odd assortment. Some names had been given Rhine by Estabrooks; others may have been suggested to him by William McDougall, for elder statesmen, social psychologists, and Harvard Ph.D.'s from McDougall's era there (1920-26) made up a large proportion of the list. As a result, *Extra-Sensory Perception* may have gone to fewer experimental psychologists and students of perception than would have been desirable. But Walter Franklin Prince was not concerned with such fine distinctions. He wrote to Rhine in May to question the appropriateness of sending the book to so many nonscientists and personal friends—he was particularly bothered by the copy sent to "a magician" (Wallace Lee)—and he noted that Rhine had neglected the physical and biological sciences as well as certain psychologists.[4] Rhine insisted firmly that his choice of recipients was based on a conscious decision to try to reach the general public, not the uncaring professional world.

> I can scarcely conceive of a better way of wasting 400 copies than to send them in perfectly wooden fashion to 400 people just because they either appear in *Who's Who* or belong to one of the four groups. I had much rather send a copy to an intelligent school teacher out in a small middle west city whom I happen to know would do something about the subject—perform some experiments, interest some students, start a working group, and perhaps write her professor in a near-by university about the subject—than to send 10 copies to the ten leading psychologists in this country who would throw them in the waste basket.... I have studied over this list of 168 names with such time as I have had and most of them are either people who are pretty certain to do something to give the book attention or are in positions where they are likely to interest someone else as a result of the book. Some of them are not prominent people. For that matter, I am not a prominent man myself and most of my fellow workers are not prominent people. None of us may ever be in *Who's Who* but surely Mrs. Wood would not have scratched us off her list.... It becomes, then, simply a matter of judgment, and I will put my judgment of an humble person I know against your judgment of a man in *Who's Who* whom *you know* nothing more about than the record therein, if I am doing the choosing. I would even rather send one to a magician or a clergyman who will do something about the subject than to the dean of American psychologists, McKean Catell, who would throw it at my head. So there we are. This is not in the least quarrelsome, it is simply my answer.[5]

Prince did not press the matter further, and the distribution of *Extra-Sensory Perception* to the psychological community remained a somewhat casual and haphazard affair.

The monograph apparently roused no strong emotion in its first psychologist readers. It was given very few reviews besides Gardner Murphy's laudatory account in the *Journal of General Psychology*—notably a very favorable one in *Character and Personality* by its editor, Robert Saudek, and a generally sympathetic one by George Dearborn in the *Journal of Abnormal and Social Psychol-*

ogy.⁶ The response of the University of Chicago, where the psychology department maintained much of the prestige it had acquired in the days of Angell and Dewey, may be indicative of the immediate professional reaction. Rhine had sent several copies of *Extra-Sensory Perception* to members of the Chicago faculty—including his one-time mentor C. A. Shull in botany, the comparative neurologist C. Judson Herrick, Henry Nelson Wieman at the Divinity School, and T. V. Smith in philosophy—but none had been sent to an experimental psychologist. Herrick and Forrest Kingsbury (the latter teaching business psychology) praised the book warmly, and the behaviorist K. S. Lashley reported on Rhine's work at a departmental luncheon, questioning only Rhine's precautions against minimal cues. There was further favorable comment within the department, and eventually Rhine's monograph was ordered for the departmental library.⁷ There is nothing in this reaction to suggest anything but a normal response of perplexity and mild curiosity. A year later a correspondent wrote Rhine that "the comment at Chicago on your own work so far is to withhold comment until other experimental work is done."⁸ The Chicago psychologists were reacting as they might have done to any unexpected experimental results that lay outside their own particular research concerns. Far from automatically denying the possibility of telepathy and clairvoyance, they considered Rhine's report with some seriousness and waited for further information.

To be sure, Joseph Jastrow dismissed *Extra-Sensory Perception* brusquely, but Jastrow was a special case. In his teaching days at Wisconsin he had always been a stringent critic of the unorthodox or the faddish in psychology; it was this position that had made him a natural choice for McDougall's 1921 Advisory Scientific Council for the American SPR and that had led him in 1924 to discourage J. B. Rhine's first exploratory inquiry about the potential value of scientific psychical research. When *Extra-Sensory Perception* appeared at the end of April 1934, Lydia Allison of the Boston SPR gave Jastrow a copy, with the idea that he might be willing to review it for a psychological journal. Jastrow looked quickly at the book and declined, explaining in a letter to Rhine that "I should be obliged to comment more disparagingly than I would care to do." He put his finger on an objection that in later years would be taken up by English psychical researchers and American psychologists alike: "The fundamental difficulty is that you have neglected to tell what the procedure is in handling the cards; and also at what stage either you or the subject was informed of the results. It is true that from stray phrases I can make a guess, but the most important section should be an account of the *procedure,* in sufficient detail so that any one could repeat the experiment under precisely the same conditions."⁹

Rhine was not disturbed by Jastrow's comments, however—the book had only been out a week, after all, and Jastrow's prejudices were well known—and he wrote back rather lightheartedly: "I feel that the difficulty that you referred to is due to your having expected to find an elaborate discussion of procedure standing right out in the front of this report. For a certain group of readers I am sure that would have been an advisable policy; perhaps it would have been the best. But for reasons which seemed to me at the time good and justifying, I distributed the description of procedure through the report and there you will find it if you have the time and patience to read through it.'"[10] The "certain group of readers" was of course the psychologists, for Rhine had had the ordinary psychical researcher in mind as his principal audience for *Extra-Sensory Perception,* as he repeatedly told his friends.

What soon changed this attitude of Rhine's was the thoughtful and careful consideration that his research received from the staff at Clark University; their exchanges with him made him aware that psychologists might be genuinely open to persuasion about his results. Only one copy of Rhine's monograph had been sent to a Clark psychologist: to Carl Murchison, who must have been somewhat nonplused when he read through it. For Rhine's work certainly posed a potential threat to some of Murchison's own plans. Murchison had for several years been trying to divert the Clark endowment for psychical research, the Smith-Battles Fund, into the hands of the psychology department; it provoked him that over ten thousand dollars should be available to subsidize psychical research, and he made various attempts to broaden the application of the terms of Battles's will. Most recently, in January 1934, Murchison had applied for a grant of three hundred dollars from the fund to subsidize the preliminary stages of research into the social behavior of chickens that he himself planned to carry on—he had asserted vaguely that "the use for which I request it lies well within the purposes for which the fund was given to the University" [!]—and his application had been granted.[11] Now, only a few months later, a book had suddenly appeared that could weaken Murchison's claims for the sterility of psychical research. Precisely how Murchison reacted we cannot be sure. But in the middle of June 1934 W. S. Hunter, Murchison's associate in the Clark psychology department for ten years, wrote a long letter to Rhine discussing the book, and it is at least reasonable to suppose that Murchison had passed his copy of *Extra-Sensory Perception* on to Hunter for examination.

Hunter was an experimental psychologist who had received his Ph.D. at Chicago in 1912, a prominent behaviorist who was not prepared to concede the reality of extra-sensory perception any more than was Murchison. As he himself explained to Rhine, "With my theoretical background, there are many questions that arise for which I should want satisfactory answers before I could accept your interpretation of the data."[12] He was willing, however, to rule out cheating and to agree that some individuals were able to score above chance in guessing Zener cards. The question for him was how this might be accomplished by normal means. Granted that the cards were arranged in a chance order, would not the subject's guesses be far from random, determined by personal reaction tendencies? and might it not then be possible that "the subject [is] guessing correctly only because one of his reaction tendencies coincides with a temporary run in the cards?" Rhine was astonished by so thoughtful a letter from a psychologist, since he had expected most psychologists to be closed-minded, and he replied in kind, thanking Hunter for his courtesy and sending him a sample deck of cards to try out.[13] After two weeks of further work Hunter wrote once again to acknowledge that "the probability mathematics are the crux of the whole question" and to say that he was turning the problem over for analysis to a junior colleague at Clark, R. R. Willoughby.[14]

The ensuing correspondence between Willoughby and the Duke parapsychologists lasted for two years, culminating in a series of critical articles in psychological journals by Willoughby with appended rebuttals by Rhine or Charles Stuart. Yet the tone of the correspondence was never acrimonious; Willoughby's letters, in fact, were regularly flippant, almost offhand, although they reveal that he was giving very serious attention to the statistical basis for Rhine's claims. This was by no means Willoughby's first exposure to psychical research. He had taken his M.A. in 1923 and his Ph.D. in 1926 from Stanford, where he had done some work with J. E. Coover. In 1928 he had been hired by Clark to teach abnormal psychology, and, curiously, in his first year there he was supported by the Smith-Battles Fund, at Carl Murchison's rather blunt suggestion: "The purposes for which the Smith-Battles fund was created lie within the general field of Abnormal Psychology, if they exist as research problems at all."[15] With this background, together with an interest in statistics, Willoughby was the obvious man at Clark to tackle the problem of *Extra-Sensory Perception*.

Willoughby went into the investigation of extra-sensory perception sharing Hunter's judgments that Rhine's results were sure to have a normal explanation and that the probability mathematics lay at the heart of the problem. On the first judgment, if not on the second, he continued to remain firm. He was willing to entertain the possibility that truly extra-chance card-guessing might take place for which no normal explanation was obvious; but, as he told Rhine, "Naturally I will not admit your clairvoyance hypothesis in [that] case, on the same grounds that I will not admit, because I cannot demonstrate how a rabbit was 'taken from a hat,' that rabbits really grow in hats."[16] Nevertheless, while he could not take Rhine's claims seriously—or perhaps *because* he could not do so—he did scrutinize his techniques and methods with extreme care. We have already discussed Willoughby's principal objections: his feeling that Rhine should have compared his ESP results with an empirical chance series and his resistance in particular to the idea that the normal distribution would describe such a chance series. Willoughby himself proposed an alternative experiment that would avoid all his statistical and methodological criticisms. Let a subject be tested in the DT condition exclusively, and let him know nothing concerning his success or failure during a series of two hundred tests. "Each of the 200 tests should consist of a target series, a guessed series (or call series), and a second shuffled series; all these data should be used in the evaluation, which should consist of the plotting of the two distributions of correspondences upon the same axes and of the determination of the significance of the differences between them by sound statistical methods."[17] A few months later Willoughby described this experiment in more detail in an article submitted to the *Journal of Applied Psychology*, labeling it a "prerequisite" for a clairvoyance hypothesis and specifying the use of the chi-square technique in evaluating the results.[18] He predicted that such an experiment would show no uncontrollable extra-chance factors at work, "and unless and until such prediction is shown to be in error, we shall not regard the concoction of hypotheses of the mechanism of ESP as a profitable investment of energy."[19]

Yet despite this disclaimer, Rhine's report continued to tease Willoughby. He had ruefully remarked in August that "this business has a great many of the characteristics of a cross-word puzzle (from my standpoint); it has no special relation to anything I regard as important, and yet I can't take it and I can't leave it alone."[20] In the fall he went on looking into ESP, trying experiments with Zener cards first on a local "psychic," then on his secretary and himself. Suddenly in November he learned that the Smith-Battles "spook fund" need not be expended on animal work (as Murchison had done), and in some excitement he wrote to Rhine suggesting that he apply for money from the fund to bring one or two of Rhine's best subjects up to Clark to be tested. Rhine replied with some irony that he had "allowed my stock of clairvoyant and telepathic subjects to run out" and proposed instead that the Smith-Battles Fund be used to support a Duke graduate student in a year's research at Clark.[21] It is a measure of the seriousness with which Rhine was being taken at Clark that Willoughby and Hunter both apparently felt the idea worth pursuing. Eventually, however, Willoughby settled for a seventy-five-dollar grant from the Smith-Battles Fund, intended to support further extra-sensory experimentation on a slightly larger scale and to subsidize a visit to Duke in April 1935.[22]

Rhine had by this time become fully convinced of the need to take psychologists seriously, and he extended himself to meet the anticipated criticism. He could not show Willoughby any successful experiments in telepathy or clairvoyance at the moment, to be sure, and chose not to fuel Willoughby's skepticism by discussing the ongoing work with psychokinesis. But he showed him the parapsychology laboratory, illustrated the methodology he had used, and then turned him over to Charles Stuart to deal with the mathematical questions he had raised. Here Stuart was several steps ahead of Willoughby—he had already prepared refutations of both of Willoughby's critical arguments—and left him with nothing to say. At Willoughby's insistence, a chi-square test was finally applied to the data—and to his chagrin the Duke results still proved to be significant. When Willoughby left Durham he commented offhandedly, "Well, I guess I'll go back and tear up my next paper."[23]

He did not, of course. Willoughby never brought himself to accept telepathy and clairvoyance, but his criticism of Rhine became indirect and was henceforth directed towards questions of method rather than of mathematics. Instead of accepting Rhine's early results and trying to deny their significance, he now attempted to check the results themselves by performing his own experiments. This indirect attack was inevitably frustrating, since the negative results Willoughby obtained proved nothing about Rhine's and were "always vulnerable to the criticism that I may not have subjects that could be expected to be sensitive."[24] He had hoped that some sort of collaboration with Rhine to settle some central experimental questions could come out of his visit to Duke and was disappointed when Rhine showed no interest in such collaboration. Consequently, when he had used up his Smith-Battles money, he published a paper summarizing his results[25] and called a halt to his experimental investigations into psychical research. Yet Willoughby retained enough curiosity about the field to supervise a Clark undergraduate named Ralph Rothera in a series of experiments on extrasensory perception, experiments that seem to have struck him as promising. He confessed to Rhine in September 1936 that he had abandoned his earlier objection to the statistics and that Rothera's work, together with reports he had heard from C. R. Carpenter, had led him to conclude "that the problem may turn out to be one in *subliminal* perception, a field in which I, at least, would feel on a little firmer ground; the general nature of these data is that uncontrolled signals, with some subjects and under some conditions (approximately the ones you describe), lead to super-chance results which progressively drop to chance as one possible cue after another is controlled; the mechanism of this is likely to be extremely intriguing, but hardly supernatural."[26] How this explanation might apply to some of Rhine's experiments, such as the Pearce-Pratt work, Willoughby did not say.

The exchanges between Clark and Duke had thus had their effect on psychologist and parapsychologist alike. Rhine had at last accepted what Prince and Gardner Murphy had felt from the beginning, that psychical researchers should properly be attentive and responsive to the judgment of academic psychologists; and he had gained some confidence that he could satisfactorily meet their objections, even though he still had no thought of merging his research into the larger field. Willoughby, conversely, had been led by his review of extra-sensory perception from a thorough skepticism to a sort of reluctant acknowledgment that Rhine's results might deserve serious consideration. In this respect it would prove to be Willoughby's guarded receptivity, not Jastrow's scornful dismissal, that would typify the reaction of most American psychologists to parapsychology.

II

During 1935 Rhine began at last to receive a series of serious inquiries about his work from the professional community. This was of course not due to a steady accumulation of data; rather it seems to have been produced by pressures outside as well as inside the scientific world. The rapidly intensifying publicity Rhine's work was receiving in newspapers and magazines kept it continuously before psychologists' attention. Increasingly they were being asked for their professional opinion on extra-sensory perception—not only by members of the broad public but by their own graduate students, eager to explore so potentially exciting a field. Most were intellectually too honest to dismiss it out of hand, as Jastrow had done, and many perhaps had some latent curiosity about the subject; consequently Rhine began to receive a growing number of requests from early 1935 on for reprints, for information as to his methods, or for sample ESP cards.

One department of psychology that expressed sympathy and encouragement was that of the University of Colorado, an open and flexible department, in some respects not unlike that at Duke. Its chairman, Karl Munzinger, was a learning theorist of some originality with views similar to those of Donald Adams. Both Munzinger and T. S. Howells at Colorado were among the psychologists who began to correspond with Rhine in early 1935. "I am much impressed by the results," Munzinger wrote, "as well as the scientific temper of your undertaking," and he criticized the "narrow orthodoxy in psychology" that made it difficult to work on such problems.[27] What interest the Colorado psychologists may have felt in Rhine's work was reinforced that fall when John Schoolland, an instructor in the department who had been working on a Ph.D. at Duke during 1934/35, returned to Boulder. In his year at Duke, Schoolland had naturally come to know something of Rhine and his work, and while it would be wrong to see him as either a convert or an eager proselytizer, he was certainly a spur to interest in ESP at Colorado when he returned in the fall of 1935. In November he wrote Rhine to announce several inquiries made of him at Boulder and to ask for information on ESP so that he could respond knowledgeably. "There seems to be considerable interest," he reported, "in response to which I have tried to contribute toward a favorable attitude."[28] Rhine urged Schoolland to try some experiments, and Schoolland was mildly agreeable to the idea; he was engrossed in finishing up his dissertation, but he promised to carry out some experiments with a volunteer—the wife of a university regent—if he could find the time. Whether or not he ever did so, he helped keep attention upon Rhine's work alive at Colorado, among physical scientists and mathematicians as well as among psychologists. Munzinger wrote Rhine at the end of 1936 to thank him for a reprint and to reiterate his belief in the impressiveness of Rhine's results and his admiration for Rhine's courage and persistence;[29] at the same time, as we have seen, he encouraged a Colorado graduate student (Dorothy Martin) in a program of parapsychological investigation.

At other schools experimentation began independently of any Duke connection. Thus in January 1935 Rhine received a letter from Theodore Karwoski, a psychologist at Dartmouth who had taken his Ph.D. at Harvard with L. T. Troland in 1928 and was working in color vision and perception. Karwoski acknowledged receipt of a reprint with thanks, and he went on to explain that he and Adelbert Ames (in physiological optics) had been considering working on extra-sensory perception, "or the extension of consciousness as we call it," for some time. They had found Rhine's original monograph very useful and hoped to begin work on a problem shortly, although they were not yet able to give the project much time. Three months later Rhine sent ESP cards to Karwoski, in case time should appear for research; apparently, however, it never did, for there was no further response.[30]

There were a dozen or more inquiries of this sort in 1935 and 1936: from J. B. Miner at Kentucky, J. E. Rauth at Catholic University, J. E. Marsh at the University of British Columbia, John Clark at Earlham, K. H. Baker at Minnesota. A few are of particular interest; for example, the response of Donald Adams's brother Eugene, who was teaching philosophy at Colgate. Gene Adams had visited Durham at Christmas, 1934, had seen Rhine's laboratory, and had gone back to New York "very much interested in the possibilities of [Rhine's] telepathy experiments," as he told his colleague G. H. Estabrooks. He undertook to try to repeat Rhine's work at Colgate and reported offers to help from all sides there: from the university president, G. B. Cutten, who gave Adams his personal "psychic library," on down to the baffled "local standard bearers of behaviorism," who promised all the time they could spare. Adams was unable to do more than a few irregular experiments that year, but in early 1936 he agreed to supervise a psychology major in a more systematic attempt at replication. A month later Adams reported at least mildly promising results.

> Covert, the student who has been working for me, reports preliminary test scores of more than fifty people, only one of which involves a significant variation from normal expectation. This particular one put on a most baffling performance one evening. At the end of the usual twenty-five preliminary tests, he had averaged about five, and Covert was ready to dismiss him. However, he insisted that he was "psychic," and Covert gave him fifteen more runs (target shooting). To the amazement of Covert and two other observers, he averaged 9.8 for his final fifteen runs. Next day, calling them from the bottom of the pack, he was back to five. I look forward with interest to seeing what he will do during the remainder of the year.[31]

Another serious attempt at replication began at Antioch College, where the chairman of the psychology department, Clarence Leuba, reported himself admiringly impressed by *Extra-Sensory Perception*. Rhine sent him a pack of ESP cards, and by November the pack had been completely worn out by testing, leading Leuba's assistant to inquire about purchasing another dozen packs. "We have not yet been working long enough on the subject to have any valid results," she reported, "but we have gotten some rather startling positive examples of telepathy with the cards, and are working very enthusiastically on checking them."[32] Antioch interest was not confined to the psychologists: Walter Kahoe in philosophy reported that he and his students, "as well as several other members of other departments," had been carrying on experiments.[33]

At Kansas, J. F. Brown (a friend of Donald Adams's from graduate school) began a series of tests in the summer of 1935. To simplify the statistical treatment of his results, Brown used a 3 × 3 rather than a 5 × 5 format, asking his subjects to guess cards from a deck of nine cards formed from three suits.[34] The results that Brown eventually amassed were, as he put it to Rhine, at least "suggestive." Brown had laboriously worked out the theoretical distribution of chance matchings from two 3 × 3 decks and found that the results of nine thousand matchings of one such shuffled deck against another conformed satisfactorily to this distribution. Yet his three subjects were able to guess consistently above chance expectation: each guessing some eleven hundred decks, they averaged 3.04, 3.07, and 3.23 successes per deck. During the summer of 1936 Brown retested his best subject, Thomas Thurman, with ESP cards, with which Thurman now could do nothing: his average for eleven hundred decks was 4.97. Attempts to get Thurman to guess the wrong card repeatedly, in series with both kinds of decks, produced results in the right direction, but not significantly so.[35]

As Brown described his feelings, he was left very much of two minds. He had originally expected his work to disprove Rhine's claims; by his own admission, indeed, he had *tried* to disprove them. Clearly he had not done so; yet his own results were by no means incontrovertibly positive either. "I am personally convinced that your results are not due to the shuffle and are not accounted for by chance," he agreed, and he allowed Rhine to make what use he chose of the data; but he was not enthusiastic enough to be willing to write up his research for publication. In October 1936 a former Duke graduate student who had come to work at the Menninger Clinic, Wally Reichenberg, gave a seminar to the clinic on ESP, and Brown agreed to chair it. At the seminar Brown admitted that while he could not give a normal explanation of the results—certainly Rhine's statistical treatment was sound, whatever Willoughby had said—he remained skeptical of ESP. Rhine heard all this from Wally Reichenberg,[36] and he pressed Brown to publish something in the forthcoming *Journal of Parapsychology,* whose first issue was less than five months away: either a summary of his own findings or, if he still chose not to do this, a statement that he was satisfied with the adequacy of Rhine's statistical methods and of his precautions against shuffling error, "as these are the areas questioned most." But Brown remained unwilling to commit himself even this far.[37]

Besides Brown, only one of Rhine's professional inquirers ever reported the results of his work back to the Duke laboratory: Kenneth H. Baker of the University of Minnesota. But unlike Brown, Baker had decided to publish his results, and he had sent off a paper to the *Journal of Experimental Psychology* in January 1937.[38] As he wrote to Rhine a month later, he was "somewhat discouraged with my findings," which were (as he put it) "equivocal," since he had not been able to identify anyone capable of producing consistently high scores. Nevertheless, the Minnesota psychology department was still intensely interested in the problem, and Baker had given a departmental seminar the day before on Rhine's work—at which seven out of fifty listeners expressed the conviction that they personally possessed extra-sensory ability. The seminar had also raised specific questions, which Baker passed on to Rhine: Why did he confine himself to geometric figures on cards? Why not try to increase the number of agents and percipients, perhaps using severe critics of the work as agents in order to have successes carry still further conviction? Some members of the colloquium seem to have pressed for further experimental data, and Baker expressed marked sympathy for Rhine's position here: "I don't blame you at all for being rather impatient with those who keep requesting that you supply more and more data to support your contention of the existence of E.S.P. I think I made that point rather clear in my report to the seminar. If psychologists in general were as critical of all experiments and observations as some people are trying to be in your case, there would undoubtedly be a lot of question marks in our text books today."[39] The tone of Baker's response, even more sympathetic than Brown's, reveals a genuine open-mindedness in the Minnesota department about the possibility of extra-sensory perception.

The unsettling effect that Rhine's work was coming to have upon American psychologists, the reluctant half-acceptance that more and more were willing to grant it, is nowhere better seen than in the case of Donald Adams, Rhine's colleague at Duke. One reason for singling out Adams as an example is of course his association with Rhine and his direct acquaintance with the details of the work; another is the jealousy clearly felt by Adams (as by the other members of the department) for Rhine's standing with McDougall and for his prominent role in departmental affairs. As a postdoctoral student in Germany in 1930, Adams had explained to his mentor, Robert Yerkes, just what sort of professional position he hoped to find: it should be in psychology broadly conceived, and it should consist mainly of experimental research.[40] The next year Adams was at Duke, and in most respects he found what he wanted there. If the material surroundings were still somewhat primitive, McDougall was using his influence in the university to keep the departmental teaching load at an unheard-of six hours, leaving plenty of time for research; moreover, Adams increasingly found McDougall a profoundly sympathetic figure. While Adams was perhaps a more thoroughgoing experimentalist than his chief, he could not help but appreciate McDougall's breadth of mind and his vision of the psychologist's wider concerns. But the exasperation with Rhine that broke out in the spring of 1934 was inevitable: Adams saw Rhine as an incompetent psychologist, a not particularly careful experimentalist, a man with limited interests who was using his influence with McDougall to split the department and was thereby destroying the hopes for the future that he and the others had in mind. Under these circumstances Adams was bound to feel suspicious of Rhine's research.

Reinforcing Adams's mistrust was his thorough skepticism of psychical phenomena in general. As an undergraduate at Pennsylvania State University and a graduate student at Harvard (M.A. 1925) and Yale (Ph.D. 1927) Adams had come up against the claims of psychical research over and over again and had never managed to take them seriously. His first psychology teacher, the man who convinced him to enter the field, had been a student of Coover's at Stanford and had come away with an open-mindedness towards psychic phenomena at which Adams could only marvel. In his year at Harvard, Adams had served as a subject in one of G. H. Estabrooks's experiments designed to test the effect of alcohol upon telepathy, but he did so only for the sake of the "excellent whiskey" Estabrooks used and never bothered to find out the results of the research. William McDougall's support for psychical research he interpreted as one more obstinacy in that defender of lost causes. The defense on scientific grounds that his fellow graduate student, Hudson Hoagland, offered for the study of Margery did trouble him, however, and he had just volunteered to sit in on some of Hoagland's experiments when Hoagland declared his belief that Margery was a fraud.

Characteristically, Adams himself provided this sketch of his early feelings, in a very personal, detailed retrospective essay written in late 1936—an essay meant, not for publication, but to try to clarify his own position in his own understanding.[41] As he described his subsequent development, he first became aware of an inconsistency between his prejudice against psychic research and his professed ideal of scientific dispassion late in his graduate career, when his admired teacher, Robert Yerkes, raised the possibility of telepathic influence upon experimental animals, much to Adams's consternation; but it was a purely intellectual awareness, and his resolve to maintain an objective attitude in psychic matters was not seriously tested until he encountered Rhine's work at Duke. "My resolve met this test admirably. I viewed his efforts with a real, if somewhat pitying, sympathy, *so long as I could defer a critical examination of his results*. When his monograph appeared, this could no longer be done, and my resolve vanished. I wanted not the truth, but to prove his positive conclusions wrong." Yet as Adams confessed, he had to stretch to do so. In part, he based his objections on an *ad hominem* dislike of Rhine and of what he perceived as his professional limitations; but he had to concede that Rhine's "statistics seemed impeccable and his gradually more rigorous conditions adequate." What the scientist in Adams found weakest was Rhine's failure to supervise closely his coworkers, who were sometimes suggestible and careless and tended to hero-worship him.

Adams was perfectly well aware that he was rationalizing his instinctive prejudices, and during 1935 he began to realize that he was not alone among psychologists in this.

> Whenever I met one at psychological meetings and elsewhere, he was almost sure to ask immediately—and hopefully—whether Rhine was crazy, duped, or crooked. When I had to reply that in my opinion he was none of these in sufficient measure appreciably to affect his results, the questioner was usually either politely contemptuous of my gullibility or plainly troubled as I was. It was noteworthy that the latter response was typical of those that I had regarded as the abler. At least two of these men went so far as to make extensive experiments, in one case including more guesses than Rhine had reported. Both found statistically significant positive deviations from chance. All this did my prejudice no good, but worse was to come.

During the summer of 1936 Adams accepted Rhine's offer of payment for a translation into English of Hans Bender's *Zum Problem der aussersinnlichen Wahrnehmung,* a monograph that Adams found enormously impressive. Bender's approach was entirely different from Rhine's, for its argument derived not from a mass of statistical results (reasoning that Adams found unconvincing) but from a careful account of certain analogies between extra-sensory and sensory perception. Adams agreed that "these analogies to well known processes of normal perception and imagery, of course, explain nothing. But they do make the bitter dose easier to swallow in that they offer at least a point of attack for the effort to incorporate extra-sensory perception into the body of ordered psychological knowledge."[42] All in all, while Adams still found the thought of psychic abilities unreasonable and distasteful, by late 1936 he had come grudgingly to conclude that the evidence could not be ignored. "There is a sort of slow inexorability about scientific method before which prejudice is silly, small-minded and futile. Nature does not seem to care in the least what we think of her." The candor of his self-examination gives us a revealing picture of the way in which even the more hostile and orthodox psychologists were coming to change their minds about parapsychology in 1935/36.

By early 1937, then, parapsychology was definitely beginning to win some limited professional recognition; it was not always rejected out of hand by psychologists because of its supposed scientific inadmissibility. A few were curious enough to attempt to replicate the Duke experiments; others, involved in well-established research programs of their own devising, were content to leave the investigation of parapsychology to Rhine and Murphy but expressed tolerance or sympathy for the work. Adams, Brown, Baker, and Munzinger (each in his own way) all testify that many members of the psychological community were by this time prepared to hear and even tentatively to accept Rhine's account of his work, were in fact content to view him as a professional colleague.

This is not to say that there was not already some criticism being voiced of "extra-sensory perception." There clearly existed a sizable group of psychologists who were on principle utter disbelievers in telepathy and clairvoyance, but they had not yet begun to do anything more than complain privately to friends about the nonsense at Duke. One such individual who remained steadily hostile was J. E. Coover at Stanford. Since 1917 Coover had paid less and less attention to psychical research—formally it was supposed to occupy one third of his time—and more and more to perfecting a kinaesthetic method of learning typing and a system of condensed writing ("notescript"). The publication of *Extra-Sensory Perception* had drawn him back to active psychical research, and he attempted to confirm Rhine's conclusions, but he obtained no results that he could consider positive. Privately he was very critical of what he felt to be the carelessness of Rhine's methods, but he made no public announcement of any sort—apparently, as his letters to President Wilbur of Stanford suggest, for fear of what "the tabloids" might do with the ensuing controversy.[43] But as long as he remained at Stanford he did all that was in his power to see to it that the university gave no support whatsoever to any representative of the Duke doctrines. In 1935 he helped quash Lucien Warner's inquiry into the possibility of working at Stanford by insisting that "his present eagerness to work on Dr. Rhine's problem... incline[s] me to think that he is too advanced (toward the occult) for us."[44] When Rhine wrote to President Wilbur of Stanford a month later, intimating that Coover's apparent inactivity suggested that the Stanford endowment was being misspent, Coover went a little further in explaining to Wilbur how he felt about the Duke claims. "If Dr. Rhine is interested, as his letter indicates, in some kind of cooperation with this department, I should be very glad to have, say, one of his students who has demonstrated the capacity working here on a regular program for his degree. What I wouldn't want, and what Dr. Rhine no doubt especially desires, is to have one of his men carry out independent research for the mere purpose of verifying Rhine's findings in another university. I have no objection to extra-sensory capacity but it's got to be extra-sensory capacity determined by methods that meet my approval."[45] Coover seems to have tried to replicate Rhine's work himself during 1934-36, but without success.[46]

In the fall of 1936 Wilbur began to look to replacing Coover, who was to retire the next spring at the age of sixty-five, and he asked Coover himself for suggestions. Coover's thoughts were certainly interesting: he proposed that the next Fellow in Psychical Research be someone who could pursue the experimental investigation of brain waves, which had always been popular among psychical researchers as a hypothetical explanation of thought-transference and which in the mid-1930s were attracting attention from psychologists and physiologists. For Coover, this area of research had the added advantage that it would automatically eliminate the Duke group from consideration. As he wrote to Wilbur,

> On Jan. 6,... Dr. Terman conferred with me upon the matter of selecting an applicant for the Psychical Research Fellowship, and the kind of work we should have done. I resisted the suggestion of an intensive repetition of Rhine's work (at Duke), not granting it adequate for scientific attention; he has statistically significant excesses over probability, but he has uncovered no factors that correlate with them, and has no right to any term to name the extra-chance causes responsible for them (such as "Extra-sensory Capacity[")].... I suggested the promise of the new technique, electroencephalograms of Brain Waves; research must be good in experimental psychology, and in physiology; applicant should have produced significant work; no fresh Ph.D. to flounder about in occultism. We need do no more rough stuff in P.R.[47]

When Terman eventually canvassed the profession for a possible appointee, however, he did not reveal any of these earlier discussions; he simply announced that Stanford was looking for a young psychologist to hold the position for from one to three years. "We hope," he wrote McDougall, "to have a succession of gifted young psychologists on this fellowship and believe that by this plan psychical research can be brought into more intimate contact with general experimental psychology."[48] The eventual appointee was John L. Kennedy, who had been a Stanford undergraduate (1934) and was just finishing up his Ph.D. that spring at Clark—not on brain waves, but on a neurophysiological subject. Coover stayed at hand and for six months worked closely with the new Fellow in devising a program of experimentation. When Coover died in February 1938, Kennedy was well into the preliminary stages of some new research.

While Coover's hostility to extra-sensory perception remained relatively covert, Joseph Jastrow's did not. Rhine sent a copy of the first issue of the *Journal of Parapsychology* to Jastrow in April 1937; Jastrow replied courteously enough but insisted upon his disagreement with Rhine over "the major issue of E.S.P." While he raised minor objections to Rhine's mathematics, he agreed that something "anti-chance" was in operation and rebuked Rhine for piling up so many guesses when 4,000-6,000 would have been enough. His principal objection, however, as in 1934, was to Rhine's methods of card-guessing, and he insisted that he would have to have the findings verified by completely independent techniques before he could accept them. His closing remarks were by no means unkind—"I must accordingly differ as sharply from your conclusions as ever; but I regard the problem as worth investigating"—but he may have taken offense at Rhine's reply that it was "the *younger* psychologists who are perhaps not as firmly settled in their views of what, *a priori,* ought to be."[49] Jastrow (who was then seventy-three) complained bitterly of the tone of the reply and shortly thereafter began to make himself heard as an almost obsessive opponent of extra-sensory perception.

Despite occasional criticism, however, by 1937 Rhine had some cause to feel contentedly that the professional establishment of parapsychology was not far removed. The criticism did not seem serious. Jastrow and Coover were men of a much earlier generation, about to disappear from psychology, who so far seemed unwilling to offer any public opposition. Willoughby, finally silenced by Stuart, had never really convinced the psychometricians. Besides Willoughby, only two men so far had actually reported failure to replicate his results, both in Carl Murchison's *Journal of Experimental Psychology:* Kenneth Baker at Minnesota and a Princeton undergraduate, William S. Cox, who had investigated 132 subjects in 1934/35, his senior year.[50] Such failures did not particularly worry Rhine. We have seen that he had come to explain his own difficulties of replication as due to some subtle change for the worse in the psychological conditions of experimentation at Duke, in the atmosphere surrounding experimenter and subject; he now interpreted the failures of Cox and Baker as due to inadequate salesmanship by carefully disinterested experimenters.[51] There were of course rumors of other researchers who had failed to replicate his work, but Rhine could cite similar rumors of successful experiments to counter them, as well as the work of Warner, Pratt, and Woodruff. He was finding some moral support, too. In May 1937 Vernon Lemmon of Washington University published an article attempting to evaluate Rhine's claims objectively, insisting that most of his professional colleagues had failed to do so. "It is interesting to note that in the natural sciences, when an unexpected or improbable result is reported, the scientists usually try to verify the results first and criticize afterward, whereas in this instance the psychologists have hastened to criticize, and almost no attempts to check Rhine's results have been reported.''[52] Though he criticized what he felt to be slack experimental technique and denied that a radiation hypothesis had been proven unsatisfactory, Lemmon accepted Rhine's claim of evidence for an extra-chance factor. The signs of professional recognition that were accumulating for Rhine by 1937—invitations to speak before learned societies; inquiries from graduate students hoping to work under him at Duke; the decision by *Psychological Abstracts* to include material from the *Journal of Parapsychology*—made it clear that many psychologists now agreed with Lemmon that Rhine might really have discovered something.

III

In the fall of 1937 the tide of professional opinion seemed about to reverse itself, as a flood of articles and letters by psychologists critical of Rhine's claims suddenly appeared in print. What set this off was the accumulation of publicity given to Rhine's research. Skeptical psychologists had originally been willing to leave extra-sensory perception to the normal process of evaluation by peers, by discussion in seminar, by individual attempts at replication, and by professional publication, a process that would eventually refute or establish parapsychology as a legitimate field of investigation. But the popularization of "extra-sensory perception" as a supposed scientific fact, by science writers and nonscientist academics, raised very different questions. Should this sort of misleading exposition be allowed to go unchallenged? In particular, E. H. Wright's supportive articles in *Harper's* in November and December 1936 struck a number of psychologists as so uncritical as to demand some sort of response. Then in September 1937 the openly commercial Zenith radio programs began, claiming Rhine's authority from the outset; simultaneously *New Frontiers of the Mind* appeared as a Book-of-the-Month-Club selection, obviously written for a popular audience, and the ESP cards appeared for sale in drugstores throughout the country, all "copyright J. B. Rhine."

It was a tendentious article in the October 1937 *Scientific Monthly* by Chester E. Kellogg of McGill University that first began to crystallize the psychologists' unease into resentment.[53] Kellogg had learned contempt for psychical research twenty-five years earlier as a graduate student under Hugo Münsterberg at Harvard, but he insisted that he had taken up extra-sensory perception in a purely dispassionate spirit. As he explained the genesis of his investigations, the graduate seminar at McGill had discussed *Extra-Sensory Perception* at length in 1934/35 and had aroused his concern over the statistical methods employed. This led to a brief critical article in the *Journal of Abnormal and Social Psychology* in

September 1936[54] and a subsequent correspondence with Stuart. "I had hoped," Kellogg wrote Rhine after the *Scientific Monthly* article appeared, "that we might reach substantial agreement on the other points at issue, in the same informal way, and that the presentation of your evidence might be modified accordingly, when the publicity, beginning with the articles in Harper's, precluded carrying on in that way and keeping the issues private."[55] Certainly there was nothing private in the *Scientific Monthly*, a magazine of popular science edited by the psychologist J. McKeen Cattell—indeed, it may very well have been Kellogg's design to confront Rhine in the area in which he had so far been supreme, the public consciousness. Cattell would certainly have been delighted with the article, for he was very much of Jastrow's opinion on psychical research, "altogether set in his mind against the validity of ESP," as one correspondent wrote confidentially to Rhine.[56] Kellogg continued to publish articles critical of the Duke mathematics (his criticisms have been discussed in detail in chapter 7, above), and his example incited a number of other psychologists to join the onslaught.

Inevitably it was Kellogg's original attack, the semipopular article in the *Scientific Monthly*, not his later articles in psychological journals, that attracted the most public attention. What was principally remarked in it was its unpleasant tone and its implication that, properly evaluated, none of Rhine's results—not even the Pearce-Pratt work—was of real statistical significance. However firmly Kellogg might profess neutrality, it seemed clear to Rhine's supporters that the article was an unfair and inaccurate attack. Waldemar Kaempffert gave the criticism more publicity by reviewing it in the *New York Times* in November.[57] The one man who did not take part in the public debate was Rhine himself. When Kellogg's article first appeared, Rhine's friends urged him to respond, and Cattell at first agreed to publish Rhine's own account of his work (though insisting it must not be controversial); but then Cattell demanded major revisions and finally declined to let any paper appear at all.[58] The effect of the publicity was certainly to awaken some doubt in the public mind about Rhine's success, but beyond that it gave unsympathetic psychologists a tangible reason for opposing parapsychology. Kellogg's critique of the mathematics established him as a champion of this school, although many—perhaps without the inclination or the ability to master the subject for themselves—took his criticisms on faith. The more sympathetic were simply bewildered by the debate pro and con. In November, describing the effect of the past few months upon the Harvard psychologists, a correspondent reported that "more than once I have heard a professor say, 'Well, after the Duke experiments I realize how dangerous it is to say anything about probability.' "[59]

Kellogg and a few of his fellow psychologists were clearly disturbed by what they took to be a danger to their profession (and the role of statistics therein). Kellogg complained that "the public is being misled, the energies of young men and women in their most vital years of professional training are being diverted into a side-issue, and funds expended that might instead support research into problems of real importance for human welfare.'"⁶⁰ Hence it is fascinating to find that a number of statisticians were equally outraged by *Kellogg,* whose arguments they saw as discrediting *their* profession. For several years Rhine had been contacting mathematicians for help with the statistics, and during 1937 he had attracted the interest of two relatively eminent ones in particular. Thornton Fry of the Bell Telephone Laboratories, author of a widely used statistical textbook, had met Rhine during the meetings of the American Mathematical Society at Durham/Chapel Hill in December 1936; Edward V. Huntington, professor of mechanics in Harvard's mathematics department, had begun an animated correspondence with Rhine after encountering the statistical side of parapsychology in Wright's *Harper's* articles and had visited the Duke laboratory in September 1937. Both these men found Kellogg's article in the *Scientific Monthly* offensive and misleading at the same time. Huntington was particularly distressed. He himself helped to resolve Kellogg's criticism that Rhine's estimates of significance were based upon an inappropriate theoretical distribution, and he prepared a broad defense of the Duke mathematics for the *American Scholar.*[61]

Fry was similarly unhappy with Kellogg's writings, but he favored a refusal (by Rhine and by the mathematics community) to dignify them with a reply. Eventually Huntington was won around. It was not that they saw any substance in Kellogg's arguments that might be difficult to handle; rather, the danger they perceived in tackling Kellogg publicly was of compounding the literate public's confusion over technical mathematical arguments—a confusion not likely to help Rhine.[62] Rather than continue the polemic, therefore, Huntington consulted with others, including Burton Camp of Wesleyan and S. S. Wilks of Princeton, and drew up a statement, which Camp, as the current president of the Institute of Mathematical Statistics, released to the press at the December meetings of the institute (held jointly with meetings of the American Mathematical Society and the American Association for the Advancement of Science at Indianapolis): "Dr. Rhine's investigations have two aspects: experimental and statistical. On the experimental side mathematicians, of course, have nothing to say. On the statistical side, however, recent mathematical work has established the fact that, assuming that the experiments have been properly performed, the statistical analysis is essentially valid. If the Rhine investigation is to be fairly attacked, it must be on other than mathematical grounds.'"[63] Among his colleagues Huntington was perfectly candid about his motives. "No mathematician and no statistician has seriously complained.... One or two psychologists have published what they hoped would be devastating refutations. I hate to see the prestige of mathematics and physics called into question. I hope that those equally resentful with me over these attacks will join with me in regard to making some statement which will be of use to Dr. Rhine.... We should speak before any more discredit is brought on us.'"[64] His remarks confirm that some statisticians perceived Kellogg's attack on Rhine's mathematics as a threat to their own emergent discipline.

For at this time statistics still had not clearly established itself as an independent professional field. Although statistical methods had been advancing since the early nineteenth century, it was only in the twentieth century that their development was rapid, in the research of such men as Karl Pearson, R. A. Fisher, and A. N. Kolmogorov. American statistics had not yet produced any individuals of this stature, though increasingly it displayed a disciplinary self-consciousness in the 1930s: the second statistical journal to be published in this country, the *Annals of Mathematical Statistics,* was founded in 1930, and in 1935 the Institute of Mathematical Statistics was established as the field's professional organization. However, the statisticians' sense of their own independence and their search for professional recognition were not generally endorsed in other fields. Worse yet, nonmathematicians had an altogether too generous view of what "statistics" included; professional statisticians like Harold Hotelling of Columbia complained that wholly unqualified individuals were regularly being appointed to teach "statistical" courses in departments like economics or psychology. In order to free themselves from the suspicion of mainstream mathematicians, the statisticians felt it necessary to disavow the amateurishness and superficiality with which their subject was being taught and applied in other fields. Huntington's campaign to discredit Kellogg and his allies in psychology is fully understandable when we recognize how concerned statisticians were at this time to protect their subject from amateurs in other fields.[65]

At almost the same moment that Kellogg was exposing Rhine's statistics to critical examination, other psychologists were beginning to attack his experimental technique. The latter was potentially the more vulnerable area of work, although—despite the weaknesses in it that English critics had already identified—it is not at all clear that Rhine felt particularly sensitive on the subject before the real torrent of objections began to flow late in 1937. The first sign of what was to come was the *Saturday Review of Literature*'s October review of *New Frontiers of the Mind,* which its editors had assigned to B. F. Skinner.[66] By 1937 Skinner had already begun to make himself known as one of the leading behaviorist psychologists. After nearly ten years as a graduate student and research fellow at Harvard, he had come in 1936 to the University of Minnesota to teach. Given that Skinner was reviewing a Book-of-the-Month-Club selection, a book written in a decidedly anecdotal and rather heavily dramatized style, his review was remarkably restrained, perhaps because it was clearly meant as a review of the work on extra-sensory perception, not merely of *New Frontiers.* While Skinner mentioned Kellogg's objections to Rhine's method of evaluating chance expectation, he concentrated upon Rhine's claims to have provided unequivocal experimental proof of telepathy and clairvoyance. He argued that Rhine had not systematically excluded all alternative hypotheses that might explain his success and indeed had been willing to introduce ad hoc hypotheses, such as the necessity of the proper psychological atmosphere, to explain his failures. "The result is that Professor Rhine is always very close to presupposing what he undertakes to prove, and his account is thereby seriously weakened for the critical reader."

Just a month later, Skinner came upon an even more devastating source of possible error in Rhine's work. Examining a deck of the commercial ESP cards, he found to his surprise that in certain lights they could easily be read from the back, due to warping caused by the drying of the imprinted ink. He startled first his wife and then the Minnesota Students' Forum with a suddenly discovered "gift" of ESP before revealing the defect in the cards; he then mailed an affidavit to Rhine testifying that he had called one hundred cards correctly in succession, adding the sardonic note: "I suppose you know about this, but we found it amusing. Do you plan to use data obtained with these cards?" Rhine acknowledged that flaws existed and explained that this made screening necessary in rigorous experiments (an announcement to this effect did in fact appear in the December *Journal of Parapsychology*). He assumed that Skinner meant to give the news to the public, but Skinner denied this, calling his discovery "only another example of the kind of thing which is responsible for the failure of many of us to take your work seriously."[67]

Nevertheless, Skinner was certainly troubled at how the public image of scientific psychology might be distorted by the claims being made for ESP. He had just been brought in as liaison officer between reporters and the American Psychological Association during the association's summer meetings, and he had learned something of the dangers and opportunities involved in publicizing science. "In my opinion," Skinner wrote the APA council in October 1937, "the profession is being needlessly damaged in the eyes of the public by its current press reports."[68] Reporters typically looked first for scoops, popular accounts of research, or opinions on sensational issues. Skinner apparently felt that the association had not yet fully recognized how important public opinion was to the support of psychological research, nor that it might be able to form public opinion through the press. If manuscripts of important papers, not merely abstracts, were made available to reporters before their actual presentation, serious science reporters—like Marjorie Van de Water of Science Service or the Associated Press's Howard Blakeslee (who was complaining that "*psychologists are the only science group . . .* which does not have regular machinery for getting at least a fair showing of important manuscripts in advance")[69]—would be willing to give them fuller and more careful treatment. Skinner urged the association to establish a committee to study its relation with the press, and when this was eventually done he served on the committee briefly. Thus in late 1937 Skinner was bound to be particularly sensitive to the damage that favorable publicity for ESP might do to orthodox psychology. He kept his word to Rhine and did not publicize his discovery of the flaws in the ESP cards, but he immediately passed it on to one colleague whom he knew to be preparing a critical assessment of the Duke work, with a comment no doubt evoked by his recent experience with the press: "The more I think of it the greater seems to be the need of counter-propaganda. Although this doesn't disprove ESP any more than Rhine's results prove ESP, it may have the same kind of effect upon the public, which cares very little for proof anyway."[70] The colleague—perhaps the most outspoken critic of Rhine's work—was Dael Wolfle of the University of Chicago.

Wolfle had taken his Ph.D. at Ohio State in 1931 and had gone to the University of Mississippi the next year to begin its psychology department; in the fall of 1936 he had joined the University of Chicago board of examiners to supervise examinations in the biological sciences. Wolfle had done work in the application of statistical methods to psychology, and perhaps for that reason had been asked to review the second (1935) printing of *Extra-Sensory Perception* for the *Journal of Educational Research*. The review, which appeared in the May 1937 issue, refused to admit that Rhine's work yet merited scientific consideration, stressing that Rhine's statistical formulas were incorrect (since a subject's calls were almost certainly interdependent rather than independent) and arguing that Rhine had not excluded the possibility of rational inference (in BT-5, which Wolfle believed was Rhine's usual experimental procedure). Later that year Wolfle agreed to prepare a comprehensive review of the work on extra-sensory perception for the *American Journal of Psychiatry*, and he began to read extensively in the field and to discuss it with those professional colleagues whom he knew to have some involvement with it: Skinner at Minnesota, Estabrooks at Colgate, G. D. Higginson at Illinois, Willoughby at Brown, and Kellogg at McGill. None expressed any belief in Rhine's work, although Willoughby surprisingly warned Wolfle to "handle the matter pretty gingerly"; apparently Rothera's work and Carpenter's recent experiment had made him feel that there might be something yet to be discovered, if only "a puzzling problem in minimal cues."[71] Wolfle had already sent the manuscript of his review article to the journal when Skinner wrote him on 16 November to announce his discovery of the flaws in the ESP cards, and after amazing the Chicago Quadrangle Club with a demonstration of his own "ESP," Wolfle added a description of the flaw at the end of his article, which appeared in January 1938.[72]

Wolfle constructed the main body of his review, not about a presentation of Rhine's various data, which he summarized cursorily in four or five sentences, but about a consideration of the alternative hypotheses that Rhine claimed to have excluded: chance, fraud, incompetence, sensory perception, and rational inference. He recapitulated the objections others had made under these headings, adding several "other possible explanations"—principally hyperaesthesia and recording error. On the whole, Wolfle's analysis gives evidence of by far the closest reading of Rhine's reports in the critical literature on the subject, and it is unfortunate that so careful a review should have been combined with a wide-ranging personal attack upon Rhine, for this brought his article down to a much more unpleasant level. Under the heading "Hypothesis of Incompetence" Wolfle assembled a paragraph of masterly innuendo:

> Rhine concludes that "the Hypothesis of Incompetence will, I think, find few adherents and no justification." There has been relatively little outside observation of Rhine's work (it was observed by Wallace Lee, a professional magician, and by W. H. Wright, Chairman of the Department of English Literature of Columbia University) and Rhine is quite right in saying that it is more important to get confirmation of results from other laboratories than it is to have a committee observe him and his subjects. Rhine writes of his training that he at one time planned to enter the ministry, but that as graduate students he and his wife "turned by common consent to the field of professional forestry. The woods seemed to offer a free and natural life, one in which we might hope to escape the fog of an increasingly dubious philosophy and work out at least a practical formula for existence." A desire to enter the field of psychic research led the Rhines to work with Professor McDougall and to an appointment at Duke. Rhine is not a member of the American Psychological Association. Only three of the 18 authors whose articles have been cited above as evidence for ESP are members.[73]

Inevitably these comments confirmed Rhine in his belief that psychologists were bound to be emotionally biased against him, and indeed it is impossible not to think that Wolfle's gratuitous remarks betray an instinctive disbelief that psychical researchers could ever be taken seriously. They were the sort of comments that some psychologists had undoubtedly been making to one another privately; Wolfle was the first to express them in print.

Wolfle's career had closely paralleled that of Harold Gulliksen: the two had studied together at Washington, roomed together at Ohio State, and had ended up together on the Chicago board of examiners—Gulliksen for the social rather than the biological sciences. They had had a common concern for psychometrics—both attended L. L. Thurstone's Wednesday evening seminars at Chicago—and had developed a close friendship and working relationship. In preparing his review of ESP, Wolfle had constantly discussed with Gulliksen the problems it posed and had repeatedly given him his manuscript for comments, so that the final version was in a real sense a joint production.[74] Late in 1937 Gulliksen wrote a similar appraisal of extra-sensory perception for the *American Journal of Sociology*, recapitulating the principal arguments in Wolfle's article without contributing anything more.[75] In the developing controversy, Wolfle and Gulliksen came to be widely quoted as the most thoughtful critics of the Duke methods. Few people recognized how close they were or that their views had developed conjointly. Their papers were misinterpreted as independent judgments of extrasensory perception and hence seemed all the more damning.

The psychologist critics of ESP found temporary reinforcement in the reports that Science Service now began to issue at the direction of Watson Davis. On 12 October 1937 the news service released to the press four articles inspired by

Rhine's book and the "Zenith program . . . plugging telepathy," articles confessedly prepared as "skeptical reporting of the parapsychological front."[76] The articles pointed out alternative explanations of apparent ESP; called attention to failures of replication by Ralph Gundlach at Washington (who had conducted radio tests in Seattle) and William Griffith at Reed College;[77] and summarized the statistical criticisms just published by Kellogg in *Scientific Monthly* (whose editor, Cattell, was a trustee of Science Service). A second salvo came on 31 December in an account of the Indianapolis meetings of the AAAS that mentioned the mathematicians' discussion of Rhine's statistics in a light tending to weaken the Duke case.

> [E. V. Huntington] finds the mathematical methods used by Dr. Rhine in analyzing his scores, subjected to attack by psychologists, are "good enough for his purposes", although his methods "cannot bear close analysis."
>
> "This doesn't mean you have demonstrated the reality of telepathy, does it?" Dr. Huntington was asked. And his answer was that it did not.

The account included for good measure a rumor that Chicago faculty had been able "to make high scores through images appearing on the backs of cards."[78]

Science Service's psychology editor, Marjorie Van de Water, had been at Indianapolis and had learned there that Dael Wolfle had more information about the transparency of ESP cards. From Wolfle she got full information about B. F. Skinner's discovery, plus copies of the Skinner-Rhine correspondence, and used this as the basis for a third debunking article, for release on 31 January. Entitled "The Camera Takes a Hand at the ESP Game," it used photographs under special lighting to demonstrate how the ESP cards could be read from the back and implied that this explained the Duke success; no account was given of the successful experiments reported where subjects could not see the back of target cards.

Many people interested in the Duke research found these reports disturbingly one-sided. E. V. Huntington reacted to the December release by complaining directly to Watson Davis about the truncation of Camp's statement and the misleading context in which it had appeared. When he received no satisfactory response, he sent a protest at the end of January to the individual members of Science Service's board of trustees.[79] One of these was C. G. Abbot, the secretary of the Smithsonian Institution, who had long been a supporter of Rhine and had himself performed successful experiments in clairvoyance; Abbot had already begun independently to put some pressure on Davis.[80] When Rhine was sent a copy of Van de Water's account of the defective cards at the end of the month, he too protested the slanted reporting. Davis in return invited Rhine to draw up a factual article "discussing general criticisms and current developments [of] your researches";[81] but by this time the trustees were deciding to step in. On 5 February their executive committee met and called Davis back to a sense of what was due the objectivity of science. W. H. Howell, chairman, wrote Huntington apologetically that the committee had "expressed a unanimous opinion that it was not in the province of Science Service to take sides in a scientific controversy, but to report as accurately as possible divergent views as they appear in the literature."[82] Abbot wrote more pungently of the attitude toward Davis: "The position of Science Service was discussed in the meeting of the trustees on Saturday, and although the others present besides myself were either exceeding or moderately sceptical of the reality of the extra-sensory perception, Davis was instructed that Science Service must maintain a colorless attitude towards this and all other controversial questions. . . . He disliked this rule very much, but the trustees were unanimous."[83] Subsequently Davis made no difficulty about accepting Rhine's account of his work for release, although he insisted on changing Rhine's manuscript from the third to the first person—apparently trying to dissociate Science Service itself from the positive report. The release, dated 4 March, struck Huntington as "first class in every respect."[84] Rhine was not entirely happy with Davis's later reporting of his work, but it was certainly much less obviously tendentious.

Unquestionably, these accumulated criticisms had some effect in shaping professional opinion about extra-sensory perception, but they do not seem to have destroyed the willingness of many psychologists to concede the possibility that it might exist. In February 1938 an M.A. candidate in psychology at Southern Methodist University, James C. Crumbaugh, sent out a questionnaire to seventy psychology departments throughout the United States asking for their opinions of the Duke studies and testing their familiarity with other work in psychical research. The answers he received—one hundred replies from forty-three schools—indicated that psychologists were in the main still open-minded about ESP. Sixty-one felt that there was some value in the work, whether positive or negative; fifty-seven agreed that the Duke research might be valid. Very few, however, expressed complete conviction on either point. To be sure, there remained a very considerable group convinced that the claims for ESP were valueless, citing the mathematical and experimental weaknesses that Kellogg and Wolfle were proposing: "Rhine confuses dependent and independent probabilities"; "subject's foreknowledge of all possibilities"; "apparent choice of the wrong probability curve"; "failure to give all pertinent data in reports"; "systematic errors of recording"; "a wish which fathers the findings"; "incompetence as regards experimental method." It is also apparent that comparatively few of the psychologists polled, sympathetic or not, knew much about psychical research; forty-four knew nothing of J. E. Coover (who had just retired at Stanford), and thirty-five had never heard of the *Journal of Parapsychology*. Nevertheless, a definite majority of Crumbaugh's sample admitted not only Rhine's right to conduct his research but the possibility that there might be something in it.[85]

IV

This pattern of tolerance exhibited by American psychologists towards Rhine's research in the years immediately following its first publication can be confirmed by tracing the reaction of the Harvard psychology department to the work, in particular the reaction of E. G. Boring. Boring had been left the one senior psychologist in a department of philosophers by the departure of the more philosophically-minded William McDougall for Duke in 1926/27—a departure that Boring apparently did not find seriously distressing. He was ambitious of the establishment of a truly professional psychology at Harvard, and when the chemist J. B. Conant became president of the university in 1933, Boring found administrative support for his plan to make psychology a department distinct from philosophy, a separation accomplished in 1936. By that time the behaviorist K. S. Lashley had been drawn to Harvard from Chicago, and in 1937 Gordon Allport was promoted to a third permanent position in the department.[86]

As we have already seen, Boring had inevitably become somewhat familiarized with the problems of psychical research during the 1920s. McDougall took him to sittings with Margery, which left him slightly bemused, and when McDougall went to Duke, Boring became de facto supervisor of the Hodgson Fund and of Estabrooks's clairvoyant investigations. The attitude towards the field that slowly formed in his mind was somewhat milder than his mentor E. B. Titchener might have liked: psychical research, he came to believe, is of little scientific value because it purports to prove a negative in that it maintains that certain phenomena—telepathy, clairvoyance, or the like—have been produced by no known normal means.[87] Boring had been growing steadily more operationalist in his view of psychology and science generally; science, he believed, should be concerned only with positive relationships and causal connections.

For this reason, it is remarkable that Boring's first recorded reaction to ESP was one of guarded respect. He wrote in 1935:

> My own feeling about psychic research is that it is made peculiarly difficult by the amount of human emotion that is aroused in connection with it, and that in general this fact would be a strong reason for leaving it alone. My more fundamental reaction to it is that the problems are ordinarily formulated as negatives, the establishment of a proof that something is inexplicable, and I do not see why the Harvard Corporation as trustees of this gift should feel justified in having money spent on problems formulated in this futile manner. I am speaking in general, for it is some of your work that constitutes one of the few exceptions to this rule.[88]

The comment was elicited in the course of a correspondence over the Hodgson Fund, for Rhine was of course very much interested in every possible source of support available for psychical research. Indeed, Rhine shortly went so far as to suggest that Harvard might assign the use of the Hodgson Fund to Duke, or at least collaborate in a research program, so that the fund might "get into proper use." This unwanted advice on how to apply Hodgson monies seems to have irritated Boring,[89] but it did not lead him to slight Rhine's experimental claims, which he remained willing to take with some seriousness. In the fall of 1936 Boring invited Rhine to Harvard to speak to the psychological colloquium—expenses and honorarium to be paid from the Hodgson income.[90]

Rhine reacted to the invitation with a mixture of satisfaction and apprehension. It was the first major sign of acceptance by the uncommitted professional community that he had received, and it marked the culmination of two years in which isolated psychologists at lesser schools had shown a growing interest in ESP. McDougall was naturally delighted at the news. Equally naturally, Rhine felt some concern about the reaction he might expect from a psychology department whose sympathy for psychical research was by no means assured. Boring went out of his way to soothe Rhine on this point: "We realize that in so controversial a subject you would naturally feel some diffidence about coming up against what may possibly be our criticism, but I do not sense any antagonism of this sort within the Department. Everyone seems to feel that a positive result has been got and that an explanation for it is needed, since to call it clairvoyance or telepathy is simply a polite way of saying that we do not know how it occurred. ... What we really want is advice about extra-sensory perception, and that is why we would like to import you."[91] The colloquium was eventually scheduled for 18 November; dinner and discussion would follow at the Harvard Faculty Club.

The result was unquestionably a triumph for parapsychology. Rhine returned to Durham with a sense of having largely won over—or having reduced to perplexed silence—a group that had been prepared to find problems with the work. Boring's correspondence with Rhine in the aftermath of the colloquium similarly reveals a certain sense that the Duke work might have something to offer psychologists. In particular, Boring was impressed by the same evidence that had struck Donald Adams in Hans Bender's paper: the revelation that extrasensory perception seemed to share certain features with ordinary sensory perception. Boring stressed three in particular that seemed important to him:

> (1) SP loses in precision when the O[bserver] is not encouraged by being given easy judgments ... [that is, in SP as in ESP] the O needs to be reassured that he is doing well or he breaks down and gets worse. (2) Subliminal SP is very difficult and is easily upset by fatigue, distractions, etc. ... (3) ... it does not take attention to make accurate judgments in psychophysics. The O may be occupied with something else and knows about his judgment best by listening to his voice talking, as if the utterance were automatic. This state is easily acquired in a long series of judgments, and might resemble what you get in ESP.[92]

These similarities, which Boring had not anticipated, made him the readier to see extra-sensory perception (whatever that might mean) as a legitimate part of psychology.

What still disturbed Boring about Rhine's conclusions were philosophical rather than metaphysical difficulties. To begin with, he still felt that Rhine was trying to prove a negative, to "reify the difference [between chance and extrachance results] into an entity called ESP."[93] The objection took Rhine by surprise, since he had long prided himself on a pragmatic avoidance of hypotheses. Hence Boring took the pains to explain a little more fully what he found unsatisfactory.

> What I am driving at is my intuition of the mental attitude of you and your colleagues in this research, and, if I can make it clear to you what I think it really is, then you can readily correct me if I am wrong. I think that you feel that when you have got a significant difference, you have got Something. It is here that my charge of reification lay. What you have got when you have this difference is just this difference. ESP is a negative concept; it states a correlation but gives no systematic implications of the correlations, so that the fact remains isolated. Most isolated facts of this sort are scientifically uninteresting, and that is why, I think, the critic of your work inevitably tries to psychoanalyze you, seeing why you take an inexplicable correlation as important when it is usually the explicable implications that are scientifically important.[94]

Moreover, Boring had always been unwilling to concede that statistical laws functioned satisfactorily as scientific explanations,[95] and now (as Willoughby had already done) he raised objections to Rhine's automatic assumption that a chance ESP score should be five out of twenty-five.

> You see you say, "If nothing intervenes but known processes of nature, an average of approximately 5 may be expected to occur." My point is that you do not know the processes of nature that make the 5 occur. You have certain conditions under which the 5 occurs, but the essential thing that makes it come up 5 you do not know, and this is why the 5 is quite as mysterious as the 9. In coin tossing the analogy is that you do not know what makes a particular throw heads instead of tails. In general my point is that you have not got along very far scientifically until you can reduce to the particular form the statistical average.[96]

Boring discussed this wider point with other colleagues in early 1937, without finding much agreement. His objection went far beyond the special case of Rhine's work, as he himself readily recognized: "It is a criticism that the statistical method never leads to a full understanding of the nature of phenomena, but I think perhaps we had better stop discussing it, since it looks as if no one is ready to accept it."[97] About the phenomena themselves, however, Boring expressed no reservations.

Indeed, whatever doubts the Harvard psychologists might have had about the good faith or the competence of the Duke researchers seem to have been dispelled by Rhine's enthusiastic report of the successes recorded elsewhere. In response to an inquiry from Rhine as to how Boring and the department would react to the prospect of seeing ESP research at Harvard, Boring replied:

> You ask about my personal attitude toward having psychic research being done here at Harvard, and you distinguish between my attitude and the Department's. The trouble

is that the Department's attitude affects mine. It had gotten worked up to the stage where it might have undertaken something if there had been only the Duke results, but, when it found so much going on at so many other places, it rather lost interest because it thought it could not do anything crucial. (And we had not heard about the Stanford Fellowship then.) If the Department were with me, I think the simplest thing would be to put somebody on some of this work that you have started, or perhaps take the psychophysical problem and see if that could be settled. We would be using the Hodgson money for just what the donors intended. However, I do not feel that I have the energy to support a move of this sort if the Department is not back of it. I have lots of other things that I am much more interested in, and I do not want to be the sole person responsible for what is done, with the constant checking of results and fussing over publication, because it is an emotional subject. Moreover, we are very short of space in the Laboratory, and there is a question about displacing other things for this in that respect. In other words, you see I am well disposed to research of this sort if it does not require too much effort on my part, too much special responsibility, too much exertion of pressure upon other people. It just does not seem to me to be worth all that.[98]

The passage is extremely revealing of the way in which the Harvard psychologists had responded to Rhine's work, treating it much as they would have any piece of research interesting but not directly related to existing research at Harvard and apparently well in hand elsewhere. It would be ironic if Rhine had been so convincing that he persuaded the Harvard audience not to look into it themselves.[99]

The Harvard psychology department would undoubtedly have been willing to leave psychical research to itself at this point if outside, nonprofessional interest had not forced them to go further. Public interest has regularly been the most important force constraining scientists to go against their instincts and spare some thought and time for what they would ordinarily find peripheral to their real interests. In this particular case it was Henry James, William James's son, who was able to push Harvard further into the support of psychical research. He was in a position to do so because his role in Harvard affairs was considerable. He had graduated from the college in 1899 and from the law school in 1904; as a young lawyer in practice he had "had a good deal to do with drafting" the letter of gift establishing the Hodgson Fund in 1912. In 1922 he had become one of the overseers of the college, and in 1936 he was selected as a fellow of the Harvard Corporation; he thus had come to have an important voice in the affairs of the university.

In the spring of 1937 James seems to have convinced Harvard's President Conant that the status of psychical research now warranted Harvard's consideration as to whether the Hodgson monies might be usefully expended in some way. Conant, more than the psychologists, was prepared to believe that the current reports of extra-sensory perception indicated the existence of a phenomenon that merited further study. To a graduate student in the psychology department he confided that as a chemist and a physical scientist he just didn't know how to cope with ESP. He believed that Rhine was honest, he believed that he was doing his best to collect data correctly. He had the assurances I guess of the American Statistical Association or something that the statistics being used were acceptable, and therefore he couldn't personally understand what was happening. So he thought we should do in psychology or parapsychology what was done in the physical sciences. When you got something you didn't understand you called it an effect, and you put the name of the discoverer on it. So he wanted to call ESP the "Rhine effect," and he said then you just act as if it were a fact and try to connect it in with the rest of science and in the meantime refer to it as the "Rhine effect."[100]

It was very much in this spirit that he supported James's proposal and encouraged the formation of an informal committee of Harvard scientists to discuss the use of the Hodgson Fund. Eventually the committee came to include Boring (chairman) and Lashley from psychology; the astronomer Harlow Shapley, who had been a friend of L. T. Troland's and had been one of Margery's investigators in the 1920s; E. B. Wilson from chemistry; the mathematician G. D. Birkhoff, dean of the Faculty of Arts and Sciences; and F. L. Hisaw from zoology.

The committee was never happy about its responsibilities. Boring himself was probably most sympathetic, and even he continued to feel that the purposes supposedly served by the fund were based on a fallacy, that of trying to prove a negative; as he expressed it to James, any successful use of the fund would be a defeat of its original purpose. James replied to Boring in a very long and constructive letter, insisting that the intent of the fund had been simply to support study of psychical phenomena whose origin "appears to be independent of the ordinary sensory channels"—it did not exclude the possibility that a physical interpretation for the phenomena studied might ultimately be forthcoming. The objections of the remainder of the committee James treated a little more sharply.

> Wouldn't it be a little half-hearted to shy away from the situation just because, as Shapley says, there have been numerous delusory phenomena in the history of biological and natural science; or because of the danger which Lashley points out that the work may tax the emotional energy of the Committee; or because, as Hisaw remarks, research men have not infrequently wasted their time on hoaxes. Such anxieties are natural enough at this stage, but don't they vanish if one can decide on a sensible and practical experimental research procedure? . . . Rhine and his associates appear to have opened a sort of experimental lead. It is about the first good experimental lead that has presented itself in the area to which the Hodgson Fund is dedicated.

Boring agreed that if a man trained to do scientific work could be found who had a strong experimental hunch to follow up, it might be worth supporting his research at this point.[101]

It was at this moment, in the spring of 1937, that Gardner Murphy inquired of Boring about the possibility of having Saul Sells, a recent Columbia Ph.D. then teaching temporarily at Teachers College, work at Harvard with Hodgson Fund support.[102] Boring rejected the idea, for the committee had not yet come to any conclusion about how Hodgson funds should be used. But by year's end it had adopted James's point of view and had begun to support Ernest Taves, a Columbia undergraduate, in a program of experimental psychical research supervised by Murphy. The experiments involved a sort of card-guessing by unselected subjects, in a rather Rhinean fashion (except that the target cards were not ESP cards), and had yielded mostly chance results by June 1938, when Murphy and Taves reported on their work. Even so, the committee agreed to subsidize the research for a second year.

It must not be thought from this account that the committee ever became markedly enthusiastic about psychical research. Sympathetic to Murphy, perhaps: "I think we all feel in the committee," Boring wrote Henry James, "that Murphy is a reliable investigator, not merely honest but superior to ordinary suggestion and prejudice, and yet a man who goes at the thing with enthusiasm and sympathy."[103] But they kept their own involvement in it carefully at a minimum. Boring was quite candid: "We are all working a little reluctantly from a sense of duty, more or less out of a devotion to the spirit of fair play and tolerance and the belief that closed minds sometimes prevent discovery. But I do not have the time or patience to read the numbers of the new Journal of Parapsychology, which is ever so much more scientific in its approach and its welcoming of critical articles than are the S.P.R. journals."[104] Nor, one may imagine, did anyone else. The chairman was repeatedly thankèd for getting Murphy to make *short* reports. The lack of enthusiasm felt by the committee was not properly understood by Murphy for a long time; he not unnaturally was overjoyed by the professional sanction and support being given to psychical research, and he made a point of keeping the Harvard committee informed of his progress. Boring had to urge him to keep his reports brief, and at one point late in 1938, when Murphy apologized for a two-week delay in sending in a report over the summer, Boring replied testily that the committee really cared very little about the matter and could quite well wait for several months: "[It] would much rather have the reports that it asks for than reports that are volunteered . . . as if you were trying to create an obligation for the Committee to read them. There is nobody at Harvard that is tremendously excited over psychic research, and it is my impression that the members of the Committee would readily resign from a not too welcome task if there is much work to do."[105] Once having been aroused by President Conant and Henry James, the committee showed itself willing to support research by a competent colleague and to leave its details very much alone.

At the same time, the two psychologist members of the committee had become aware that in some respects certain real concerns of mainstream psychologists—perception and behavior—might overlap with those of psychical researchers and that the Hodgson funds might reasonably be employed in these cases. Carl Murchison had had similar thoughts in requesting Smith-Battles money for his chicken work at Clark ten years before, but with far less justification, for as it happened both Boring and Lashley *did* actually have an interest in psychological problems analogous to those being raised in parapsychology. Twenty-odd years before, Lashley had done some research with John B. Watson on the homing of birds, and during 1937/38 he had begun to advise a graduate student in biology, Donald R. Griffin, on a thesis concerned with the homing of Leach's petrels.[106] There was surely no one in the psychology department less likely to take extra-sensory perception seriously than Lashley, "an adamant behaviorist who argued that even mental events and processes were coextensive with physiology and behavior." Nevertheless, as Lashley realized, in the terms of the Hodgson deed of gift, as clarified by Henry James himself, homing should be a legitimate study for the psychical researcher; "it seems to be almost extra sensory in the sense that no one can find out what sensory cues are used." The words are Boring's, seeking Henry James's endorsement, and James raised no objections.[107] Consequently, during the summer of 1938 the committee awarded Griffin two hundred dollars to support a field trip to Grand Manan Island to study Leach's petrels. Griffin, like Lashley, remained unimpressed by the evidence for extra-sensory perception, and in his 1942 dissertation he did not even include the parapsychological among those explanations of homing behavior that merited consideration.[108] But in whatever spirit the work was actually done, it was certainly research that parapsychologists would readily have considered a legitimate and important topic in their own field. Indeed, in the 1950s Gaither Pratt carried on his own research (with inconclusive results) into the possibility that telepathy might be used to explain the homing of pigeons.[109]

Boring, like Lashley, was becoming conscious that some of his concerns bordered those of parapsychology, and he too began to inquire into using psychical-research funds to support research in these borderline regions. In 1937/38, James Grier Miller, an M.A. candidate in psychology who was already something of a protégé of Boring's, had launched upon a program of experiments upon the nature of unconscious behavior that would eventually culminate in his Ph.D. dissertation; he began by testing whether subjects could discriminate visually between target figures without being aware that they were doing so.[110] Apparently at President Conant's suggestion, Boring and Miller applied for and received a grant from a minor, almost unused psychic-research fund to help with costs of the study of unconscious perception. When Boring raised the question of Hodgson support with Henry James, however, James was discouraging, saying that since subliminal perception could be shown to take place through normal sensory channels, it could not be a legitimate problem for the Hodgson Fund. While Boring protested mildly that subliminal perception could *appear* to be independent of the normal senses, he did not carry the question further.[111]

The very form of Miller's experimentation shows that he and Boring were willing to apply the techniques of parapsychology in contexts devised to meet their own particular needs. Miller arranged his test for subliminal perception by telling his subjects that they were being tested for clairvoyance and asking them to visualize the target card—taken from an ESP deck—as if it were displayed on the milk-glass screen facing them. Actually, he was reflecting images of the ESP cards upon the translucent screen at variable strengths of illumination below the limen of awareness to see whether as illumination increased, the number of correct guesses would also increase, even though the subjects remained unaware that an image was being presented to their sight.

Miller was careful to begin with "a long run of the cards . . . without any illumination at all to be sure that S was not gifted with ESP"; then with increasing illumination increasing success at guessing did take place until the subjects suddenly recognized with surprise that the screen actually *did* hold a visible image. The use of the ESP cards and the experimental exclusion of clairvoyance were described perfectly matter-of-factly, with no embarrassed hint that this required any sort of defense from a psychologist; ostensibly, at least, Miller's language suggested a perfectly straightforward borrowing from another area of normal research.[112] Zener's selection of target figures was proving of some utility for the experimental psychologist.

It would obviously be absurd to suppose that Lashley and Boring thought they were really engaged in psychical research; no doubt they joked privately about their supposed involvement within the field. But however little they believed its theories or shared its goals, they could recognize in it a remote but genuine relationship with their own activity. Many of their professional colleagues in the mid-1930s seem to have felt the same way. They did not dismiss the subject out of hand, as the previous generation of psychologists had done; rather, they reserved judgment until further information was available.

CHAPTER TEN
Towards Professional Acknowledgment

I

It would be unrealistic to expect that subjects as far removed from normal experiences and beliefs as was psychical research could ever move directly from rejection to acceptance by the scientific world. Even under the best of circumstances, we might more reasonably anticipate a period of transition during which scientists soften outspoken criticism of the new subject and concede the legitimacy of its study, without actually being willing to accept its conclusions. Eventually, after further development of the subject, integration into mainstream science would come more easily. Whatever the difficulties with this model, it does to some degree fit the case of parapsychology at the end of the 1930s. There is considerable evidence to suggest that a widening discussion of the field was convincing many American psychologists that it needed to be taken seriously and that it should be given at least auxiliary status in their profession until further experimentation had taken place.

Much of this discussion was stimulated by a small number of insistent critics of parapsychology, among whom perhaps the most visible was John L. Kennedy, who had replaced J. E. Coover in the fall of 1937 as the Fellow in Psychical Research at Stanford.[1] Kennedy, who had just completed his Ph.D. at Brown, had had no previous exposure to psychical research. Having planned since junior college to become a "scientific" psychologist, he had taken a heavy undergraduate load in biology and mathematics as well as in psychology in order to prepare himself. In his two final undergraduate years, which he spent at Stanford, he was research assistant to Paul Farnsworth and Calvin Stone and was generally "indistinguishable from a graduate student." Coover never crossed his horizon there. When he went on to Brown as a graduate student, he was put to work for his dissertation on a study of sensory functioning in which Leonard Carmichael and Carl Smith were interested, and he applied for the Stanford fellowship as one possible option for Depression employment. When he learned to his surprise that he had won the appointment—to be sure, he was not precisely an unknown quantity to Lewis Terman and the Stanford department—he quickly began to read in the literature of psychical research: Coover, Carl Murchison's *Case For and Against Psychical Belief,* and Rhine.

The Stanford psychologists made it clear to the new Fellow that they would like him to concentrate his efforts on the study of Rhine's ESP; in the main, like

Terman, they were skeptical of the Duke results but puzzled by them as well, and they were anxious for the puzzle to be cleared up. Kennedy found this attitude entirely natural, and he too approached ESP as a scientific anomaly calling out for resolution: How is it that certain researchers have claimed positive results from an experimental situation that for other experimenters has yielded only chance results? It was, as he himself has put it, a chance to demonstrate his scientific muscle. For the first time, he met Coover: Coover was "circumspect" in his relations with Kennedy, but they had many discussions of the problems, and Coover steered him to the appropriate literature. Inevitably, sensitized by Coover, the department, and his own training, Kennedy looked for ways to reduce the abnormal to normal, to explain the occasional extra-chance results— that is, to identify sources of error. He reported his first results in December 1937 to the meetings of the AAAS at Indianapolis: of one hundred college students tested for clairvoyance, only one scored significantly beyond chance expectation, and *he* proved to be using subtle sensory cues. "Just what these cues are," Kennedy concluded, "and how they became effective remains to be determined." In a paper published in the spring of 1938, Kennedy verified photographically Skinner's claim that the backs of the commercially manufactured cards bore visual cues.[2]

By this time, however, Kennedy's research had taken a new tack, for it occurred to him that another explanation of strikingly positive card-guessing results might lie in regular errors made in favor of ESP by experimenters anxious to prove its existence; such errors, he argued, were most likely to occur precisely in those test situations in which reported success had been greatest (for example, open matching). Coover reacted enthusiastically to this hypothesis and encouraged Kennedy to pursue it experimentally. In February, just after Coover's death, Kennedy circulated a preliminary report of his findings.[3] He had counted the recording errors made by three different agents in his own ESP tests: two, he admitted, made practically no or very few mistakes, but in 4,600 trials the third agent made sixteen recording errors favoring ESP and only one against. An average of .087 errors in every run of twenty-five cards was of course still not significant, but Kennedy concluded somewhat precipitately that "increasing the number of trials, assuming the error rate is constant, will eventually cause the excess to be significantly above that expected from the laws of chance. These results are not striking but they do illustrate the possibility of finding selected agents who may make these unnoticed errors consistently." Charles Stuart later commented teasingly: "It is enlightening to imagine what would be said if an ESP investigator reported the results of tests of three subjects in the form: 'One subject chance, one slightly above chance, a third with an average of 5.087 for 184 runs. Since the latter is an experimentally-determined rate for ESP, in 10,000 runs the scores would be highly significant. These not striking results do, however, illustrate the possibility of selecting consistent ESP subjects!' "[4]

At the meetings of the Western Psychological Association at Eugene, Oregon, on 17-18 June 1938, Kennedy delivered several talks (of varying degrees of formality) dealing with his investigation of ESP: he treated his attempt to replicate Rhine's results, his work on the recording-error hypothesis, and a study of "unconscious whispering" as the factor possibly explaining marked cases of success. Nor was Kennedy the only critic of parapsychology to speak at these meetings. He was joined by Knight Dunlap of U.C.L.A., an old crony of Coover's. Dunlap had thirty-five years more experience in psychology, and the additional years gave Dunlap a very different perspective on Rhine's claims. Having studied Margery closely with Henry McComas in 1926, Dunlap was entirely familiar with the extravagant claims that some psychical researchers could make. He took Rhine's popular appeal to be one more manifestation of the immemorial public fascination with the occult. What worried him more was the professional recognition that ESP had begun to claim, a disturbingly new phenomenon that to him pointed up the problems of contemporary psychology. For the Duke work itself—"the Rhenish combobbery," as he contemptuously put it—he had only sarcasm (although his summary of Rhine's procedure is inaccurate enough to suggest that he had not really read the parapsychological literature with any care). From the now-standard list of "errors" in statistical and experimental procedure he singled out Rhine's supposed selection of data as most fatal to a properly conducted plan of research.

How was it that, despite so obvious a procedural failing, not only Rhine but a certain number of other psychologists had accepted the validity of the Duke experiments?

> We can readily understand the attitudes of the two or three middle aged psychologists who have rushed into print with acclaims for Rhine, and confessions of long standing beliefs which they now feel it safe to proclaim. These men have in other regards been open to suspicion, and considered on the fringe of psychological respectability. Furthermore, I believe, they have all been, in the past and at impressionable ages, apprentices of McDougall, and came under the spell of his attractive personality... ... These men, we can account for. But what about the larger number of men of the younger generation, who... are converts to the Rhenish faith, repeating naively Rhine's experiments with his fallacies? Presumably, these men have had graduate training in psychology, but it has done them little good. Further, it seems true, that many persons who have had one or more undergraduate courses in psychology have from them received no protection against the occult racket, and no basis for evaluating evidence in the field of psychology. Something seems to be wrong with our systems and methods of instruction and training.[5]

Selection of data, he said, is not confined to parapsychologists; it may be found in biology and psychology as well, and professional scientists must learn to take the requirements of experimental method more seriously.

This was in fact the real burden of Dunlap's paper. The success of ESP was actually symptomatic of the decline of tough-minded experimental psychology, which had once ruled at the heart of the profession. For this decline there were a number of contributing reasons. The "hoary superstitions revived by Freud" had helped to blind students to the strict demands of psychological problem-solving; certain ancillary fields, like mental testing, had begun to infringe upon the name of psychology, so that it had become possible, Dunlap snorted, "to get a doctor's degree in psychology by devising and scoring a new intelligence test or personality test"; other disciplines—zoology, sociology, education—had begun to dismember psychology at certain schools by laying claim to the animal or clinical research originally done by experimental psychologists. In the present confused state of the profession it was thus perfectly possible, if humiliating, for a man trained as a plant physiologist to win acceptance as a "psychologist" and all the easier for one more occultist fad like parapsychology to gain public support. Dunlap suggested a sound test for extra-sensory perception and sarcastically expressed his willingness to study any telepath sent to him from Duke, but his concern was less to discredit the parapsychologists than to recreate and reestablish psychology itself.

The meeting at Eugene was not the only professional meeting that spring at which ESP was brought under discussion. At the meetings of the eastern branch of the APA, at New York University on 1-2 April, three different speakers had attacked Rhine's research. F. H. Lund of Temple had merely reported that in strict testing of 596 subjects for extra-sensory perception he had found no single consistent high scorer. Steuart Britt of George Washington University and Hyman Rogosin, a graduate student, had been much more caustic. Rogosin attacked Rhine for selecting his data and suggested that his position under an eccentric like the Lamarckian McDougall made his work all the more suspect; besides, Rogosin argued to the audience, astronomical extra-chance figures could never prove the existence of ESP. Britt went still further in his attack upon the work, raising the possibility of fraud on the part of Rhine's subjects and making much of the fact that the ESP cards could be read from the back. Some of his hearers, indeed, thought that Britt had gone too far and complained that his dogmatism was unscientific.[6] Perhaps sensing a shift in mood among his colleagues, Joseph Jastrow now sent out a questionnaire to canvass them for their "expression of opinion of Dr. Rhine's work advanced in support of E.S.P." He inquired specifically about three issues:

1. In re: the possession of special or unusual psychic powers by certain individuals, absent in the ordinary endowment.
2. In re: the probably or demonstrably suspicious factors or sources of error in the experimental procedures employed in the Rhine results, including adequate control and selection of data.
3. In re: the effect of publicizing these findings without indication of their acceptability to professional psychologists.[7]

Jastrow's characteristic interest in discrediting psychical research is perfectly apparent in his phraseology. Indeed, he may have selected his respondents to that end: no questionnaire was sent to Rhine until he requested one, nor, apparently, to anyone else on the Duke staff. What sort of response his loaded questions drew, however, is impossible to say. Some months later Jastrow claimed he had received 160 generally negative replies, a summary of which he promised to publish in a chapter of a forthcoming book, but they seem never to have appeared in print.[8] Psychologists may have shown themselves more tolerant of psychical research than Jastrow cared ultimately to reveal.

The feeling expressed by men like Wolfle and Dunlap that Rhine did not deserve to be treated like a professional psychologist was to some extent reinforced by what they perceived to be Rhine's own attitude. In *New Frontiers of the Mind,* for example, he seemed to be speaking somewhat disparagingly about the weaknesses of the larger field.

> In justice to my fellow psychologists (whom I have perhaps adopted with greater equanimity than they can master in adopting me) it needs to be said that they have problems to solve and conditions to contend with which are incomparably more complicated and perplexing than those of the biological and physical sciences .... But the science of psychology (if it may be so described) suffers from the peculiar difficulties of its problems and the reluctance of those who are exploring them to abandon either the old guardrails of philosophy or the alluring analogies of other and more advanced sciences.
>
> Whatever may be the reason, our psychology, our knowledge of the human mind, has not yet reached the point where we can make precise statements of general principles in the field.[9]

Moreover, some nonpsychologists like Kaempffert published similar remarks, alienating in the process psychologists who might otherwise have remained neutral in the dispute.[10]

In February 1938 Donald Adams wrote from Swarthmore (where he was spending his sabbatical year) to warn Rhine of the anger his language was arousing and to argue testily that "you can hardly expect people who know something about the field to hasten to accept as a 'fellow psychologist' any one so ill informed as to make such a statement." In a subsequent letter Adams explained himself at greater length.

> I do not believe that psychology should be or can be Quantitatively precise in the sense of physics, or for that matter that it will (or can) even use the same or any other metric geometry. But because a kind of measurement is possible in one subject matter (physics) that is patently not possible in another (psychology) is no reason to sabotage (with the damning *ad hominem,* "Physicalistic!") all efforts at precision in the ways that *are* appropriate to the latter. Yet that is just when you have been doing for years, and then turn round and upbraid the people who are actually working at the matter, for want of precision! I am quite aware that the sabotage is not malicious (as in some quarters it is) but proceeds simply and solely from want of interest to lead you to look into what psychology is up against and trying to do. Nevertheless—oh! I had better quit before I get really worked up.[11]

Adams's reaction was no doubt colored by his resentment of Rhine, but it captures well enough the difficulty that psychical researchers had always had in reconciling an admiration for the precise methods of the physical sciences with an antimaterialistic philosophy. There was bound to be some discontent at the prospect of merging parapsychology into psychology: either it would lose its scientific character, or it would be tied to some materialist, physicalist basis.

Rhine himself, then, saw no urgent need to associate the fields closely. It made no real difference that others were less aloof—as we have seen, Gardner Murphy was actually appointed to the council of directors of the American Psychological Association in 1937, and Gaither Pratt had been accepted as an associate member of the APA. But both these men had had a training in psychology and viewed themselves as straddling both fields. Murphy, moreover, was convinced that psychical research would turn out to be a segment of the larger field and therefore that psychologists could be both used and instructed; Rhine was far less confident of this. Though Murphy pressed him to apply to the APA for membership, he did not do so, and only in February 1938 did he agree to apply for associate status.[12]

However, this professional criticism of the Duke research in fact had very little success in turning professional opinion against parapsychology. This is apparent from the results of yet another questionnaire meant to appraise attitudes toward ESP, this one sent out from New York University by Lucien Warner and C. C. Clark to the 603 full members of the APA in July 1938. Three hundred fifty-two of those addressed returned the completed form (half of them signed), which attempted to survey the attitudes of psychologists to extra-sensory perception, the basis for their judgment, and their criticisms of experimental parapsychology.[13] Only five accepted ESP as "an established fact": Lucien Warner, Gardner Murphy, Barbara Burks, C. B. Bliss, and John Gray Peatman.[14] But nearly half the remainder conceded that it might exist. Twenty-six admitted that it was "a likely possibility," including Robert Yerkes and W. S. Hunter. One hundred twenty-eight called it "a remote possibility," among them Lewis Terman, R. R. Willoughby, Harry Helson, R. S. Woodworth, Edward Tolman, and Leonard Carmichael. To be sure, many of those who called ESP a "remote possibility" heavily stressed the word *remote;* but only 51 insisted that it was an impossibility, and 142 took the easy option and labeled it "merely an unknown." Thus only one in seven admitted to the adamant opposition parapsychologists anticipated from the psychological community in general (thirty of the fifty-one acknowledged that their hostility was based on "purely a priori grounds"). More than half the respondents said that their opinions had been formed by reading "J. B. Rhine's book" (presumably most had in mind the recent *New Frontiers of the Mind,* not *Extra-Sensory Perception*), yet nearly half said, too, that they had studied experimental research reported in the *Journal of Parapsychology*. Most encouraging of all, perhaps, 90 percent of the respondents agreed that the investigation of extra-sensory perception was a legitimate scientific undertaking, and the great majority of these accepted it as "within the province of academic psychology."

The fact that so many of the questionnaires were returned signed permits a few further interesting prosopographical generalizations about their authors. It does not seem to be true, for example, that younger psychologists were more sympathetic to ESP than their elders: the average age of those who accepted extra-sensory perception as likely was about fifty-one; of those who rejected it as impossible, fifty. What differentiated the groups was their professional subfield. Rhine found a number of his supporters among the clinicians and abnormal psychologists, many of whom had had congruent experiences with their patients. But virtually all of the experimenters save W. S. Hunter and Christian Ruckmick rejected Rhine's conclusions, complaining about what they took to be inadequate precautions and weak scientific methodology in the published work. To this extent, at least, the attacks of C. E. Kellogg and his successors had begun to have their effect.

Still more interesting than these generalizations, however, are the suggestions of individual writers, for their often quite pungent comments on the questionnaire highlight the issues that in the summer of 1938 were most closely identified with ESP in psychologists' minds. Some show that a few psychologists automatically classed parapsychology with the study of the supernatural, as extra- or para-scientific: "[It] belongs with alchemy; inheritance of acquired characteristics, with witchcraft, demonology, etc." (Karl Dallenbach of Cornell). Others raised questions about Rhine himself. Many respondents expressed distaste and even contempt for Rhine's personal involvement with the subject, and a feeling that his uncritical devotion to it made his results immediately suspect and discredited psychology. "If [such experimentation is] done at all," urged John Dashiell, Rhine's neighbor at the University of North Carolina and president of the APA for the year, "[it] should be done by others than those also interested in Telepathic horses, Divining rods, Haunted houses, Lamarckian heredity." Rhine's association with McDougall was clearly not a wholly unmixed blessing. Even more widely resented was the publicity and commercialism that had become associated with ESP and with Rhine: "The use of radio ballyhoo and similar publicity justly subjects the whole business to suspicion and discredit. (This is a very mild way of putting it)" (unsigned, from Memphis).

Beside this emotional, almost instinctive hostility, however, the questionnaires reveal in the majority of respondents a serious concern for the substantive problems raised by the field. For those who reacted emotionally against ESP, there were almost as many who recognized and regretted the same tendency in themselves or their colleagues: "I deprecate ... the disposition among certain psychologists (see Jastrow's recent biased inquiry) to take a dogmatic and even vindictive attitude toward those who are interested" (John Fletcher, Tulane). Many writers took the trouble to suggest not merely respects in which experimental controls might be tightened up but types of research whose possible value had been indicated by their own professional experience: experiments using more significant material than geometrical figures, checking the possibility of suggestion or subliminal perception, and studying the state of mind and attitudes of experimental subjects.

One matter in particular that clearly perplexed these psychologists was what Rhine's data and terminology really meant. For some, "extra-sensory perception" was inescapably a paradox. The questionnaire had offered two alternative definitions of the phrase: "a) the acquisition of perceptual knowledge without the aid of the recognized senses" and "b) the extra-chance coincidence of 'stimulus' to 'response' when the former is presented under such conditions that it can effect no known sensory endogran [endorgan]." Some respondents were willing enough in principle to believe that other senses or sense organs might remain to be identified, but they doubted that that was what Rhine really meant by "extra-sensory"; and if he did not, they found his language meaningless. Esther Uhrbrock wrote, "From a biological viewpoint I cannot conceive of any perception which is 'extra-sensory' if one means by sensory all degrees and combinations of sensitivity...." Others found the two definitions offered quite inconsistent, and their choice of which to favor reflected their psychological orientation. Some denied that *b* could be a definition at all—it was "like defining intelligence in terms of mental tests," someone wrote scornfully from Urbana. Just as many, however, rejected *a* as "impossible," "unscientific," or simply "the bunk!" and maintained that *b* was more factually descriptive. This second, rather positivistic attitude was undoubtedly strengthened by the operationism then gaining influence in psychology, a methodological program that held that psychological concepts were properly described only in terms of the operations used to test and reproduce them. In any case, there was no accord on what Rhine's phrase might mean. Carney Landis of Columbia spoke for a few in commenting that "in my opinion Rhine has found something but incorrectly explained his finding. The analogy is the discovery of hypnotism by Mesmer—who explained it as animal magnetism."[15] For these respondents, the data remained a legitimate subject for future interpretation, however unsatisfactory Rhine's explanatory language might seem.

There is, finally, one theme recurrent in these comments that suggests the answer to a puzzling question implicit in the scarcity of psychological reaction to *Extra-Sensory Perception* between 1934 and 1937—that is, why psychologists (with few exceptions) before C. E. Kellogg had not taken up Rhine's challenge. As Dorothea Crook of Skidmore succinctly put it: "Suppose ESP has been firmly established by the positive results so far obtained—So What?" From reading the apologies of psychical researchers or the credos of men like Joseph Jastrow, it is easy today to assume that ESP was a subject that deeply *mattered* to anyone who was concerned with the activity of human understanding. Yet this is not so. Many psychologists simply found extra-sensory perception—true or false—an uninteresting or unimportant problem in comparison with the current frontiers of psychological research.

It is my feeling that psychologists should not waste their time on this type of thing when there are so many other important and worthwhile problems to solve. Life is too short and time is too precious to be fritted away on this type of problem. [Andrew W. Brown, Illinois]

Too many worthwhile topics are clamoring for the psychologists' attention. [Luton Ackerson, Institute for Juvenile Research, Chicago]

It had taken the public furor over parapsychology to make the profession think about taking a stand on the field (even so, many frankly admitted they were entirely uninformed), but now most would have agreed with Willard Olson of Michigan "that the whole business is improbable but that someone must satisfactorily explain [away] the data."[16]

As secretary of the APA, Olson was in a position to act for the profession, and he had already arranged for a session to consider ESP at the September meeting of the association by the time the Clark-Warner questionnaire was sent out. To organize it he had appealed to John Kennedy,[17] Stanford's new Fellow in Psychical Research, who had taken up the field with energy. As Kennedy planned the session, it was not to judge overtly the validity of the Duke results or of ESP but to consider what had come under recent attack, that is, the methods used in ESP research. As finally worked out it was in fact entitled "Experimental Methods in ESP Research." It was to begin with six ten-minute papers, more or less paired opposing and defending methodology, and was to conclude with an hour's informal discussion of the subject.[18]

The event was obviously crucial, and most of the staff of the parapsychology laboratory—Rhine, Gaither Pratt, and Greenwood—went up to Columbus, under some apprehension as to their reception. The ESP session was not until the final day, but included among the papers on the first day, 7 September, was one that awakened all their suspicions: Clarence Leuba of Antioch, who had written so hopefully to Rhine in 1935, spoke to emphasize how favorable cases could be unintentionally selected to give higher than chance results and insisted on the need to include all the data from all the subjects tested in reporting any experiment. Rhine was in the audience, and from the floor he insisted that he had always done so, defending himself against S. H. Britt as well; but, successful defense or no, Leuba's paper seemed to suggest what was to come.[19] Greenwood's turn under the gun came the next afternoon, when he was introduced into the midst of an informal critical discussion of methods of statistical evaluation and put in the position of defending the Duke work against questioners.[20] When the ESP session took place on Friday afternoon, Rhine appeared to be anxious and tense.[21]

Yet after all things did not go badly. The session included the following papers: L. D. Goodfellow, "The Zenith Foundation Telepathy Broadcast"; T. N. E. Greville, "The Application of Mathematics to ESP Problems"; H. O. Gulliksen, "The Use of Sensory Cues in ESP Methods"; J. B. Rhine, "The Exclusion of Sensory Cues in ESP Research"; J. L. Kennedy, "Recording Errors in ESP Methods"; and G. Murphy, "Safeguards in Recording and Handling ESP Data."[22] Goodfellow's paper argued that the Zenith results, while statistically significant, were explicable in terms of guessing patterns in the radio audience; it was by no means an attack on the Duke experiments. Thus defenders outnumbered critics by 3 to 2. Moreover, as arranged, Rhine and Murphy were ideally placed to blunt critical comments. Gulliksen once again threw scorn on the defective commercial cards, pointing out that this was only the most flagrant example of the possibility of sensory cues in ESP situations; but he granted that under certain conditions—screening or long-distance experiments—sensory cues of this sort could be ruled out. Rhine happily agreed to this and insisted upon the work in which all possible sensory cues had been eliminated. Kennedy's paper raised a more novel point (which he had first raised at the Eugene meeting earlier that summer): the possibility that when the experimenter recorded both the object card and the subject's guess at the same time, he might unconsciously tend to change one to agree with the other. He reported that in an experiment of his own, he had found four hundred errors in eighty thousand trials, which had the effect of raising the score by a small amount. Murphy had been in California during the summer and had talked with Kennedy about his paper; he was therefore able to put together a carefully reasoned reply showing that Dorothy Martin's work or the Pearce-Pratt experiment could not be explained away by recording errors of this sort and citing Bernard Riess's recent success at Hunter College as a further example of the same.

The subsequent open discussion was immediately dominated by familiar critics of ESP—Hyman Rogosin, Knight Dunlap, S. H. Britt—who reiterated their own objections to the field.[23] Britt, for example, criticized Rhine for having rushed into print in 1934 before his evidence was airtight. Since Rhine had been led to change his experimental conditions repeatedly since then, Britt argued, the original data should never have been claimed to "prove" extra-sensory perception: "The fundamental proposition in any field of science is that we should have all our results reported, know all the implications of our data and be sure that all experimental problems are fully worked out before making any claims." The audience was not entirely sympathetic to the critics. Herbert Sanborn wrote Rhine after the meeting, in effect to apologize for his colleagues even though he classed himself as "an inveterate sceptic."

> [Rogosin's talk] irritated me so with its glittering generalities and its sophomoric, cavalier attitude that I felt that his bluff should be called. If I am not quite open-minded, I do at least feel that I want to try to retain my scientific integrity to some extent; and I feel that some of my friends have made some objections that do not seem quite fair. . . .
> Perhaps it will do no harm for me to tell you of a conversation that I had one evening of the session with [G. M.] Stratton [Berkeley, emeritus] and McComas [Johns Hopkins] in the course of which we came to discuss ESP. Stratton said among other things, "I don't like the attitude of _____ or _____ in their criticism and I do

approve the patient, courteous behavior of Rhine in continually accepting criticism
and in attempting continually to improve his methods in the light of such criticism'';
and with this we all agreed, without any of us venturing to express an opinion of the
validity of the results or conclusions. I think, too, that Britt's suggestion you should
have held up publication until you had established everything... is silly and was
adequately answered by the ridicule of a later speaker.... A remark like this seems
too naive for a scientist.[24]

The answer to Britt mentioned by Sanborn was from Pratt, who in the course of his remarks alluded to the gap that remained between critics and proponents.[25] There was a middle ground, however, which rested in the possession of psychologists who, while dubious, were still willing to give Rhine a serious hearing on the basis of his best evidence.

On balance, the parapsychologists could be—and were—content with the symposium. Murphy spoke of it gloomily as "drops of water against the Rock of Gibraltar" and yet admitted that the leaders of the APA were by no means hostile.

It was very evident that at least two members of the council, namely Carmichael and
Jones, were considerably impressed. There was clear evidence in the case of Carmichael that he regretted raising his voice against you and some tentative gropings
by Jones in the same direction. I think that McGeogh and Olson were slightly but not
so profoundly affected. Allport is I guess pretty near the middle of the seesaw. Among
the younger people particularly those in the late twenties and early thirties a much
greater influence was exerted, as is of course to be expected.[26]

Rhine, in a memorandum to President Few, spoke glowingly of the professional response. Where he had expected hissing, he said, he had received pronounced applause, and there was "no evidence whatever of hostility at any time." "To indicate the temper of the audience, spontaneous applause followed the reply which one of the ESP investigators [Pratt] made to the most outspoken critic [Britt]. I do not know of anything that has happened in the history of our public relations over ESP that has been more gratifying than that manifestation. I interpreted it to mean that even the psychological world is a more tolerant one than we have expected."[27] This was not mere outward reassurance. Rhine himself was growing aware that psychologists would be ready to take parapsychologists more seriously if they were given a dispassionate account of the facts.

II

The APA meetings had found Murphy in the middle of the experimentation being supported at Columbia by the Hodgson Fund, a series of experiments which seemed to be yielding positive and quite interesting results. Rhine was looking forward to seeing Murphy's eventual report in the *Journal of Parapsychology*, and he was dismayed when he learned in October 1938 that Murphy felt it would not be "wise" to publish it there, intending it for a major psychological review instead. To Murphy's mind, the paper he was planning would be too long and too technical for the journal as it was presently designed.[28] For Rhine and Murphy continued to disagree over the proper editorial policy of the journal: Rhine was eager that experimental papers of all sorts should continue to fill it, while Murphy felt that its articles were in general too superficial to attract psychologists and regretted the time that Rhine had had to take from experimental work to give to editorial activity. In response, at first tentatively and then more seriously, Rhine suggested that Murphy might want to take over editorial control of the journal. All sorts of reasons encouraged the idea. Rhine had had to bear more and more of the editorial responsibilities at Duke, for of the other editors, Charles Stuart was unwell, and William McDougall was obviously fatally ill. To allow the responsibility to pass to Murphy would be only sensible, for Murphy could share the work with Riess and Taves and could institute the editorial policies for which he had so long argued. "Furthermore," he told Murphy, "I believe it would be a good policy for me to completely relinquish editorial connections in the interest of furthering the interest which psychologists have begun to show in this *Journal*."[29]

Murphy was far from eager to take over the publication, and he was besides "sick with overwork" (he was already editor in chief of *Sociometry*), but he was persuaded by the thought that he might free Rhine to recreate his original experimental successes. By January 1939 he had agreed, and the details had been worked out. He and Riess would serve as editors, with Taves as managing editor; the journal would appear semiannually, rather than quarterly, and would aim at publishing long and fully descriptive monographs.[30] Nothing more was said of publishing his own current research elsewhere.

Murphy had been encouraged in his plans for a "new" *Journal of Parapsychology* by arrangements that had developed out of the Columbus meetings for academic psychologists to play a role in reviewing and criticizing its contents. One of those who had taken part in the concluding discussion at the Columbus symposium was Saul Sells, who as a graduate student at Columbia had shown some interest in psychical research and had won Murphy's high recommendation. Towards the end of the discussion, Sells had raised the question whether the *Journal of Parapsychology* could set aside a certain amount of space to respond to criticism of published papers, and apparently on the spur of the moment Rhine proposed that Sells serve as chairman of a committee of psychologists that would review past and future papers submitted to the journal.[31] To Murphy's mind, this was an inspired suggestion (though he thought it came from Sells, not Rhine), and he had enthusiastically followed up the idea with his colleagues when the session ended. "I feel . . . that Dr. Sells' suggestion was made in absolutely good faith and that if fully carried out it would mean a great deal to the advancement of our subject. After the presidential address Friday night I talked to Hilgard who was utterly sympathetic and immediately agreed when I asked if he would be willing to serve on such a committee. He, as you know, is among the most powerful figures on the west coast and for the first time became aware at the symposium that there was a lot more to this thing than he had ever heard from Kennedy."[32] Murphy went on to propose the organization of regional subcommittees: Hilgard, Kennedy, and R. C. Tryon ("the honestest soul alive") on the West Coast; John Volkmann, Sells, and Murphy in New York; and perhaps a third subcommittee in the Chicago area. Sells discussed Murphy's ideas with him, and the committee began to take shape in November very much along these lines: a New York subcommittee of Murphy, Sells, Irving Lorge from Teachers College, Louis Long from the College of Physicians and Surgeons, and Volkmann; a New England subcommittee composed of J. J. Gibson at Smith, John McGarvey at Mt. Holyoke, and R. R. Willoughby at Brown; and a West Coast group of Hilgard, Kennedy, and Tryon. When Murphy assumed the direction of the journal the next year, therefore, he was able to look forward to constructive criticism from a panel of not unsympathetic psychologists (to which he had appointed Lillian Dick of Sarah Lawrence in his own place).[33]

The first issues of the reorganized journal give the impression that it was indeed on the verge of becoming an instrument of communication between psychologists and parapsychologists. For one thing, the former were beginning to address the latter directly in the pages of the *Journal of Parapsychology*. Back to back in the June 1939 issue came two important articles from critics of the field who demonstrated a willingness to discuss their objections quite dispassionately. One of these was J. E. Coover's last contribution to psychical research, a response to critics—F.C.S. Schiller and R. H. Thouless—of the manner in which he had analyzed his data in his 1917 monograph.[34] The paper had been virtually complete at Coover's death in 1938, and J. L. Kennedy had given it a final polish and then decided to submit it to Rhine. This paper was followed by an article contributed by a group of experimental psychologists led by Knight Dunlap that attempted to define the criteria they felt necessary for adequate experimental procedure in studying ESP.

To be sure, the way in which this latter paper had come into existence reveals that the psychologists' interest in communication was still one-way. The talk that Knight Dunlap had given to the Western Psychological Association in June 1938 had been so sarcastic in tone that Rhine was told it had been a personal attack. Rhine wrote Dunlap for a copy of the paper, insisting that he knew it must have been scientific rather than personal criticism, and Dunlap agreed: "Many psychologists, unfortunately, take attacks on methods and conclusions as personal attacks: Titchener did; McDougall did at one time; I am glad to know that you are not so disposed."[35] But he did *not* send a copy of his paper. Perhaps he would now have been embarrassed to do so. For in August, Dunlap had agreed to take on the responsibility of consulting with colleagues to draw up for parapsychologists a complete brief for a methodologically satisfactory test of extrasensory perception.[36] Eventually Henry McComas (Johns Hopkins), Karl Dallenbach (Cornell), Samuel Fernberger (Pennsylvania), and Harvey Johnson (Tulane) joined him in collaboration: Dunlap selected them as "meticulous experimenters." He presented them with an outline for their collective discussion at the Columbus meetings, then sent them a draft for further comments in October, more questions in November, and a final version at the end of December.[37] It described a variation of a screened matching experiment whose every aspect was narrowly defined. Control of visual cues, method of scoring and recording results, uniformity of working conditions, selection and supervision of subjects (and experimenters), protection of records against tampering, elaboration of data—all were rigidly specified, far more rigidly than in any ordinary psychology experiment.[38]

The eventual paper was largely Dunlap's, but all the other collaborators (including Percy Cobb, who was drawn into the project later) had given it prolonged and serious criticism. Karl Dallenbach, for example, who had classed parapsychology with alchemy and witchcraft on the Clark-Warner questionnaire, had initially joined Dunlap's enterprise in a rather gloomy mood: "I shall gladly coöperate with you, but I am sorry that you were drawn into coöperating with Dr. Ryan [sic]. I haven't any confidence in his ability to follow out a rigorous scientific program."[39] Yet by November Dallenbach had become much more optimistic about communicating with parapsychologists. Dunlap had thought at first of publishing the article in the *American Journal of Psychology,* but while Dallenbach (as editor) was willing to accept it, he felt it only proper to send it to the *Journal of Parapsychology* first.

> I agree with you, we should reach the *proper* readers; and who are they? I take it that they are the ones interested in psychical research; the ones who believe or are inclined to believe the ESP experiments. Where will we reach them? In the *American Journal?* I think not. Indeed, I think it would be buried there as far as reaching the readers we hope for. In *Science* or the *Scientific Monthly?* Again, I think not. The particular audience we seek isn't among the readers of either of those journals. Indeed, I believe that the chemists and physicists who bulk large among the readers of those journals will fail to appreciate the need for the requirements we stipulate. The readers of *Parapsychology* are the very ones we wish to reach; and we as experimental psychologists wish to show them the procedure that we regard as necessary for proof and demonstration of their faith. The requirements we specify will stand there in bold relief among the exhibits of the kind of "research" against which we are inveighing. My chief concern is that the editor of that journal will not accept it, but if he doesn't we have a "card in the hole."[40]

The article was submitted to the *Journal of Parapsychology,* and Rhine received it with pleasure, forwarding it to Murphy and Taves, the new editors.[41] And in some measure his satisfaction was justifiable. If the Dunlap paper was less indicative of open-mindedness than perhaps he believed, it did at least indicate that psychologists thought readers of the journal were objective enough to appreciate the force of an argument on scientific method.

The major article in the June journal, however, was from Gardner Murphy and Ernest Taves, a report on their past eighteen months' work.[42] They had encountered some unanticipated statistical difficulties and had had to cut their paper back sharply; nevertheless, Murphy was still able to tell Rhine that "after enormous labor I feel fairly well satisfied with our own report although it contains much less definite data and a much less significant contribution than many already published."[43] Murphy and Taves had begun with the hypothesis that traces of ESP ability might reveal themselves by success in two or more experimental tasks carried on simultaneously. They therefore asked a variety of subjects to perform a quadruple task (QT) in a single session: to guess (1) fifty-two playing cards; (2) fifty ESP cards; (3) fifty cards bearing either a circle or a cross; and (4) fifty cards bearing a circle or blank. The experimenter shuffled and prepared the target cards in a cubicle in Murphy's office, locking the cubicle before admitting a subject into the main room and asking him to guess the ordering of the various sets of cards. The scores in the various tasks were then compared to see if any statistically significant correlation of high scores was present.

Although the authors reported only limited success, their results were still as positive as any Murphy had ever obtained, with the possible exception of his work with René Warcollier fifteen years before. Their analysis of covariance revealed that while there was no statistical significance in the individual scores for any of the four tasks alone, there was a statistically significant correlation between the scores obtained on playing-card color and the combined scores obtained on circle-cross and circle-blank guesses; that is, high scores in guessing the color of playing cards were associated to a significant degree with high scores in tasks 3 and 4 combined. The effect was found only in the five most intensively studied subjects (including the authors), and it tended to disappear once the subject had learned of the significant correlation. Murphy and Taves drew only the most cautious of conclusions from this statistical anomaly: "Either (1) the covariance effect may be a genuine result with certain subjects working under certain conditions, or (2) it may be an extremely unusual and at present obscure statistical artifact."[44] The authors urged that other experimenters try to replicate their results, warning them only to try to maintain their subjects in ignorance of the effect being sought; they recognized explicitly that their own findings were simply "markedly suggestive."

Still, in the aftermath of the APA symposium, what was impressive about the paper was not so much its findings as its detailed account of the experimental methodology that had been employed, an account so full as to anticipate some of the criteria in Dunlap's article. The detail was clearly designed to "normalize" an ESP experiment, that is, to establish a neutral experimental situation with enough precision to permit any other experimenter to replicate it. Exactly how the target cards had been prepared and set out in the cubicle, as well as how the scores had been recorded and checked, was carefully described—in the latter regard the authors acknowledged their debt to John Kennedy's criticism of experimental procedure. A floor plan of Murphy's office and the cubicle and photographs of the office and of the table with the target cards set out accompanied the paper. A very full analysis of the data was provided, and an appendix even provided the scores of all the QT's that had been recorded—an even thousand—in case a reader wanted to carry out his own analysis of the results.

Of all the parapsychological papers published in the *Journal of Parapsychology* in 1939, the one most fully and explicitly responsive to the issues raised by psychologists was that contributed by J. G. Pratt and J. L. Woodruff, which appeared in the December issue.[45] Their experimental procedure was there set out with painful care, and specific discrepancies from the requirements laid down by Dunlap in the previous issue of the journal were acknowledged and explained. Indeed, the authors had already altered their procedures once, midway in their research, to accommodate them to the criticisms made at the APA symposium. Even so, the authors still reported the occurrence of ESP in their tests, and their conclusions extended the parapsychological paradigm a little further. The Pratt-Woodruff report thus has some claim to represent experimental parapsychology at its most highly developed stage to this point and merits careful description; the length of the description that will be necessary is a measure of the growth of the methodological sensitivity of the field.

Pratt and Woodruff had begun (in "Series A," 1,468 runs, March–August 1938) to test for ESP by screened touch matching; the one variation in the now traditional STM procedure was the installation of a second, backup screen placed on the experimenter's side of the main screen in order to shield more completely the experimenter's card deck from the subject. In "Series B" (2,400 runs, October 1938–March 1939) a number of important procedural refinements were introduced, designed to safeguard against sensory cues and against errors or collusion in recording the data. Instead of a separate backup screen on the experimenter's side, a sloping screen was permanently attached to the main screen on that side below the aperture, slanted up toward the experimenter. Furthermore, the target cards were moved from their normal position under the aperture of the main screen (where they had been in full view of both experimenter and subject) to pegs located on the subject's side of the main screen; in this way only the subject could see what the cards were. And in the final four hundred runs the target cards were placed with their backs outward on these pegs so that their identity was concealed even from the subject. Five blank cards were placed under the aperture of the main screen, each one just below one of the suspended target cards. To indicate his choice, the subject pointed to one of these blank cards.

Series B was carried out with Woodruff serving as experimenter and Pratt as witness, the latter stationed about six feet behind the subject. The procedure was as follows: at the start of each run, Woodruff shuffled and cut the experimenter's

Joseph Woodruff (right) testing a subject by the STM ("screened touch matching") procedure in the Pratt-Woodruff experiment. (Courtesy of Burke M. Smith.)

deck of ESP cards while Pratt gave the five target cards to the subject. The subject was supposed to change their order from that of the previous run and then place them on their pegs. Pratt then took up his observation position behind the subject, and the experiment commenced. The subject pointed to one or another of the blank cards under the aperture (corresponding to the target card he was guessing), and Woodruff dealt out his ESP cards into five piles opposite these blanks (but under his sloping screen) in accordance with the subject's indications. In the last 830 runs, Pratt also shuffled Woodruff's deck. In the final 400 runs, in which the target cards were placed so as to be unidentifiable even to the subject, Pratt also rearranged and repositioned them on their pegs.

An elaborate procedure was worked out for recording the data. At the end of each run, Woodruff noted down the distribution of the cards of his deck in each of the five piles into which he had placed them on a score sheet, at the top of which he wrote the name of the subject, date, type of test, and his initials. On another score sheet, Pratt recorded the same identifying information together with the order of the target cards. The two score sheets had, moreover, been serially numbered with the same number in advance. When each had completed his recording, the two sheets were clipped together and deposited in a box that was kept locked. Then, as a double-check, Pratt sorted out the "hits" in each of Woodruff's piles of cards, and both Pratt and Woodruff recorded the number of hits in their laboratory notebooks. The clipped sheets were subsequently checked by an independent secretarial assistant, who juxtaposed Pratt's and Woodruff's complementary records. Finally, the score sheets' tally was compared with the scores recorded by Pratt and Woodruff in their laboratory notebooks. Where there were discrepancies, the *lower* score was taken.

Besides attempting to demonstrate ESP, the authors were hoping to pursue a specific problem squarely in the Duke research tradition—whether variation in the size of ESP symbols affected the rate of scoring in ESP tests—and hence they used as many as four different sizes of symbols as the targets for guessing. Two other questions also closely related to earlier Duke research were taken up in the course of the work: (1) Was there any inverse correlation between positive scoring rate and "experience" in doing ESP tests—that is, was there evidence of anything resembling a decline curve? (2) Was there any direct correlation between positive scores and novelty of the ESP task (introduction of a new symbol size or, in series B, of a procedural modification)? Both these subsidiary questions were related to the issue that had become so important to Rhine and his group: the psychological context of ESP successes.

Pratt and Woodruff reported significantly positive results in both series and a CR for the two combined of 7.80. More definitely than Murphy and Taves, they felt able to claim to have demonstrated the occurrence of ESP. Beyond this, however, they seemed to have learned something about the particular questions they had set out to investigate. ESP evidently did not depend on symbol size; however, ESP scores *did* seem to vary with what might be called position effects in the course of the experiments. Most notably, new test material did appear to influence the scoring rate positively. Subjects tended to do best in the first few sessions after a new size of symbol was introduced; subsequently their scores on this symbol would decline, suggesting the medium- to long-range decline effects Rhine had noted earlier. Moreover, within each session, patterns already established in the literature also appeared: an initial period of adjustment (whether or not the test material was new) would be followed by an episode of higher than average scoring, which in turn declined fairly precipitously in the course of the session if the material was old but remained more consistently above average for new material, at least in the first few sessions in which it was used. Conscious knowledge that a new symbol was being used seemed to enhance the positive effect.

This last conclusion, expressed guardedly, fitted in well enough with the conviction that had developed within the Duke laboratory that ESP subjects were prone to "general psychological satiation" with a particular stimulus and that only cordial subject-experimenter interaction could assure consistently successful work. Pratt and Woodruff had taken pains to encourage precisely such interaction:

The experimenters always showed the subjects the new sizes before beginning the daily session in which they were first used. This was frequently done in a manner which challenged the subject to do better with the new sizes before beginning the daily session in which they were first used. During experimental sessions the experimenters, particularly Woodruff, frequently encouraged the subject between runs in the same challenging manner. The introduction of a new size of stimulus was made a special "talking point" between the experimenters and the subject. This probably resulted in a greater interest of the subjects in the new material for a few sessions.[46]

Inevitably, therefore, no attempt was made to isolate physically the experimenter from the subject during the actual test situation. Knight Dunlap and the other psychologists had not objected to this practice in their June paper meant to describe the ideal experimental situation; hence, in their systematic account of how possible sources of sensory cues had been excluded, Pratt and Woodruff did not even think it necessary to apologize for the joint presence of subject and experimenter at the test.

Even so, C.E.M. Hansel has argued quite recently that this makes the Pratt-Woodruff success automatically suspect. He has identified a nonrandom feature in the scoring pattern of their best subjects and has suggested that this could have been due to a surreptitious manipulation by the experimenter of his deck of ESP cards so as to conform them to his subjects' guesses.[47] While Hansel cannot be said to have proven his charge, he has certainly found an unsettling anomaly in the data, one whose impact would have been less had the protagonists been isolated from one another during the experiment or had a standard shuffling routine been instituted for the target cards.

Nevertheless, current judgments as to the soundness of the Pratt-Woodruff experiments are irrelevant to an assessment of their significance for parapsychology in 1939. Hansel's intransigence and determination to treat *any* positive results as suspect goes far beyond anything that the psychologists on Murphy's advisory committee felt when the Pratt-Woodruff paper was submitted to them for judgment. Their report to the journal's editors expressed real satisfaction with what they saw. "The members of the Committee have been impressed with the thoroughness with which the experimental work has been conducted and the report treated. From the standpoint of 'repeatability' the report is very satisfactory. The procedure has been described in complete detail. Every step is explicitly written up.... It is hoped that future writers on ESP will use the Pratt-Woodruff manuscript as a model for careful reporting of experimental results."[48] For, as the Murphy-Taves paper had already indicated, the issue of "proving" psychical phenomena had momentarily become subordinate to the task of normalizing psychical research in experimental psychology—by establishing rigorous, standardized experimental situations and a systematic pattern of reportage and evaluation. What the Pratt-Woodruff experiment represented, in effect, was not so much a proof of ESP as a generally successful attempt by parapsychologists to accede to the standards demanded by academic psychologists, without compromising their own concern for the delicate psychological atmosphere in which psychical phenomena could be expected to occur. A serious and constructive exchange between the two communities had at least been initiated.

As if in confirmation of this willingness for discourse between psychologists and parapsychologists, the APA council of directors chose in that summer of 1939 to approve Rhine's application for associate membership, which had been deferred the previous year. Narrowly, to be sure: five to four, Murphy reported, "Muenzinger, Allport, Valentine, Freeman, Murphy voting yes. As far as I know (and it was again a full discussion) Murphy *alone* accepts ESP, the others feeling that your work is *psychological work* and letting it go at that."[49] But that was the critical concession from the profession. With its most prominent exponent, parapsychology had been acknowledged by psychology.

III

The plans Rhine was making to renew research at last were interrupted early in 1939 by an inquiry as to his future publication plans from Bill Sloane, a friend who worked for the publishers Henry Holt, which set him to thinking of writing a technical book on ESP that would systematically explain and justify the methods and conclusions of his laboratory. The entire Duke team spent most of the spring and summer on the project. Though the editors at Holt recommended that he aim such a work at a university press, Rhine sent the finished manuscript to the publisher of *New Frontiers of the Mind,* John Farrar, for consideration. When Farrar declined to publish the book, Holt agreed to do so, with the understanding that Rhine's future trade books would be given to them as well. Bill Sloane gave the manuscript its title, *Extra-Sensory Perception After Sixty Years,* and the book appeared in June 1940.[50]

Though the book began by briefly setting recent research in historical context, it carefully examined only the research of the last ten years. Much of the work tried to establish the occurrence of ESP on the basis of a few unexceptionable tests. A very full account of the various mathematical and experimental techniques that had been developed was followed by an attempt to summarize the objections raised by psychologist critics—the authors distinguished thirty-two such objections—to account for the data by normal means. Then they discussed the half-dozen experiments they considered best (including the Pearce-Pratt, Riess, Pratt-Woodruff, and Murphy-Taves studies) in order to show that these experiments could not be explained away by any of the objections, singly or in combination. They invited leading critics of parapsychology to comment on these arguments in turn and printed their replies (although only three, most notably C. E. Kellogg, took advantage of the opportunity), concluding by expressing the conviction that ESP had been experimentally demonstrated. The remainder of *ESP-60* accepted the phenomena as established and turned therefore to the mass of evidence collected under a variety of looser conditions during the previous decade to try to decide what could be said about the nature of ESP, before ending with a survey of "unsolved problems" awaiting the attention of students of the field.

The form and scope of *ESP-60* cannot help but invite comparison with *Extra-Sensory Perception,* Rhine's first monograph, published six years before. Formally, as a matter of fact, the two were entirely dissimilar. Nothing like the systematic analysis of methodology found in the newer book had been given in *Extra-Sensory Perception.* The earlier book's entirely historical approach had not always allowed the reader to define the precise conditions under which the pre-1934 research had been performed, and therefore in *ESP-60* the demonstration of extra-sensory perception was referred strictly to the most carefully described of experiments. Two pages on statistical method in the first book had been replaced by a twenty-seven-page chapter (and seventeen statistical appendices), yet Rhine's original delight in reiterating high p-values had disappeared, while the very style of *ESP-60* was more subdued. The emphasis in the 1934 monograph on the metaphysical implications of telepathy and clairvoyance had become much more restrained (though it was not entirely absent). Whereas *Extra-Sensory Perception* had been still in the amateur psychical-research tradition, *ESP-60* was attempting to meet the needs of psychologists as well.

Yet beyond the manner and detail of presentation, there was little in *ESP-60* that would have been unfamiliar to Rhine's readers of 1934—a more sophisticated basis for evaluation of data, certainly, and some new variations of guessing techniques (OM and STM), but nothing fundamentally different about the underlying experimental situation. Moreover, the conclusions that *ESP-60* drew about the nature of extra-sensory perception were not markedly different from those of the earlier work, with the exception of its discussion of the recently described subject-experimenter effect. Indeed, they were perhaps slightly more restrained; though the authors still suggested that the faculty might be demonstrable in 20 percent of the population, they did not now suggest that it might be a species-wide endowment. ESP, the book concluded, was possessed by both sexes and all ages, was not particularly marked in psychics or subjects in trance, was more easily demonstrated when motivation was high, and could be affected by drugs. ESP performance was unaffected by size, shape, or distance of the targets from the percipient; it was unconscious, erratic, and unstable, incapable of development through use. In these last judgments the optimism of 1934 was again somewhat tempered, as in Rhine's opening warning for the experimenter not to be discouraged by an inability to obtain evidence of the phenomenon. But in ending with suggestions of topics needing further investigation, *ESP-60* deliberately maintained the original picture of a science susceptible of systematic articulation.

Gardner Murphy had learned about Rhine's plans to write a technical survey only after the manuscript for *ESP-60* was well advanced, and he had not been happy with the news. To his mind, the work Rhine was planning was not the sort of thing likely to impress psychologists, and he made his reservations known.[51] Rhine replied in terms that help explain whom he hoped to reach with the book. "I have come to the view that we dare not concern ourselves with trying to 'impress the psychologists,' in the way you mean. We hope to interest young psychologists actively and to provide opportunity for them to work. We hope to have a number of older men open-minded and neutral, and that hope is being realized as rapidly as could be reasonably expected."[52] Rhine was less ready than Murphy to believe that trained psychologists could be won over by any evidence he could provide, and he was concerned instead to ensure the entrance of new, young scientists into the field.

In the end, however, both audiences were reached through the distribution policy that Rhine worked out with his publishers. In lieu of some of his royalties, he was enabled to offer a free copy of the monograph to any psychologist who accepted his invitation, with only the suggestion that the volume "be brought, by way of a review, to the attention of psychologists—students or staff—of your college community, and any critical comments passed on to us."[53] No fewer than two hundred psychologists accepted the invitation. Not only is the offer itself a measure of how Rhine had reoriented his search for an audience since 1934, when he had aimed primarily at the sympathetic general public, but the response of the psychologists is an indication of how they in turn had become ready to examine the scientific issue for themselves.

The professional response to *ESP-60,* as it made its way back to Rhine in private correspondence, was still mixed, but it was perhaps slightly more favorable than that expressed in the APA survey conducted two years before. Only about 10 percent of the recipients did in fact return comments to Rhine after reading the book. Murphy, in a volte-face, wrote glowingly of it as a "superb contribution,"[54] and a few other hitherto uncommitted academics also expressed the feeling that Rhine had now proved his case.[55] Others wrote to agree that extra-sensory perception at the least merited further investigation.[56] Edmund Conklin of Indiana, who expressed his regret that he had *not* been able to agree that ESP had been demonstrated, took pains to explain his objections fully and ended his long letter with a suggestion that was far from a blanket rejection of the Duke research as in any sense "unprofessional": "It seems to me beyond doubt that variations far above chance scores are and can be obtained. The problem then should be to explain why those variations occur. Is it not quite within the range of possibility that some other explanation than ESP may be discoverable? It would seem to me so, and it further impresses me that you are exactly the person to seek to make the discovery."[57] None of his professional correspondents evinced the unwillingness to consider his evidence that Rhine had once expected to encounter from psychologists.

Another indication that serious professional consideration was now being given to parapsychology was the number of reviews given *ESP-60* in psychological journals; *Extra-Sensory Perception* had only received three, and one of those had been volunteered by Murphy. The reviews were, again, mixed, ranging from favorable (James Snyder, in *Character and Personality*) to hostile (anonymous, in the *Psychoanalytic Review*).[58] The spectrum of opinion undoubtedly reflects the varying orientations of the subgroups to whom they were addressed, as well as those of the authors themselves. The most thorough and thoughtful review, written by Henry J. Garrett of Columbia, appeared in the *American Journal of Psychology*.[59] Garrett emphasized the improvements in technique and presentation made by the Duke researchers since 1934 but explained that, even so, he remained dubious: "It must be admitted that the data which [the Duke workers] present, and the arguments deduced therefrom, strike one as extremely plausible. I must confess, however, that I am still unconvinced that the evidence so far produced is really crucial...." Garrett readily conceded the legitimacy of the statistical techniques, but he argued that extra-chance guessing still could only define ESP as a negative, "namely, something other than the factors which have been controlled." Further, he explained, the evidence presented seemed to him to show that with increasingly tight experimental controls ESP guessing fell closer and closer to chance levels. "I have found very little evidence of [ESP]," he concluded; "I rather wish I could." This not unsympathetic appraisal could not have satisfied Rhine, who had tried to meet objections like Garrett's in *ESP-60* itself; even so, it shows very well how far psychologists had come round to a sense that the work of contemporary psychical researchers deserved to be read carefully and treated seriously. Very few psychologists of 1940—always excepting men like Jastrow[60]—were any longer expressing the belief that psychical research had no claim to be considered a legitimate scientific activity, whether they believed in telepathy and clairvoyance or not.

We have repeatedly used the psychologists at Harvard as a guide to the professional reaction of psychical research, and it will be useful to do so once more. The response of the department to *ESP-60* was all that Murphy or Rhine could have hoped for. Gordon Allport, who had taught social psychology at Harvard since 1930 and had voted for Rhine's APA membership in 1939, wrote him an extremely positive letter upon reading the monograph over the summer of 1940: he praised the manner of presentation enthusiastically, suggested leads to pursue from the standpoint of a psychologist in a different field, and urged Rhine to continue his work.[61]

Though certainly cooler in expression than Allport's, E. G. Boring's response was even more encouraging for the future of the new field.

> You will be surprised to hear that we have been considering seriously giving this book—or perhaps a part of it—to Psychology A as required reading in January when the course goes on to problems of scientific method. In doing that there would be no primary intention of treating the book either positively or negatively. It obviously illustrates the difficulty of analyzing a situation and the various sorts of things that can

be done, and all that is positive. It also shows the difficulty of a frontal attack upon a phenomenon which has a name but which does not have definitely known conditions as determinants.[62]

And indeed in the fall semester of the academic year 1940/41 the bulk of *ESP-60* was made assigned reading for students in the introductory psychology course at Harvard.[63] Whatever the terms in which it was treated in discussion sections or referred to in lectures, the result would still have been to expose beginning students to the best case that could be made for experimental psychical research—perhaps the first time this had ever been done outside Duke. When we consider that the Hodgson committee had agreed in 1939 to support Murphy and Taves for a third year and was preparing to make a grant to J. L. Woodruff for 1940/41, it is clear that the Harvard psychologists were quite ready to allow parapsychology to prove itself within the wider scientific forum.

The response of psychical researchers to *ESP-60* was for obvious reasons bound to be quite different from that of psychologists. In America, most of the tiny community of scientific psychical researchers was in fact at Duke and had had a hand in the composition of the work. There remained the constituency of the amateur societies. In New York, Murphy had recently begun an attempt at a rapprochement with the ASPR, which late in the 1930s had shown signs of moving away from its once credulous and even antiscientific position, and a measure of its movement was the decision by the ASPR *Journal* to ask Murphy's young associate, Ernest Taves (who was about to finish a Ph.D. at Columbia under John Volkmann), to review *ESP-60*.[64] It was the English community, however, whose response to *ESP-60* was bound to be most telling, for there an initial welcome of extra-sensory perception and of Rhine's attack upon psychical-research problems had given way to a widespread disbelief that card-guessing experimentation with unselected subjects would ever give consistently positive results if carried out with appropriate precautions. Whately Carington's enthusiastic review of *ESP-60* in the SPR *Proceedings* was not necessarily a sound indicator of English feeling,[65] for Carington had always been one of Rhine's strongest English advocates. Other members of the society might send Rhine polite congratulations,[66] but in a way the crucial judgment was bound to be S. G. Soal's, since it was Soal whose unremittingly unsuccessful experiments had convinced the English that the card-guessing approach was bound to be fruitless. As it happened, Soal had just begun to achieve positive experimental results with a variation on Rhine's methods, and he announced to Rhine his high opinion of *ESP-60*, destined, he predicted, to become a standard experimental textbook for psychical research; he declared that it had given him as much pleasure as anything he had ever studied.[67] It was an astonishing concession from someone who only two years before had been a bitter and unpleasant enemy of Rhine's design for psychical research—although privately Soal expressed the reservation that an experimental approach revealed less about the psychical processes involved than did the study of spontaneous cases.[68] As word began to spread in England of Soal's change of heart and of the success that this intensely skeptical researcher was beginning to have, experimental investigation of the Duke type into telepathy and clairvoyance was bound to seem, much more surely than before, to be the long-sought route by which psychical research could gain recognition as a true science.

Soal's allusion to Rhine's book as likely to be "the standard textbook on experimental E.S.P." reminds us of Oliver Lodge's assessment of Charles Richet's hopes for his *Traité de métapsychique,* published only eighteen years before. The *Traité* had not been able to serve as a text for psychical research for a variety of reasons. In part it had been due to Richet's inability to convince his audience of psychical researchers that his openly materialist, antisurvivalist assumptions were adequate foundation for the would-be science—which in turn had been due to the psychical researchers' own inability to agree upon that foundation. In part, however, it had also been due to the character of the *Traité* itself, a diffuse, often anecdotal account of the whole range of reported psychic phenomena, described with little attention to detail—often so loosely as to seem to scientists largely uncritical.

Soal was right: *ESP-60* held some promise of serving as a textbook for a science, far more than Richet's work. We have already suggested that Rhine's first book offered a general paradigm for investigation that could hope to attract new students to the subjects of clairvoyance and telepathy. His work of 1940 reinforced the paradigm by embedding it in a context of procedural and mathematical detail and in effect gave explicit guidelines to those who wished to try to confirm and extend his conclusions, suited not only to the amateur but to the skeptical scientist as well. Rhine sent a copy of *ESP-60* to Oliver Lodge, whose writings had first led him to look into psychical research, and the eighty-nine-year-old Lodge—almost the last survivor of the early SPR—acknowledged the gift in a scrawled note two months before his death. Lodge's letter can stand as an indication of how far experimental psychical research seemed to have come.

> I had heard so much about your experiments in telepathy that I rejoiced to get an authoritative account, and especially to know that a University Professor of Psychology was taking up the subject. And now I find that you were aware of my own work in the same direction, although it was carried on in a back-stairs manner and had no University status. At the same time I was personally convinced of the reality of what you have rechristened E.S.P. & desire no more evidence; only now the subject is on the way to becoming respectable, treated in a handsome volume, published by Henry Holt, and vouched for by several Professors as a branch of Psychology.[69]

Epilogue

In 1940 Oliver Lodge could reasonably take comfort in the thought that psychical research seemed to be on the threshold of establishing itself among the sciences; the difficulties of the past quarter-century had finally been resolved. English investigators had at last begun to replicate the striking card-guessing successes reported in American experiments. The most serious objections of psychologists had apparently been met, and both in Great Britain and in the United States there were signs of widening academic tolerance and even a certain willingness to support the subject. Yet parapsychology remained poised on that threshold; scientific respectability and full incorporation into the academic community continued to elude the field. Parapsychology simply did not proceed to develop as one might expect an emergent scientific specialty to do.

To be sure, this assessment requires one major qualification. Parapsychologists could continue to feel during the 1940s that their discipline was growing together around a dominant experimental-statistical research program, and they could point with some confidence to evidence of its further progress. The institutional coalescence was epitomized by a "palace revolution" in the ASPR in January 1941 in which Jocelyn Pierson was a leading figure: the Margery group was removed from control of the society and replaced by individuals—most notably Gardner Murphy—concerned once again to insist on scientific standards, though interested still in the study of a broad range of psychical phenomena. It was a stunning reversal of the policies that had been maintained by the ASPR since William McDougall's ouster just twenty years before. Another rift was healed, symbolically, when the remnants of the Boston SPR merged back into the reformed American society. At the same time, English psychical research (though much diminished in the war years) centered increasingly upon the possibilities of the Rhinean experimental program. During 1941–43 S. G. Soal and K. M. Goldney carried out a variety of experiments with Basil Shackleton and Gloria Stewart, designed to confirm the apparent ability to guess cards precognitively that Soal had happened to find in their data in 1940. These experiments, which did indeed provide further evidence of both precognitive and retrocognitive telepathic guessing under apparently rigorous conditions, were variously witnessed by members of the English psychical-research community as diverse as C. D. Broad and Ina Jephson, E. J. Dingwall and Robert Thouless, and ensured that the SPR would at last concede the importance of

Rhine's original results.[1] Thouless's presidential address to the Society in 1942—in which he positivistically coined the neutral term "psi phenomena" to subsume ESP and more—marked something of a watershed in the orientation of English investigations:

> Existing researchers have not merely proved the existence of the psi phenomena; they have also found out some odd and unexplained things about them that are a challenge to further research. Let us now give up the task of trying to prove again to the satisfaction of the sceptical that the psi effect really exists, and try instead to devote ourselves to the task of finding out all we can about it. With fuller knowledge of its nature, the difficulties of believing in its existence may appear less formidable than they do now.[2]

Within only a few years, there were several signs that Thouless's hopes were being realized. For one thing, the range of possible psi phenomena brought under scientific study was extended to psychokinesis, or PK (the ability to influence the physical environment by mental activity), in papers published both in the United States and in England. The American work, in which a subject tried to affect the fall of dice, had been carried out in the mid-1930s, but publication had been withheld until 1943 for fear of its effect on the acceptance of parapsychology. Interestingly, the re-survey of the mass of the PK data led to the discovery of a pattern that appeared to be general and consistent: a decline in the PK scoring rate both within individual runs and also from run to run in a PK experiment, which became known as the "QD [quarter distribution] effect."[3] This effect was not unlike the position effects that had been discovered in the earlier experimental work, such as the "decline effect" and the "U curve." In its apparent lawfulness, the QD effect might be taken as a criterion for authenticity of PK data regardless of the variability of conditions under which they had been compiled. To Rhine, the discovery of the unforeseen QD effect was a most heartening substantiation of the reliability and meaningfulness of much of his PK data, and therefore also of PK itself.

The English research on PK was performed in direct imitation of the Duke experiments; it was on the whole much less successful. Nevertheless, given the initial English failure but subsequent success in replicating Rhine's ESP results, their first PK failures did not seem quite so critical; moreover, R. H. Thouless himself seemed actually to have had some slight success in PK tests.[4]

The publication of the PK work was unquestionably a dramatic milestone in parapsychology, but perhaps closer to Thouless's prescription for research was the appearance in the mid-1940s of a number of studies of the psychological context of ESP manifestation. Rhine had already suggested in 1934 that personality type might be linked to ESP ability, although he had not pursued the idea. The first researcher to take up the idea vigorously was Gertrude Schmeidler, a young clinical psychologist who in the summer of 1942 happened to attend a special Harvard seminar in psychical research offered by Gardner Murphy—Murphy had been invited by E. G. Boring to give the class in commemoration of the centenary of William James's birth. That fall Schmeidler began a series of experiments under Murphy's guidance at Harvard, continued when she moved to New York, in which she tested the hypothesis that subjects' initial attitudes towards psychical phenomena might help predetermine the direction of their scoring in ESP tests.[5] Preliminary questioning divided test subjects into "sheep," those willing to accept the reality of psi, and "goats," those adamantly skeptical; and, as she reported in the 1943 ASPR *Journal,* "sheep" consistently scored above mean chance expectation and significantly better than "goats" in ESP tests. Very similar studies were soon being carried out by investigators at Duke, most importantly by Betty Humphrey, who had entered the graduate program there in 1941 to work with Rhine. Humphrey's Ph.D. dissertation, completed in 1946, attempted to correlate personality types with ESP ability and argued that, by various personality measures, the well-adjusted, stable, extroverted, self-confident individuals did better in ESP tests than the maladjusted. Schmeidler and Humphrey both continued to work along these lines until the early 1950s, developing this as a normal program in behavioral research. Parapsychology could thus reasonably claim to have progressed scientifically during the 1940s.

Nevertheless, the internal progress that its adherents could claim did not coincide with a widening acceptance of parapsychology by the scientific community; the small number of institutions and localities prepared to support experimental psychical research remained very much the same. Even the situation at Duke was at best static. Rhine's sources of private funds had begun to shrink in the prewar years, and this had forced him to let some of his staff go. The war effort took still others—J. G. Pratt entered war work in Detroit, for example—though this eased financial pressures somewhat, to be sure, and Pratt did return to Duke after the war. A ray of optimism in an otherwise distressing situation was provided by the award to Charles Stuart of the psychical-research fellowship at Stanford in 1942. But any hope that this would lead to a permanent incumbency of parapsychology there was dashed by Stanford's failure to renew Stuart's fellowship beyond 1944.[6]

In New York, the other American center of academic activity, Gardner Murphy remained the natural leader. In the 1940s Murphy made use of both his longstanding connection with psychical-research societies—in this case, the reformed ASPR—and his own new professional position as chairman of the psychology department at C.C.N.Y. (1940) to knit together a small group of talented researchers who operated from one or both of these bases, including at C.C.N.Y. both J. L. Woodruff and Gertrude Schmeidler.[7] Murphy also retained the support of Hodgson money during the 1940s, Schmeidler succeeding Woodruff (who had succeeded Ernest Taves) as recipient. The complex network of affiliations that Murphy built up in the New York area was the product of his involvement with the psychical-research societies as well as with academic psychology, while Rhine's tighter-knit organization at Duke reflected a more intense focus upon parapsychology per se. Murphy's style had always been one of low visibility vis-à-vis academic science; he believed in the cautious, pragmatic approach of slow insinuation. But C.C.N.Y. and Duke remained the only academic institutions hospitable to parapsychologists in the 1940s. Looking back on this period, J. B. Rhine wrote later that "the period ended with relations between parapsychology and other professional fields even less certain than before. Despite all the progress psi research had made, it was a worse time for trying to start a new research center in any university psychology department in the country than when the beginning had been made at Duke...."[8] In Britain, the situation was similar, despite the promise opened up by the Perrott studentship at the start of the decade; the only reasonable opportunity for more than part-time involvement remained the post of research officer of the SPR.[9] In 1952 Robert Thouless could only reiterate that career opportunities in the field were sorely lacking.[10]

For its critics, there is no particular difficulty in explaining why parapsychology failed to cross the line into scientific establishment. They would claim that the subject deals with nonexistent phenomena and is consequently devoid of scientific content; in their most charitable view, parapsychologists have been deluded by their own will to believe. Conversely, ardent advocates of the field see its rebuff by the scientific community as yet another case of intransigence and intolerance by the established authorities in the face of pioneering and radically innovative research—as in the case of Galileo and his Aristotelian opponents or, more controversially, Franz Anton Mesmer and the French Academy of Sciences.

In our view, neither of these positions can begin to do justice to the question of why parapsychology failed to develop professionally after 1940; nor is a definitive explanation really feasible without detailed study of the later period. Nevertheless, we feel obliged to attempt a general though inevitably speculative answer to the problem. There is no question that parapsychology has faced severe difficulties in attempting to deal with phenomena that do not appear to be common in mankind to any striking degree, have unsettling consequences for scientific theory, and are difficult to control and to replicate. Yet these difficulties with the methodological and evidential basis of the new science, central though they be, need to be given consideration in a much wider context, professional as well as intellectual, than either critics or advocates usually do. In particular, the question of "replication" must be viewed as a function of the experimenter's attitude towards the claims of parapsychology and of his expectations, for committed psychical researchers and obdurate cynics have always had different standards by which to assess the evidence.[11] It is especially critical for our assessment to try to understand the expectations of the broad middle ground of experimental psychologists, many of whom *were* by 1939 or 1940 expressing a willingness to consider the claims of parapsychology and a few of whom were even willing to try experiments of their own. The continued interest and growing involvement of this group would have made possible the expansion of the new field.

The promise that parapsychology held for scientists in the late 1930s arose mainly out of Rhine's first research, set out in *Extra-Sensory Perception* and publicized in *New Frontiers of the Mind*. We have seen how the unprecedented claims of this research afforded a new experimentalist focus to the disarrayed psychical-research communities and how a decade later the Anglo-American communities had largely adopted this new focus. The same claims were offered to scientists outside psychical research: that telepathy and clairvoyance could be demonstrated by simple methods in as many as one in five subjects in a presumably average population of college students and that results obviously much higher than the chance norm could be thus obtained. It was in this climate of expectation that some interested psychologists had taken up parapsychological experiments; results similar to those Rhine had reported would have constituted replication and at least encouraged them to carry on further work.

By 1940 psychologists in perhaps fifty colleges and universities had experimented with ESP cards. Most of them had failed to obtain evidence for ESP, although one or two reported results with individual test subjects comparable to Rhine's. This did not necessarily lead psychologists to conclude that extra-sensory perception was a fiction, but it did effectively discourage them from continuing their research or from making room for parapsychology at their institutions. Harold DeWolf of Boston University, who for some time had been an interested correspondent of Rhine's, explained late in 1938 why he and his colleagues were bringing their experimentation to an end: "We have been convinced that all the conditions which have been specified as favorable to ESP were present during most of our experiments here. There seems so little probability, therefore, that additional tests here would produce anything but more negative evidence, which, of course, must always be inconclusive, that we are turning our attention now to other problems. I shall continue to be interested in the results of ESP investigations elsewhere." DeWolf did not state precisely what his negative results had been, but presumably they were negative in the sense that they failed to confirm the expectations aroused by *Extra-Sensory Perception*.[12]

Marginal successes of a sort that Schmeidler or Humphrey would later be able to see as replicatory, therefore, might well have seemed negative to DeWolf and other interested psychologists. Scientists with an a priori commitment to psychical research could always account for a lowering of scoring levels by appealing to inhibiting conditions in the test situation or to a lack of rapport between experimenter and subject, even to the inherent inability of certain experimenters to achieve results. Sometimes their commitment was metaphysically based; sometimes it arose from personal psychic experiences or from witnessing a strikingly positive experiment. Given such a commitment, whatever its origin, an experimenter will not easily be discouraged by a difficulty in obtaining further experimental evidence. But for those coming *de novo* to parapsychology, with no prior commitment, the situation was very different. For them positive results would have been less important as evidence for extra-sensory perception than as an incentive to continue and to deepen their involvement in a program of experimental research.

That parapsychology failed to entrench itself further in the academic setting during the 1940s is thus not necessarily an indication of a pervasive disbelief in extra-sensory perception. The failure hinged instead upon the limited energy that psychologists, however willing to accept the legitimacy of parapsychological research, were ultimately prepared to invest in the field. Telepathy and clairvoyance *might* be true—but psychologists saw little sense in spending time trying to study them themselves or in creating positions for parapsychologists as long as experimental results in the subject seemed so hard to come by, especially when there was so much else in psychological research competing for time and money. Tolerance would be shown to the committed parapsychologist, but only so long as this did not involve many demands upon professional resources— witness the case of perhaps the most open-minded of the leaders of American psychology, E. G. Boring, who was unwilling to afford a niche for parapsychological research at Harvard, preferring to leave Murphy to manage the Hodgson money *away* from Harvard. This still left it possible for the parapsychologist to stay in contact with psychology, but the initiative for communication came to rest entirely with the former. Parapsychologists were becoming painfully aware that intellectual cohesiveness and claims of experimental research would not of themselves guarantee students a professional future.[13]

Unfortunately, certain other factors worked to impede parapsychology's communication with psychologists and to isolate it from the scientific world. Ironically, the *Journal of Parapsychology* functioned in this way. At the time it was founded, in 1937, a journal had seemed to its editors the necessary concomitant of a new science. Inescapably, however, the *Journal of Parapsychology* spoke almost entirely to parapsychologists alone; very few other scientists read it or subscribed to it, even after 1938, despite Gardner Murphy's hopes for the effect of the advisory committee. Conversely, since it (together with, later, the ASPR *Journal*) absorbed the experimental papers of American psychical research, psychologists encountered almost no writings on the new field in the journals they did read—and those few were consistently unsympathetic. Nor did parapsychologists continue after 1938 to present their work to APA sessions. The sole paper on the topic read at the 1939 meetings was a critical one delivered by John Kennedy,[14] and for many years thereafter the field was not represented at all on the annual program.

To some extent the policies of J. B. Rhine in the 1940s reinforced the isolation of parapsychology from mainstream science. He wrote to his sister-in-law at the end of 1940: "Just at the moment I have the feeling that one of the shifts that will occur next year is a shift back to our own real clientele, that is, those really interested in the field of parapsychology . . . away from the rank and file of the A.P.A. After all, it is mainly our friend Murphy who has been trying to sell the A.P.A. on ESP."[15] He had always believed that existing science was unwarrantably constrained by a rigid materialism, and had felt that a new scientific world view was inevitable at the same time that he insisted on the fundamental importance of the experimental method. Now, starting in the 1940s, Rhine began to give increasing emphasis to the antimaterialist implications of parapsychology. The PK work, for example, was associated with the renewal of interest in the post-mortem–survival issue.[16] Murphy and the ASPR joined Rhine in this broadening of the focus of parapsychology to include not just experimental inquiries but the study of spontaneous material (including mediumistic, survival-oriented work) that had characterized early psychical research. From the viewpoint of the traditional psychical researcher, this move to broaden the field no doubt seemed to be a positive enrichment of the serious interest of the leaders of parapsychology. But it was bound to cut parapsychology off still further from accommodation with academic science. Indeed, Gardner Murphy, who by the mid-1940s stood at the pinnacle of professional success in psychology (he was elected president of the APA for 1944/45), had by now come to share Rhine's belief that the time was not really ripe for parapsychology's assimilation into psychology.[17] And, as had been true in earlier decades, psychical research broadly conceived held greater appeal for the general public than did the more austere experimental approach. *ESP-60*, designed for technical readers, would have no successor for a long time.

At Duke, of course, parapsychology was still bound to have some sort of professional hearing. Although President Few died in 1940, Rhine continued to enjoy administrative support and to play an important role in the affairs of the psychology department. By all accounts, he continued to be an effective and even charismatic teacher for both undergraduate and graduate students. He continued to plan, however, for an independent institute of parapsychology free of departmental ties. His sources of financial support, at a low ebb in the early 1940s, rose again during World War II, as donors responded to the broadening ambit of his expressed interests and perhaps to personal bereavements as well. By 1947 he had obtained both a considerable pledged endowment and the sanction of the university for a separate research institution devoted to psychical research. The trustees agreed in November "that the Parapsychology Laboratory, now nominally a part of the University's Department of Psychology, be severed from this Department and set up as a separate and independent agency to be known as the Parapsychology Laboratory of Duke University with Dr. J. B. Rhine named as its Director, and with the understanding that he retain his professional status and continue his faculty rank and standing."[18]

Independent institutional status, like the establishment of a disciplinary journal, is ordinarily an indicator of the existence—the arrival—of a new scientific specialty. But both are accurate indicators only when a suitably broad professional foundation has been established; in the case of parapsychology, one must conclude that they may have come prematurely. The creation of the Parapsychology Laboratory at Duke cut Rhine and much of American parapsychology off from the possibility of direct entree into the academic world through training in psychology. The last graduate degrees awarded by Duke for parapsychological research were Betty Humphrey's Ph.D. of 1946 and John Bevan's M.A. of 1948—the latter coming just fifteen years after John Thomas's pioneering dissertation; thereafter Duke did not offer professional training to would-be graduate students in parapsychology.

Considered as a prospective new discipline, then, parapsychology in the 1940s was unable to develop the professional structure and opportunities necessary for its sustained growth within science. It still could not assure careers to its students, and it was not attracting and training growing numbers of new personnel. More tellingly, it had been unable to increase the range and warmth of its support among psychologists. In 1952 Lucien Warner polled a sample of the APA membership along the lines of his 1938 questionnaire, to survey their attitudes towards ESP. They proved to have changed very little in fourteen years: there was still general agreement that the study of ESP was "a legitimate scientific undertaking," falling within academic psychology, but there was still only one in six respondents who conceded that extra-sensory perception was as much as "a likely possibility" (the figure had been one in twelve in 1938), and the percentage of those who admitted never having read the *Journal of Parapsychology* had increased sharply.[19] The survey results indicate clearly how little real progress the field had been able to make in winning the interest of uncommitted academics.

Nevertheless, while they did not lead either to general scientific acceptance or to a secure academic niche for parapsychology, the events of the 1920s and 1930s had marked the field unalterably. For one thing, they had brought it at least to the edge of acceptance, and hope for that acceptance has remained strong within the field. The figure of one in six psychologists prepared to accept ESP as at least likely is not high, but a survey taken twenty-five years before would certainly have shown even less support for the subject. This increased plausibility was in turn due to the changed form in which technical psychical research was being carried out and reported in the 1930s. The quantitative-experimental approach, stressed first by a few French and English researchers and then by Rhine, was becoming central to the discipline; and today, despite a return of interest to broader issues such as survival or reincarnation, it still furnishes the field's principal claim to scientific credibility. Equally important, it was in this same period that the psychical researchers themselves came finally to agree that their subject should be developed as a science. It was not only the Richets, the Caringtons, and the Rhines who felt this way, but a part of the survivalist wing of the field as well. At the end of 1951 the medium Eileen Garrett established the Parapsychology Foundation to support the study of psychic phenomena and abilities. It is particularly interesting to note that in naming her foundation Mrs. Garrett chose to identify her broad interests with the quantitative-experimental model of the Duke research. In the sense that its participants had generally agreed on the centrality of a certain mass of experimental-scientific work and in the process had gained a unifying disciplinary self-consciousness, psychical research had indeed passed into parapsychology by the 1940s.

Afterword by J.B. and L.E. Rhine

It was generous of Drs. Mauskopf and McVaugh to permit us to make these comments on their book, a history of parapsychology. It is the field to which we have devoted our scientific lives and to which these men have now given nearly a decade of part-time research, part-time since they have been full-time university teachers as well.

As they indicate in their preface, they have chosen to cover a period of about twenty-five years. As they also state, they had no knowledge of parapsychology when they began. But writing a history is somewhat comparable to undertaking a research, in that preparatory acquaintance with the field is an almost necessary first step. To undertake either one without such preparation would be exceptionally daring in any field, especially in one like parapsychology—still new, exploratory, and controversial. The authors did know about the controversial aspect of it, however. They state in the preface that it was a field "marginal" to the scientific mainstream in much the same way that alchemy, for instance, already was in the seventeenth century. "*Unlike* alchemy, however," they add, "it is by no means clear that parapsychology is fated to remain a marginal science; this will be an issue for the future to decide. And the history of science shows that whichever way that decision goes, it will still be important to have a thorough historical understanding of the field in the various stages of its evolution."

In these statements the authors appear to be asking two questions. One is whether the field is fated to remain a marginal science. The other question is, What have been "the various stages of [the] evolution" of the field? Now, what criteria should be considered in answering these two questions? In answering the first question, the criterion to be considered would be the measure of recognition that the field has achieved in the world in general and in the sciences, psychology in particular. The criterion for answering the second question would be the degree of progress that was made towards the answering of the main issue: Does an extrasensory mode of communication—a psychic ability—exist?

How adequately have these questions been answered in this text? First, did this historical survey yield evidence that parapsychology, like alchemy, is fated to remain a marginal science? Alchemy is so classed first of all because its main objective proved to be impossible. Baser metals cannot be transmuted into gold, and therefore any methods used to try to do so had to fail.

But this historical analysis found no such situation in parapsychology. Even though much of the scholarly world may have been convinced that the parapsychological objective is just as impossible as was the transmutation of metals of the alchemists, the historians instead have arrived at a different answer. They so testify, in effect, when they say in their preface, "We are prepared to believe that many of the significant results reported . . . are real and not simply artifacts of fraud or poor experimental techniques." In other words, they found factual evidence they could not discount that bears on the parapsychological objective. This has to mean that the objective is not "impossible," as that of alchemy proved to be.

The historians, however, are then unable to go farther. They say that they can make no judgment as to what these significant results mean. They add, "Whether they are . . . freaks of random distributions or the product of some as yet indescribable 'para-normal faculty' is, as far as we can see, not yet decidable."

Does this mean, in terms of the original question, that parapsychology is destined to take its place in the scientific mainstream or that its fate is to remain "marginal," a pseudo-science, like alchemy? Although the authors give no definitive answer to this question, certainly any field of investigation in which significant experimental results have been obtained, even if their meaning is not yet decidable, can hardly remain marginal; nor can it be classed as pseudo-scientific if proper methods are used. Just as nature abhors a vacuum, so science abhors facts the meaning of which is not yet decidable. Instead, situations like this are the very subject matter of science, which above all is a system for the discovery of their meaning. In this instance parapsychology embodies that system, and it is devoted to the end of finding the meaning of its significant experimental results. It may still be a beginning science with many hurdles yet to overcome, but as long as its findings, its facts, are "real," it carries within itself the impetus, the need, to continue until the meaning of those facts has been disclosed.

Turning now to the question of the various stages of the evolution of parapsychology, the first observation must be that the twenty-five-year period chosen is too short to permit a generalization about the evolution of the field, for it covers mainly the early experimental period, which involved especially the work of the Duke Parapsychology Laboratory. This was the beginning period there, when the research problems were first being shaped and tested. It was a highly productive period, and it gave strong evidence of progress toward the establishment of the reality of psi ability, but it was not a period from which a final evaluation of parapsychology today can be made. However, in this historical portrayal, any adequate account of the actual content and progress of the field in this period is lacking. Although it introduces the reader to the history-of-science approach, which in many ways is interesting, especially in its blend with the sociology of science, still the book as a whole is in need of reinforcement on the side of the content and spirit of the field itself, which would include, too, a greater emphasis on its objective. In the space necessarily given to the historical background and the traditional objectives of psychical research, that objective as

Afterword

felt anew by JBR and instilled by him into his student subjects tends to be diminished. The result comes out a bit like a description of a car without an engine, so that the account given (mainly in chapters 4 and 5) is far too flat and pedestrian to reflect the reality. This early period at the Duke Parapsychology Laboratory was inspired by a great ideal, one as fresh and shining to these workers as if it had been original with them. That ideal was to discover the *whole* nature of man and to find out whether something more than just his physical, sensorimotor side exists.

The first simple concealed-card tests suggested (because from the start of the experimental program the results they yielded were at least a little better than flat chance) that the investigation was worth continuing. At the same time, it was necessary to work out methods of procedure to suit the ramifications of the new problems as they presented themselves.

Although the specific objective of the beginning inquiry was to find out whether telepathy really occurs, it was soon discovered that the conditions necessary to test it were more complicated than those necessary to test clairvoyance. And so the simpler, clairvoyance test was the first to be carried to a tentative conclusion. But success in that test raised new questions, which in turn called for new techniques. Thus the ultimate process was never a set or static one. Each encouraging result opened other avenues of conjecture, for example, Is this like other mental processes, affected by the person's state of health? by stimulants, drugs? or is it something quite distinct from these? Is it affected by distance? by time? Each result led to other questions, each new one calling for the devising of new and appropriate methods of testing.

The process over these early years (and ever since) thus was a continuing and developing one, and gradually those who were fully informed about the test results and their reliability came to recognize that the evidence was increasingly strong, ultimately, it seemed to them, compelling—in spite of the fact that most of the learned outside world was critical, skeptical, and finding it easier to question the integrity, the intelligence, and the competence of the researchers than to accept the results as even suggestive, much less valid.

And so it is today that to most parapsychologists themselves, the rate of the acceptance of parapsychology by psychology or any other science is quite secondary. But the reader does need to know in no uncertain terms what the field was actually about—what problems were addressed and why, what advances in methods were made, and to what extent they led to progress toward the solution of the main question. With a thoroughgoing presentation of all this, a more adequate and representative picture of the field would have been given and the reader would be better prepared to judge the answer to the other question, the extent to which parapsychology is like other sciences and unlike the pseudo sciences. Besides that, he would then get a glimpse of the lure of the field, of what it is that stirs us in parapsychology today and keeps us going with or without more general recognition.

But it is necessary to remember that these authors recognized that they were beginners in parapsychology. Being such, it was perhaps inevitable that the treatment they would give the topic of the content and development of the field would be inadequate. However, in terms of the historical viewpoint, they did work hard to draw a comprehensive picture. And while they did not hesitate to say they did not know which way the field was heading, they were convinced that significant results secured under adequate conditions of control had been found. If, like everybody else who gets deep into parapsychology, these men, too, are surprised at its complexity, they probably would concede that this is not the only scientific field in which the same is true. Accordingly, it is not the only field in which progress may necessarily be slow. But in their impatience these men also think this one is on the decline, basing their judgment largely on its rate of acceptance by psychologists and other scientists. To that the shortest reply is that for the period of the 1940s, they did not fully assess the havoc of World War II or realize the extent of the interruption those years caused in all intellectual pursuits. Nor do they know sufficiently well the signs of viability and progress the field has shown in the more recent years.

While such signs in one way depend on subjective judgments, some objective criteria can be cited. For instance, the number of American universities in which study and research in parapsychology is carried out has increased considerably in recent years. Besides the department at the University of Virginia, which has been in existence for some time, some of the new additions to the list are located at Syracuse, Princeton, Washington University (St. Louis), and two separate branches of the University of California. Also, the membership of the Parapsychological Association has recently increased to about three hundred from an original number of one hundred. But since parapsychologists themselves know well this recent history, they will know how to interpret this book's gloomy forecast. And they will appreciate its testimony that even to these historians, obviously trying to look with unbiased eyes, some of parapsychology's significant results are ''real.''

APPENDIX
Glossary of Parapsychological Techniques

This list of the technical terms coined by the Duke parapsychologists to denote their procedures has been extracted from the glossary published in *JP* 1 (1937): 306-7.

BM (BLIND MATCHING): The technique in which the subject matches a deck of ESP cards to five key cards which are laid out face down before him in an unknown order. Unless otherwise stated, the order is also unknown to the experimenter.

BT (BEFORE TOUCHING): The technique in which the top card of the face-down deck is called and, after being called, is laid aside for checking at the end of the run. Each card in the deck is treated in the same way.

CR (CRITICAL RATIO): The observed deviation divided by the standard deviation.

DT (DOWN THROUGH): The technique in which the cards are called down through the deck before any are removed or checked.

GESP = GENERAL (OR UNDIFFERENTIATED) EXTRA-SENSORY PERCEPTION: A technique designed to test the occurrence of extra-sensory perception permitting either telepathy or clairvoyance or both to operate.

KEY CARD: One of the five cards (where there are five suits) against which the cards of the test deck (i.e. target cards) in the matching tests (OM, BM, TM, etc.) are matched.

MATCHING: A form of calling in which a target card is placed opposite the key card which the subject selects to identify it.

OM (OPEN MATCHING): The technique in which the subject matches a deck of ESP cards to five key cards which are face up before him.

PT (PURE TELEPATHY) = TELEPATHY: The word "pure" emphasizes the exclusion of clairvoyance.

STM (SCREENED TOUCH MATCHING): The touch matching technique with an upright screen preventing vision by the subject of the handling of the cards by the experimenter.

TARGET CARD: The card which the percipient is attempting to perceive (i.e. to call, or otherwise indicate a knowledge of).

TOUCH MATCHING: The technique in which the subject indicates his call by tapping or touching one of the five key cards while the experimenter places the target card thus called in front of the key card indicated.

Notes

KEY

AHAP	Archives for the History of American Psychology, University of Akron, Akron, Ohio
ASPR	Archives, American Society for Psychical Research, New York, New York
CUA	Archives, Clark University, Worcester, Massachusetts
DUA	Archives, Duke University, Durham, North Carolina
EGB	Edwin G. Boring Papers, Harvard University Archives, Cambridge, Massachusetts (quotations by permission of the Harvard University Archives)
HPL	Harry Price Library, Senate House, University of London, London, England
HUA	Archives, Harvard University, Cambridge, Massachusetts (quotations by permission of the Harvard University Archives)
JASPR	*Journal of the American Society for Psychical Research* (during the period 1928-31 this journal was published under the name *Psychic Research,* and references to it in this period are made under the latter name)
JBR	J. B. Rhine Papers, Duke University Library, Durham, North Carolina
JP	*Journal of Parapsychology*
JSPR	*Journal of the Society for Psychical Research*
PSPR	*Proceedings of the Society for Psychical Research*
RM	*Revue métapsychique*
SPR	Archives, Society for Psychical Research, London, England
SUA	Archives, Stanford University, Stanford, California

Chapter One

1. Charles Richet, *Traité de métapsychique* (Paris: Félix Alcan, 1922); a second edition appeared in 1923. The work was translated into English by Stanley De Brath and published as *Thirty Years of Psychical Research* (New York: Macmillan, 1923). Subsequent references in this book will cite pages in both the first French edition (*TM*) and the English translation (*TY*); many translations given here, however, are our own. For a detailed and careful account of Richet's accomplishments and career in physiology, see the *Dictionary of Scientific Biography,* s. v. "Richet, Charles Robert."
2. See Pierre Janet, "Note sur quelques phénomènes de somnambulisme," *Revue philosophique* 21 (1886): 190-98; and Charles Richet, "Expériences sur le sommeil à distance," *ibid.,* 25 (1888): 435-49. See also below, n. 44.
3. Charles Richet, "La Métapsychique," *PSPR* 19 (1907): 2-49.
4. "A Text-Book of Metapsychics. Review and Critique by Sir Oliver Lodge," *PSPR* 34 (1924): 71-72.
5. *TM*, p. 756; *TY*, p. 595.
6. *TM*, pp. 16-42; *TY*, pp. 15-38.
7. *TM*, p. 5; *TY*, p. 6.
8. *TM*, p. 74; *TY*, p. 64.
9. *TM*, p. ii; *TY*, p. vii.
10. Alan Gauld, *The Founders of Psychical Research* (London: Routledge & Kegan Paul, 1968), chap. 8.
11. A convenient if brief discussion of this material may be found in Brian Inglis, *Natural and Supernatural* (London: Hodder and Stoughton, 1977), pp. 414-18.

12. Oliver Lodge, *Raymond: or, Life and Death* (London: Methuen & Co., 1916).
13. *TM*, p. 771; *TY*, p. 607.
14. *TM*, p. 776; *TY*, p. 612.
15. *TM*, p. 783; *TY*, p. 618.
16. Charles Richet, "Un Dernier Mot sur la cryptesthésie, lucidité: réponse à M. E. Bozzano," *RM*, 1922, p. 384.
17. We attempt here to distinguish between "spiritists," those who accepted post-mortem survival as an hypothesis capable of explaining certain psychical phenomena, and "spiritualists," those for whom the hypothesis passed into belief and became the core of a quasi religion.
18. Oliver Lodge, "En quoi l'hypothèse spirite est-elle justifiée par les faits?" *RM*, 1922, p. 67. It is ironic to see Lodge defending his own rather vague psychical theories some years later in the very terms he so strongly attacks here: "If Newton had declined to consider gravitational astronomy till he understood the nature of a gravitational field, we should still be in the Dark Ages of science" (Oliver Lodge, *My Philosophy* [London: Ernest Benn, 1933], p. 296).
19. Lodge, "En quoi l'hypothèse spirite est-elle justifiée," p. 68.
20. Ernest Bozzano, "L'hypothèse spirite et la 'cryptesthésie,'" *RM*, 1922, pp. 236-46.
21. Richet, "Un Dernier Mot," p. 383. The continuing debate between Richet and Lodge may be followed in Charles Richet, "L'hypothèse spirite: réponse à Sir Oliver Lodge," *RM*, 1922, p. 156; and in Oliver Lodge, "Sequel to the Review of Professor Richet's 'Traité de Métapsychique,'" *PSPR* 34 (1924): 99-102.
22. Gauld, *Founders*, chaps. 10-11.
23. Gustave Geley, "L'ectoplasmie," *RM*, 1920-21, pp. 355-61. For a recent, skeptical account of the experimentation with Eva C., see Rudolf Lambert, "Dr Geley's Report on the Medium Eva C.," *JSPR* 37 (November 1954): 380-86.
24. *TM*, *TY*, bk. 2, chap. 2.
25. Edmund Gurney, Frederic W. H. Myers, and Frank Podmore, *Phantasms of the Living*, 2 vols. (London: Trübner, 1886), vol. 1, p. lxv. Myers had tried to introduce the term "telesthesia" as an alternative to "clairvoyance," but the term did not catch on. He later contrasted the two in the following terms: "telepathy" he defined as "the communication of impressions of any kind from one mind to another, independently of the recognized channels of sense," while "telesthesia" he applied to "any direct sensation or perception of objects or conditions independently of the recognised channels of sense, and also under such circumstances that no known mind external to the percipient's can be suggested as the source of the knowledge thus gained" (F.W.H. Myers, "Glossary of Terms Used in Psychical Research," *PSPR* 12 [1896-97]: 166-74).
26. *TM*, pp. 76-77; *TY*, pp. 65-66.
27. T. W. Mitchell, "Presidential Address," *PSPR* 33 (1923): 11.
28. "It is not surprising that [Richet] finds a difficulty about telepathy. If telepathy means direct reading and interpreting the molecular configuration in another person's brain, ... such reading is I admit frankly incredible.... The point largely turns upon the question whether mind ever acts on mind directly without the customary modes of bodily and sensory signalling, and without the unlikely and unsupported hypothesis of brain acting on brain.... Richet is naturally so impressed, through a life-long occupation with Physiology, with the material and cerebral aspect of orthodox psychic phenomena in general, that he does not feel as if he were theorising in the least when he assumes that throughout every mental action, in origin, in transmission, and in reproduction, there must be a physical concomitant at every stage" (Lodge, "Text-Book," pp. 80-81).
29. "The aim of the Society will be to approach these various problems without prejudice or prepossession of any kind, and in the same spirit of exact and unimpassioned inquiry which has enabled Science to solve so many problems, once not less obscure nor less hotly debated" ("Objects of the Society," *PSPR* 1 [1882-83]: 4).
30. For an example of this strain of thought, see the remarks by Charles Richet quoted by Eugène Osty in "Un Savant—une oeuvre," *RM*, 1925, pp. 201-14, and Osty's comments thereupon.
31. See below, chap. 2, for an illustration of this concern.
32. See, for example, Gustave Geley, "Enquête expérimentale sur la lucidité," *RM*, 1920-21, pp. 3-17: "La seule perspective d'être soumis à une enquête, de se trouver pour ainsi dire en présence d'un juge, était de nature à les paralyser."
33. René Sudre, *Introduction à la métapsychique humaine* (Paris: Payot, 1926), p. 70.
34. See below, chap. 2.

35. H.I.F.W. Brugmans, "Une Communication sur des expériences télépathiques au laboratoire de psychologie à Groningue faites par M. Heymans, Docteur Weinberg et Docteur H.I.F.W. Brugmans," *Le Compte rendu officiel du Premier Congrès International des Recherches Psychiques à Copenhague* (Copenhagen, 1922), pp. 396-408.
36. Sudre, *Introduction*, p. 133; see also the very favorable judgment by Charles Richet, *Notre Sixième Sens* (Paris: Editions Montaigne, [1928]), pp. 153-56. We have not been able to see a copy of the original edition, published in 1927. The work was translated by Fred Rothwell as *Our Sixth Sense* (London: Rider, [1929]).
37. J. G. Piddington, "Presidential Address," *PSPR* 34 (1924): 138-39.
38. William McDougall, "Presidential Address," *PSPR* 31 (1921): 107.
39. In his own presidential address in 1922, T. W. Mitchell implied his own skepticism of this altruistic view, insisting that economic dependency and "the incubus of authority and tradition" were at work in keeping physicians like himself from the field (*PSPR* 33 [1923]: 18-20).
40. See below, chap. 6; see also Paul Forman, "Weimar Culture, Causality, and Quantum Theory, 1918-1927: Adaption by German Physicists and Mathematicians to a Hostile Intellectual Environment," *Historical Studies in the Physical Sciences*, vol. 3 (Philadelphia: University of Pennsylvania Press, 1971), pp. 1-115, and *idem*, "The Reception of an Acausal Quantum Mechanics in Germany and Britain," in *The Reception of Unconventional Science by the Scientific Community*, ed. Seymour H. Mauskopf, AAAS Selected Symposium 25 (Boulder: Westview, 1979), pp. 11-50.
41. Review of *Traité de métapsychique*, in *L'Année psychologique* 23 (1922): 602.
42. L. Lapicque *et al.*, "Rapport sur des expériences de controle relatives aux phénomènes dits ectoplasmiques," *L'Année psychologique* 23 (1922): 604-11. For Richet's reply, see Gustave Geley, "A propos des Expériences de la Sorbonne," *RM*, 1922, pp. 225-29, and Richet's letter attached, pp. 229-30; see also Charles Richet, "A propos des ectoplasmes," *ibid.*, pp. 281-83.
43. "A propos de la métapsychique," *Revue philosophique* 96 (1923): 5-32.
44. See above, n. 2. An account of Janet's successful attempts to hypnotize a patient at distances of up to a kilometer is given by Bert S. Kopell in "Pierre Janet's Description of Hypnotic Sleep Provoked from a Distance," *J. Hist. Behav. Sci.* 4 (1968): 119-23; Kopell has also translated Janet's two papers on his experimental work in *ibid.*, pp. 124-31 and 258-67. Janet himself was unhappy with the assumption that his work demonstrated the existence of an "unknown faculty" of the human mind. For a full summary of Janet's life and thought, see Henri F. Ellenberger, *The Discovery of the Unconscious* (New York: Basic Books, 1970), pp. 331-417.
45. Janet, "A propos de la métapsychique," pp. 26-27.
46. *Ibid.*, p. 28.
47. *Ibid.*, pp. 31-32.
48. Lodge, "Text-Book," p. 72.
49. Charles Richet, "Réponse à M. P. Janet. A propos de métapsychique," *Revue philosophique* 96 (1923): 466-67.
50. Ellenberger, *Discovery of the Unconscious*, pp. 85, 90.
51. McDougall's involvement with psychical research after 1920 is treated below, *passim*; a collection of his writings on the subject is Raymond Van Over and Laura Oteri, eds., *William McDougall: Explorer of the Mind* (New York: Helix Press, 1967). On Schiller and psychical research, see Reuben Abel, *The Pragmatic Humanism of F.C.S. Schiller* (New York: Columbia University Press, 1955), pp. 140-44; for a concrete expression of his views, see his article, "Some Logical Aspects of Psychical Research," in *The Case For and Against Psychical Belief*, ed. Carl Murchison (Worcester, Mass.: Clark University, 1927), pp. 215-26. A biography of Lodge that discusses his career in psychical research is W. P. Jolly, *Sir Oliver Lodge* (Rutherford, Madison, and Teaneck: Fairleigh Dickinson University Press, 1975). For studies of the relation of Lodge's physics to his psychical research, see D. B. Wilson, "The Thought of the Late Victorian Physicists: Oliver Lodge's Ethereal Body," *Victorian Studies* 15 (1971): 29-48; and Charles Seth Landefeld, "Science and Spiritism in the Psychical Research of Oliver Lodge" (Senior honors thesis, Harvard University, 1974). Doyle's biographers have neglected the details of his role in postwar psychical research; some material will be found below in chapter 2.
52 Rudolf Tischner, *Geschichte der Parapsychologie* (Tittmoning/Obb: Walter Pustet, 1960), bk. 2, pp. 215-16, 343. For an appraisal of Schrenck-Notzing's achievement by a French investigator, see the obituary notice by René Sudre in *Psychic Research* 23 (1929): 250-55.

53. For a rather negative appraisal of Hyslop's career in psychical research, see R. Laurence Moore, *In Search of White Crows* (New York: Oxford University Press, 1977), pp. 156-68; see also *JASPR* 14 (1920): 425-528, for a variety of contemporary views.

54. A collection of appreciations is Friends and Colleagues in Psychical Research, *Walter Franklin Prince: A Tribute to His Memory* (Boston: Boston Society for Psychic Research, 1935). For a more recent appraisal, see Thomas R. Tietze, "Ursa Major: An Impressionistic Appreciation of Walter Franklin Prince," *JASPR* 70 (1976): 1-34.

55. Quoted from Carl Vett's announcement of the congress, 25 May 1921 (ASPR).

56. McDougall to Prince, 10 June 1921. Letters from Dawson to Prince of 5 July and 13 July summarize the financial details (ASPR).

57. Prince to Sudre, 18 May 1922 (ASPR).

58. See *RM*, 1923, pp. 277, 289; 1927, p. 315.

59. Elwood Worcester to Prince, 29 November 1921 (ASPR).

60. McDougall to Prince, 16 May 1923 (ASPR). McDougall sent the letter, written shortly after his removal as president of the society, to all the members of the Advisory Scientific Council to explain his view of what had happened in the ASPR.

61. Dawson to Tubby, 7 May 1921 (ASPR).

62. See below, chap. 3.

63. *JASPR* 16 (1922): 1-2.

64. Jastrow to Prince, 18 September 1922 (ASPR).

65. Worcester to Prince, 24 April 1923 (ASPR).

66. Edwards to members of the [Advisory Scientific] Council, 28 May 1923 (ASPR); see also *JASPR* 17 (1923): 321-23. Edwards, originally an Episcopal clergyman in Detroit, had developed strong spiritualist leanings after the death of his son in the war. Many of Edwards's papers and his journal are in the Duke University Library.

67. Miss Tubby's approach to psychical research, which heavily emphasized the contributions of J. H. Hyslop, may be studied in her book, *Psychics and Mediums: A Manual and Bibliography for Students* (Boston: Marshall Jones, 1935).

68. Prince to Worcester, 28 May 1923 (ASPR). Looking back some years later, Prince insisted that "Edwards . . . owed to her [Miss Tubby's] admirable powers as a politician, his job" (Prince to Hyslop, 3 October 1930 [ASPR]).

69. Prince to Worcester, 13 June 1923 (ASPR).

70. Hyslop's reaction to the attempt to make psychical research a subject of laboratory investigation may be studied in James H. Hyslop, "Psychic Research in Harvard University," *JASPR* 13 (1919): 355-60; and *idem*, "Leland Stanford University in Psychic Research," *ibid.*, 12 (1918): 529-44. Cf. Miles M. Dawson, "Professor Hyslop's Engrossing Interest in Psychical Research," *ibid.*, 14 (1920): 452-56.

71. Henry Holt, "A Review of Richet," *ibid.*, 16 (1922): 655-70.

72. *Ibid.*, 17 (1923): 11.

73. George E. Wright, "Professor McDougall and Spiritualism," *ibid.*, p. 215.

74. F[rederick] E[dwards], "Leakage," *ibid.*, p. 267. Edwards noted, a little sarcastically, that the ASPR had had only three psychologists as subscribers at the beginning of 1923: McDougall, Gardner Murphy, and H. Norman Gardiner of Smith College.

75. Frederick Edwards, "Sir Arthur Conan Doyle," *ibid.*, p. 272.

76. Undated typescript with heading in W. F. Prince's hand, "Marks of Scientific Degeneracy in Journal of A.S.P.R. 1923-1924" (ASPR).

77. Undated (but early 1924) reply to president and board of trustees of ASPR from Margaret Deland. On 29 January 1924, Walter Franklin Prince wrote Elwood Worcester that "there is a mad rush on to increase membership. Thousands of invitations are being sent out, the Society is fairly on its knees begging people to join" (both letters in ASPR).

78. McDougall's resignation was announced in *JASPR* 17 (1923): 390; Mrs. Deland's resignation was accepted at a board meeting of the society on 18 July 1923; and Waldemar Kaempffert and Charles M. Baldwin were elected to succeed McDougall and Mrs. Deland (*ibid.*, p. 582).

79. The imminent "note of remonstrance," with its fifty-three signatures, was announced by Elwood Worcester to W. F. Prince in a letter of 20 December 1923 (ASPR). So far we have not been able to find a copy of the remonstrance, but its substance is summarized in Mrs. Deland's reply to the board of trustees of the ASPR cited in note 77 above.

80. Prince to Worcester, 29 January 1924 (ASPR).
81. *Scientific American* 17 (1922): 389.
82. See, for example, J. Malcolm Bird, "Our Psychic Investigations in Europe—II," *ibid.*, 128 (1923): 379-80, 428-29.
83. For an introduction to the career of this medium, see Thomas R. Tietze, *Margery* (New York: Harper & Row, 1973).
84. The phrase is used in a letter from Margery's husband, Dr. LeRoi Crandon, to Sir Arthur Conan Doyle, dated 23 July 1926 (HPL).
85. In a letter to Harry Price dated 23 December 1924, Edwards commented as follows about Margery: "Private. I have never seen or sat with the Crandons and have nothing to say because I know nothing at first hand. The woman may be a fraud; or conditions may have been spoiled; or there may be a mixture of both. But I have heard this. She is Dr. Crandon's second wife,—I think; and was a nurse or secretary—I think; he was a complete unbeliever—I think; she converted him—I think; he dotes on her and the whole business—I think; and insists on sitting next to her all the time—I think; the reasons adduced for this are many—I think. One of the most fruitful causes of superstition is mediumship in the family; especially a second and perhaps ambitious wife who can pull the wool over the eyes of an elderly man.... Perhaps Mrs. C.—did not realize that her husband would give her quite so much publicity" (HPL).
86. Edwards to Price, 27 February 1925 (HPL).
87. Bird to Prince, 12 January 1931. A copy of this letter was sent to J. B. Rhine by Prince and is filed in the Rhine Papers (JBR). See also Tietze, *Margery,* chap. 15.
88. "The Boston Society is very careful indeed whom it admits, consequently, with a very few exceptions, its members are all inclined to respect scientific method and caution. It is evident that you have discovered, or at least you realize, that the real *work* must be done by the very few, and that all notions of making scientific research democratic are a fallacy and a snare" (Prince to Price, 9 September 1926 [ASPR]).

Chapter Two

1. J. G. Piddington, "Presidential Address," *PSPR* 34 (1924): 145-46. Piddington was perhaps thinking of William McDougall rather than Oliver Lodge, for the distinction is a prominent feature of McDougall's presidential address of 1920 (*PSPR* 31[1921]: 108).
2. The biography of Harry Price by Paul Tabori is untrustworthy and should be corrected by Trevor H. Hall, *The Search for Harry Price* (London: Duckworth, 1978). The laboratory, its equipment, and its projects are described in Harry Price, *Fifty Years of Psychical Research* (London and New York: Longmans, Green, 1939), pp. 317-26 (app. B).
3. Dingwall to Price, 27 November 1924 (HPL). As Dingwall pointed out in print, the laboratory occupied the top floor of the premises of the London Spiritualist Alliance (*JSPR* 23 [June 1926]: 96).
4. Anita Gregory, "Anatomy of a Fraud: Harry Price and the Medium Rudi Schneider," *Annals of Science* 34 (1977): 481-82; for a vivid character portrait of Harry Price, see pp. 470-72.
5. The announcement that Dingwall's appointment had been terminated came in *JSPR* 24 (April 1927): 51.
6. E. J. Dingwall, "The Crisis in Psychical Research," *Realist* 1 (1929): 68-81. For a critique from the opposite end of the spectrum, see The Research Officer [J. Malcolm Bird], "The Crisis in Psychical Research," *Psychic Research* 23 (1929): 323-36.
7. Theodore Besterman, "A Critical Estimate of the Present Status of Psychical Research," *Transactions of the Fourth International Congress for Psychical Research* (London: Society for Psychical Research, 1930), p. 128.
8. *JSPR* 29 (March 1935): 48.
9. "... the superior airs of the intolerable Bestermann" (Ethel Robertson to Price, 16 November 1930); "that young swine Bestermann" (Price to H. Dennis Bradley, 31 December 1930). For an extreme example, introducing a strain of anti-Semitism, see Charles Röthy to Price, 9 December 1930 (all letters in HPL).
10. Röthy complained angrily about Besterman to Oliver Lodge, as well as to Price. Lodge's reply of 4 March 1931, while recognizing that Besterman was "rather smitten with exaggerated scepticism," insisted that Besterman's attitude was certainly in the traditions of the SPR and con-

cluded that "he is a well meaning youth, who is likely to improve and who takes such rebukes as I have had occasion to administer very well, and shows a wish to learn." Price's comment to Röthy (9 April 1931) was that "Sir Oliver did really (in a mild way) rebuke Bestermann. But of course nothing will come of it. Salter and Bestermann are, at the moment, paramount in the Society..." (all in HPL).

11. Besterman's review was of Gwendolyn Kelley Hack, *Modern Psychic Mysteries;* it appeared in *JSPR* 26 (January 1930): 10-14. Doyle's resignation and the responses to it were printed as "Sir Arthur Conan Doyle's Resignation," ibid. (March 1930): 45-52.

12. *Ibid.*, p. 47.
13. *Ibid.*, p. 50.

14. Price did not reveal that in 1929 he had made the same offer privately to the IMI, which Eugène Osty had reluctantly declined due to lack of space with which to house Price's collections (see Gregory, "Anatomy," pp. 494-95).

15. Price sent a circular letter announcing his intention to members of the SPR council on 30 October 1930; and on 12 November, while the matter was still before the council, he sent a second circular to all members of the society asking for their views on the amalgamation. The council declined his offer before responses to his second letter had had a chance to come in—218 for amalgamation and 34 against, according to yet another circular letter sent to SPR members on 19 December (SPR). At the general meeting, Price's proposal was somewhat overshadowed by the lively debate over the previous resolution, moved (unsuccessfully) by E. J. Dingwall, asking for a change in the administration of the society and criticizing Besterman's role (Report of the annual general meeting held 26 February 1931, in *JSPR* 27 [April 1931]: 52-59; see also Gregory, "Anatomy," pp. 499-500).

16. "Annual Report of the Council for the Year 1930," in *JSPR* 27 [February 1931]: 24. F.C.S. Schiller, who was at the time a vice-president of the SPR and a member of its council, wrote to Harry Price on 3 November 1930 in terms that suggest that not all the society felt as confident as this statement would imply. "I'm afraid there is 0 to be done with Psychical Research at the SPR under present conditions & it is best to wait till they change (which they may do, if for no other reason than that members are resigning in large numbers)" (HPL).

17. London: C. Vernon and Sons, [1931]. A slightly shorter version of Bradley's indictment is printed in his book... *And After* (London: T. Werner Laurie, 1931), pp. 198-216.

18. René Warcollier, *La Télépathie; recherches expérimentales* (Paris: Alcan, 1921). On its prehistory, see pp. xvii-xix.

19. *Ibid.*, pp. v-vi.
20. *Ibid.*, chaps. 1-3.
21. *Ibid.*, pp. 151-53.
22. *Ibid.*, p. 153.

23. *Ibid.*, pp. 155-61. On the whole, however, Warcollier resisted the use of objective material. His collaborator Gardner Murphy went so far as to make up a set of some ten thousand stimuli that could be scored precisely—targets distinguished by geometric form, color, size, and shade—and sent them to Paris. Warcollier replied regretfully that he could not use them, that "the French mind was not so exact" and could not accommodate to them (interview, Gardner Murphy, 5 December 1973).

24. Warcollier, *La Télépathie*, p. 256.

25. "Just as pathology has rendered the greatest services to physiology, it is not impossible that all these errors of supernormal perception—so like those of our normal perception—and all these failures, perhaps even more valuable than successes, might contribute powerfully to their own explanation. It is in their will-o'-the-wisp light that we must search for the key to telepathy" (*ibid.*, p. 326).

26. *Ibid.*, pp. 316-22.

27. Warcollier did not assume the need for materialist explanations, and he was willing to leave open the possibility of a spiritist answer for some phenomena; but he insisted that "for me, telepathy is quite adequate, since I believe that it is constant and continuous in all living beings, both awake and asleep" (*ibid.*, p. 337).

28. "We must praise the effort that Mr. W. has made to isolate the phenomenon of thought-transference from the complex metapsychic mixture. But in psychology we unfortunately cannot hope to discover the "simple bodies" that are the triumph of chemistry; [its] elements do not always

explain the compound and indeed sometimes are themselves explained by the compound. Thought-transference is not a pure psychological operation that can be studied outside of the broad problem of lucidity" (*RM,* 1922, p. 53).

29. "A note in the February, 1925 number of *Psychica* contains the information that La Société d' Etudes Télépathiques founded by M. R. Warcollier in 1922, and with which transatlantic tests in telepathy were conducted by Dr. Gardner Murphy, has undergone dissolution, its committee having accepted the generous offer of Dr. Osty of becoming a department of the Institut Métapsychique International" (*JASPR* 19 [1925]: 601). The report was composed by J. B. Rhine, who at this point had just been given the responsibility of excerpting European journals for the readership of the *JASPR* (see below, chap. 4).

30. René Warcollier, *Experimental Telepathy* (Boston: Boston Society for Psychic Research, 1938), pp. 56-73. This is a collection of Warcollier's writings drawn together from *La Télépathie* and from his later articles in the *Revue métapsychique,* edited by Gardner Murphy and translated into English.

31. See, for example, Murphy's defense of Warcollier's approach in his review of the *Transactions of the Fourth International Congress for Psychical Research,* in *PSPR* 40 (1931-32): 99-104; or his defense of Warcollier against the critical remarks of S. G. Soal (on which see below), in *JSPR* 27 (February 1932): 207-13.

32. See, for example, Charles Richet, *Our Sixth Sense,* trans. Fred Rothwell (London: Rider, [1929]), pp. 122-26; E. Osty, "La Télépathie expérimentale," *RM,* 1925, pp. 5-28; and *idem,* "Comment déceler, développer et ne pas pervertir la faculté de connaissance supra-normale," *RM,* 1931, pp. 93-118.

33. Reviewed in *JSPR* 21 (July 1923): 124-26.

34. "Concerning Monsieur Warcollier's Suggested Experiments in Thought-Transference," *ibid.* (December 1923): 177. The mathematician was F.J.M. Stratton.

35. "Annual Report of the Council for 1923," in *ibid.* (March 1924): 234.

36. Ina Jephson, "Evidence for Clairvoyance in Card-Guessing," *PSPR* 38 (1928-29): 227-28.

37. R. Warcollier, "La Télépathie active et passive," *RM,* 1924, p. 359. The article was translated in Warcollier, *Experimental Telepathy,* pp. 93-100.

38. R. Warcollier, "L'Accord télépathique," *RM,* 1928, pp. 286-306; translated in *Experimental Telepathy,* pp. 74-92.

39. "I may add here that I think the difficulty sometimes felt in admitting this arises partly from a tendency to limit the scope of telepathy which has been fostered by the use of the terms 'agent' and 'percipient' to denote respectively the mind which is the source of the information transferred and the mind which receives it. These technical terms, introduced in the early days of the S.P.R., had their origin in thought-transference experiments in which the person from whom the idea is to be transferred tries to concentrate his mind on it with a view to transferring it. But the term agent is not very happily chosen, since it assumes that this effort of concentration is an effective part of the process of transference, whereas for anything we know it may have no effect except that of ear-marking the idea it is desired to transfer. It is possible that the so-called agent's part is purely passive, while the percipient has to play the active part and extract the idea or combination of ideas from the 'agent's' mind" (Mrs. Henry Sidgwick, "A Contribution to the Study of the Psychology of Mrs. Piper's Trance Phenomena," *PSPR* 28 [1915]: 319).

40. Mrs. Henry Sidgwick, "Phantasms of the Living," *PSPR* 33 (1923): 39-40.

41. "Contact or transfusion is a physical idea just as much as transference or transmission, and probably represents what actually occurs almost, if not quite, as imperfectly. But it has the merit of eliminating the idea of distance and false analogies depending on that.... The mental contact analogy seems, in fact, more elastic than the transference analogy, and lends itself better to any case where the relation of agent and percipient is not simple and obvious" (*ibid.,* pp. 422-23).

42. Jephson, "Evidence," p. 230.

43. R. A. Fisher, "A Method of Scoring Coincidences in Tests with Playing Cards," *PSPR* 34 (1924): 181-85.

44. "A Request for Assistance," *JSPR* 23 (June 1926): 87.

45. Jephson, "Evidence," pp. 253-64. On Estabrooks's *A Contribution to Experimental Telepathy,* see below, chap. 3.

46. Jephson, "Evidence," p. 243.

47. *Ibid.*, pp. 256-57.
48. The letter is in SPR:

Chère Miss Jephson,
  Avec un intérêt passionné je viens de lire votre important article. Je me permets de vous faire remettre par M. Bestermann, mon livre sur le sixième sens.
  Dans ma prochaine édition, une grande, très grande, place sera donnée à vos recherches.
  Comme vous avez raison de considérer la télépathie comme un cas particulier de clairvoyance.
  Merci mille fois pour cette contribution excellente que vous avez donnée à notre chère Métapsychique, et croyez à mes sentiments de respectueuse gratitude.
                                                                                                                Charles Richet
21 dec. 1928

49. Saltmarsh to Jephson, 7 January 1929 (SPR).
50. Smith to Jephson, 25 March 1929 (SPR). Smith (who later changed his name to Carington) had been interested in the measurement of emotion as both a psychological and a psychical-research technique for several years and continued to be so (see below, chap. 8).
51. V. J. Woolley, "The Broadcasting Experiment in Mass-Telepathy," *PSPR* 38 (1928-29): 1-9.
52. Price to Lawrence J. Jones, 12 May 1928 (HPL).
53. Woolley, "Broadcasting Experiment," p. 8.
54. One quite clear indication of this positive interest in the new experimental work being undertaken was the SPR's decision to coopt to the council two of the most active experimentalists, Ina Jephson and S. G. Soal, together with Theodore Besterman, in June 1928.
55. These details are taken from an autobiographical sketch of Soal's early life now in the possession of Mrs. K. M. Goldney.
56. S. G. Soal, "A Report on Some Communications Received through Mrs. Blanche Cooper," *PSPR* 35 (1925): 471-594. Late in 1929 and into the new year, in a series of letters in the SPR *Journal*, Soal explained his position more fully. Mere assertion of the spirit hypothesis was a sterile dead end; only a scientific approach, conducting experiments designed to bring the necessary conditions under control, could advance the subject, and he recommended that the SPR enlist trained psychologists, who (unlike the membership at large) would do such experiments (*JSPR* 25 [October 1929]: 145-46 and [December 1929]: 176-79; 26 [April 1930]: 68-69).
57. S. G. Soal, "Experiments in Supernormal Perception at a Distance," *PSPR* 40 (1931-32): 170. A much fuller statement of Soal's criticisms of Warcollier's experimental methodology may be found in "Methods in Experimental Telepathy," *JSPR* 27 (October 1931): 130-35 and (April 1932): 236-41.
58. Soal, "Experiments," pp. 171-210.
59. *Ibid.*, pp. 211-362.
60. Theodore Besterman, S. G. Soal, and Ina Jephson, "Report of a Series of Experiments in Clairvoyance Conducted at a Distance Under Approximately Fraud-Proof Conditions," *PSPR* 39 (1930-31): 386.
61. Fisher's suggestion, as well as the statistical grounds upon which he based it, are discussed in Smith to Jephson, 25 March 1929 (SPR).
62. Besterman, Soal, and Jephson, "Report," p. 375.
63. Untitled manuscript in Soal's hand, beginning, "The main experiments in Long-Distance Telepathy come to an end on March 28th . . ." (p. 4 [SPR]).
64. Besterman, Soal, and Jephson, "Report," pp. 401-2.
65. *Ibid.*, pp. 412-14. A subsequent experiment carried out by Miss Jephson to test whether used playing cards *did* carry associations that could be supernormally perceived also failed to achieve positive results (Ina Jephson, "A Behaviourist Experiment in Clairvoyance," *PSPR* 41 [1932-33]: 99-114).
66. Soal, "Experiments," pp. 165-362; the paper comprised the results of both Soal's first (1927-28) and second (1928-29) experimental series.
67. *Ibid.*, pp. 270-362.
68. Preface to René Warcollier, *Mind to Mind*, ed. Emanuel K. Schwartz (New York: Creative Age Press, 1948), p. xvi.

69. William Brown, "Psychology and Psychical Research," *PSPR* 41 (1932-33): 81. Brown, who had studied with McDougall and had followed him as Wilde Professor of Moral Philosophy at Oxford, had been associated with Harry Price's National Laboratory as well as with the SPR. He went on to argue that in fact these negative results were quite important, since if telepathy were indeed proved to be a rare phenomenon, critics would have difficulty explaining away evidence for survival in terms of telepathy.
70. E. Osty, "Télépathie spontanée et transmission de pensée expérimentale," *RM*, 1932, pp. 234-56, 305-24, 377-405; 1933, pp. 1-40. Osty quoted the dismayed reaction of César de Vesme, formerly editor-in-chief of the *Annales des sciences psychiques:* "For my part, I admit that after having learned of M. Soal's report, I feel myself less ready to attribute to a mysterious intercommunication of human thoughts certain psychic phenomena that I had hitherto believed easy to explain in this way."
71. *Ibid.*, 1932, p. 244.
72. *Ibid.*, pp. 311-15. Throughout the 1920s experimenters reported that Ossowiecki was able to describe words and pictures drawn on papers hidden from his sight, sealed into opaque envelopes or even leaden tubes. For some of Ossowiecki's early work, see Gustave Geley, *Clairvoyance and Materialisation,* trans. Stanley De Brath (New York: Doran, 1927), pp. 30-93. A remarkable effort a decade later is described by Theodore Besterman in "An Experiment in 'Clairvoyance' with M. Stefan Ossowiecki," *PSPR* 41 (1932-33): 345-51.
73. "What is transmitted in telepathy has a dynamic power quite different from that of a design, a playing card, etc. . . . It is the vibrant life of a human *psychisme* with its hopes, its desires, its worries, its anguishes, its distresses, in its manifestations in the midst of beings and things" (Osty, "Télépathie spontanée," p. 317). Soal reviewed Osty's article very favorably for the SPR (*JSPR* 28 [November 1933]: 139-44), remarking that "he concludes with perfect justice that the power to receive telepathically a thought which an agent tries voluntarily to transfer is not possessed by the generality of mankind." He made no response or indeed reference to any of Osty's judgments about his own work.
74. A copy of Soal's manuscript is in the SPR archives. Soal had originally proposed the article to Price (10 April 1934) as a way to expose the weakness of the SPR's evidence for telepathy and clairvoyance. Price however found Soal's criticism of the SPR's replacement of Dingwall by Besterman as verging on the libellous; although Soal revised the paper, it was never published—much to his disgust (Soal to Joad, 2 March 1937; both letters in HPL). A copy of Soal's manuscript is in SPR.
75. "Experimental Telepathy and Clairvoyance in England, 1881-1933," pp. 22-23 (HPL).
76. *Ibid.*, p. 34; and see Soal to Price, 8 May 1934 (HPL).
77. *Ibid.*, p. 35.
78. Charles Richet, "Lucidité et probabilité," *RM*, 1934, pp. 256-60. An earlier paper by the engineer Robert Desoille—"De Quelques conditions auxquelles il faut satisfaire pour réussir des expériences de télépathie provoquée," *RM*, 1932, pp. 410-17—had made much the same point.

Chapter Three

1. Harlow Gale, "Psychical Research in American Universities," *PSPR* 13 (1898): 583-87.
2. James H. Hyslop, "Psychic Research in American Universities," *JASPR* 11 (1917): 444-58.
3. Frederick C. Dommeyer, "Psychical Research at Stanford University," *JP* 39 (1975): 173-205.
4. Hyslop, "Psychic Research," p. 453. An unsigned letter (apparently from the chairman of the philosophy department) in the Harvard University Archives dated 9 March 1916 proposes Troland's name for an instructorship in psychology and explains: "The income to pay for this Instructorship would be given by Miss Theodate Pope of Farmington, Connecticut. Dr. Troland and I visited her in the Christmas vacation. A provisional agreement was made for $1500, but my expectation is that this would be increased to $2000. . . . If this appointment be made the University would have an opportunity to spend the income of the Richard Hodgson Memorial Fund by appropriating it to Dr. Troland to pay the expenses of his experiments."
Apparently Harvard decided not to take advantage of a still larger gift. More than twenty years later, Miss Pope (then Mrs. Riddle) recalled: "You may possibly have known of my having offered to support a chair at Harvard in 1913, for the study of this subject. President Lowell sent one of the

professors to Farmington to see me in regard to it. I wisely refrained from making any commitment beyond the first year. The University asked to have half of the annual sum given to the Department of Normal Psychology which was done, but when I discovered that they would not mention the subject of Psychical-Research in their year book, I turned my attention to founding a school for boys..., which I did at a cost of Five Million Dollars. That was the sum that I intended to eventually give to Harvard for the study of Psychic Phenomena" (Theodate Pope Riddle to J. B. Rhine, 30 October 1936 [JBR]).

5. Joseph Jastrow, *Fact and Fable in Psychology* (Boston and New York: Houghton Mifflin Co., 1900), pp. 54–55.

6. John Edgar Coover, *Experiments in Psychical Research at Leland Stanford Junior University*, Psychical Research Monograph no. 1 (Stanford: Leland Stanford Junior University Publications, 1917), p. xxii.

7. G. Stanley Hall, "Psychological Literature Review," *Amer. J. Psychol.* 7 (1895): 139.

8. For a further perspective on the response of these psychologists to psychical research, see R. Laurence Moore, *In Search of White Crows* (New York: Oxford University Press, 1977), 133–56.

9. In *Amer. J. Psychol.* 1 (1888): 128–46. This and other manifestations of Hall's response to psychical research are discussed in Dorothy Ross, *G. Stanley Hall: The Psychologist As Prophet* (Chicago and London: University of Chicago Press, 1972), pp. 162–64, 174–75.

10. *Amer. J. Psychol.* 1 (1888): 141.

11. "We rejected every claimed fact in which the psychological facts were without a physical substratum, as in the case of departed spirits and those in which psychical facts influenced one another without physical intermediation, as in telepathy" (Hugo Münsterberg, "Psychology and Mysticism," *Atlantic Monthly* 83 [1899]: 82). Münsterberg apparently tried to keep Harvard from accepting the Hodgson funds, and when that failed, he attempted to hinder the research carried on with Hodgson money by L. T. Troland (Bruce Kuklick, *The Rise of American Philosophy, Cambridge, Massachusetts, 1860-1930* [New Haven and London: Yale University Press, 1977], p. 420n; Kuklick gives a good general conspectus of Münsterberg's psychology on pp. 196–214).

12. Jastrow, *Fact and Fable*, pp. 76–77. Margaret Washburn indirectly confirms the widespread existence of this public image of psychology in "Some Recollections," in *A History of Psychology in Autobiography*, vol. 2, ed. Carl Murchison (Worcester, Mass.: Clark University Press, 1930). Describing being interviewed about graduate work at Columbia by James McKeen Cattell sometime in the 1890s, she wrote, "I blessed the hours I had spent on Wundt's article [on psychological methods]; instead of speaking as I am sure I was expected to do, of hypnotism, telepathy and spirtism, I referred to reaction-time, complication experiments, and work on the limens and Weber's Law, and was rewarded with the remark that I seemed to have some knowledge of the matter" (p.339).

13. *William James on Psychical Research*, ed. Gardner Murphy and R. O. Ballou (New York: Viking Press, 1960).

14. James Rowland Angell, *Chapters from Modern Psychology* (1912; reprint ed., New York: Longmans, Green, and Co., 1921), p. 147.

15. Angell to Coover, 31 January 1918 (excerpted on sheets titled "Opinions on the Monograph from Letters of Acknowledgement") (Coover Correspondence [4512], SUA).

16. The quotation is from Battles's will, filed for probate 15 January 1907. (The University voted to accept the bequest a year later.)

17. H. Addington Bruce began the series with a talk on "Modern Theories Regarding Ghosts"; then Carrington spoke on "Is Psychical Research a Science?"; then Henri V. Baril, on "Some Tricks and Illusions of Seances Shown and Explained, and Some Others Not Explained"; and finally Prince, on "Spiritism and Subconscious Activities." Hall commented in December 1917 that "the lectures certainly attracted marked interest and were well attended" (letter to H. A. Willis, 13 December 1917 [CUA]).

Hereward Carrington had assisted Hyslop in the revived ASPR in 1907 and 1908 and had then left to carry on independent investigation of psychic phenomena. He won a certain reputation from regular publication during the ensuing years and acted as delegate to the First International Congress of Psychical Research in Copenhagen in 1921. In the same year, he founded his own American Psychical Institute, but it functioned only until 1923—it was reestablished, however, in 1932 (see below, chap. 8, n. 77).

18. Hall to Cross, 24 May 1916 (CUA).

19. Coover to T. W. Stanford, 28 June 1917 (Coover correspondence [4512], SUA).
20. Coover to R. L. Wilbur, 10 April 1918 (Coover correspondence [4512], SUA).
21. J. E. Coover, "Science vs. Psychical Research," *Homiletic Review* 80 (1920): 435–40.
22. Coover's laboratory notebooks survive and are incorporated among the Rhine Papers (JBR).
23. Coover, *Experiments in Psychical Research*, p. 50.
24. *Ibid.*, p. 54.
25. *Ibid.*, pp. 51–54.
26. *Ibid.*, p. 124.
27. *Ibid.*, p. 52.
28. *Ibid.*, p. 69.
29. *Ibid.*, p. 124. It must be stressed that Coover's study went far beyond straightforward card-guessing. In the card-guessing tests themselves he attempted to correlate degree of success with variation in distance between tester and reagent, and in time interval between the action of the tester and the guess of the reagent, even correlating the card drawn by the tester with the succeeding guess of the reagent (*ibid.*, pp. 77–79). Coover also tested nonpsychical conditions that might influence the test outcome—subliminal and peripheral impressions, corneal reflections, whispering, and habitual card preferences.
30. *Ibid.*, p. 501.
31. Coover to Wilbur, 31 December 1917 (Coover Correspondence [4512], SUA).
32. Titchener to Coover, 5 March 1918 (excerpted from sheets entitled "Opinions on the Monograph from Letters of Acknowledgement") (Coover Correspondence [4512], SUA).
33. It did, however, serve as the model for one other (much less ambitious) experiment from a psychological laboratory. In 1924 Hulsey Cason, a recent Columbia Ph.D. then teaching at Syracuse University, submitted to the SPR a brief experimental test (5,440 trials) for "thought-transference" in unselected subjects using four different geometrical figures drawn on cards as the targets; he cited Coover as the standard authority, used Coover's terminology ("experimenter" and "reagent"), and, like Coover, evaluated successful guessing in the group of eleven subjects considered as a whole, not individually. Also like Coover, he reported entirely negative results, concluding condescendingly that "the members of the various psychical research societies show a peculiar aversion to the scientific method of experiment" (Hulsey Cason, "A Simple Test for Thought-Transference," *JSPR* 21 [October 1924]: 314–19; the report is prefaced by some critical remarks from the *Journal*'s editor).
34. Harvard University comptroller's records. It will be remembered that Troland's salary also was to be paid with funds supplied by a private donor interested in psychical research (see above, n. 4).
35. *J. Abnorm. Psychol.* 8 (1913–14): 405–28.
36. Leonard Thompson Troland, "Paraphysical Monism," *Philos. Rev.* 27 (1918): 39–62; see also *idem*, "The Significance of Psychical Monism for Psychological Theory," *Psychol. Rev.* 29 (1922): 201–11, and *The Mystery of Mind* (New York: D. Van Nostrand Co., 1926), chap. 15. As a graduate student in psychology, Troland had already offered what now seems a prescient theory of genetic reproduction (see Arnold W. Ravin, "The Gene as Catalyst; The Gene as Organism," in *Studies in History of Biology, 1,* ed. William Coleman and Camille Limoges [Baltimore: The Johns Hopkins University Press, 1977], pp. 19–32. Ravin points out that in this time, 1914–15, Troland was attacking the "new vitalism" of Driesch and Bergson; but clearly it would be a mistake to see him as a simple materialist).
37. "The modern scientific psychologist believes that all of the activities of living organisms, i.e., all behavior, either of the lower animals or of human beings, and no matter how complex, can ultimately be explained in these general terms. It is only in the thought of a limited number of psychologists, however, that this belief involves a rejection of a consciousness which coexists with the physical processes, although it does not influence them" (Leonard Thompson Troland, *A Technique for the Study of Telepathy and Other Alleged Clairvoyant Processes* [Albany: Brandow Printing Co., 1917], p. 4).
38. *Ibid.*, p. 24.
39. *Ibid.*, pp. 10–18. Troland provided drawings and a wiring diagram to make the character of his machine as clear as possible.
40. *Ibid.*, p. 25.
41. *Ibid.*, p. 23.

42. "The subjects were seated in very comfortable Morris chairs on opposite sides of a small table, the distance separating their heads being about six feet. Both subjects faced in the same direction. The experiments were carried out in complete darkness and a screen placed in the center of the table also served to shut off their view of each other. . . . No conversation was permitted between the subjects during a sitting, and no other persons were present" (*ibid.*, pp. 21-22). Coover's conditions were just as casual (see the description of experimental conditions in Coover, *Experiments in Psychical Research*, pp. 35, 54).

43. Troland, *A Technique*, pp. 25-26. On Troland's attitude, see also Hyslop, "Psychic Research," p. 456.

44. F.C.S. Schiller, review of *A Technique*..., *PSPR* 31 (1921): 223.

45. "William McDougall," in *A History of Psychology in Autobiography*, vol. 1, ed. Carl Murchison (Worcester, Mass.: Clark University Press, 1930), p. 199. The entire autobiographical statement is invaluable for details on McDougall's life and career, as well as for his rather despairing mood.

46. See W. M'Dougall, "In Memory of William James," *PSPR* 25 (1911): 11-29.

47. McDougall, "William McDougall," p. 203.

48. We have in mind particularly his *Introduction to Social Psychology* (1908) and his *Body and Mind* (1911).

49. A discussion of McDougall's accomplishments at Oxford may be found in R. C. Oldfield, "Psychology in Oxford—1898-1949," *Bull. Brit. Psychol. Soc.* 1 (1950): 345-54.

50. For a general discussion, see Edna Heidbreder, "William McDougall and Social Psychology," *J. Abnorm. Soc. Psychol.* 34 (1939): 150-60; and Harold G. McCurdy, "William McDougall," in *Historical Roots of Contemporary Psychology*, ed. Benjamin B. Wolman (New York: Harper and Row, 1968), pp. 111-30.

51. This attitude developed early in his career. Of his postgraduate studies, he later wrote: "At Göttingen I followed Müller's lectures on psychophysics and on the experimental investigation of memory. They were admirably thorough and detailed. Yet I felt sure that these were not the main lines of progress for psychology" (McDougall, "William McDougall," p. 204).

52. William McDougall, *Body and Mind: A History and Defense of Animism*, 7th ed. (London: Methuen & Co., 1928), pp. 220-21.

53. David L. Krantz and David Allen, "The Rise and Fall of McDougall's Instinct Doctrine," *J. Hist. Behav. Sci.* 3 (1967): 326-38.

54. See, for example, his contribution in John B. Watson and William McDougall, *The Battle of Behaviourism: An Exposition and an Exposure* (London: Kegan Paul, Trench, Trubner, 1928). Grace Adams, *Psychology: Science or Superstition* (New York: Covici Priede, 1931), pp. 231-41, exemplifies the popular view of McDougall as primarily an opponent of Watsonianism.

55. *Body and Mind*, p. 349.

56. "The Need for Psychical Research," in *William McDougall: Explorer of the Mind*, ed. Raymond Van Over and Laura Oteri (New York: Helix Press, 1967), p. 46; the address was originally delivered 25 May 1922 and may also be found in *JASPR* 17 (1923): 4-14.

57. *Ibid.*, p. 47.

58. "Presidential Address to the Society for Psychical Research," in *ibid.*, p. 53; the address was delivered 19 July 1920 and may also be found in *PSPR* 31 (1921): 105-23.

59. "Psychical Research as a University Study," in Van Over and Oteri, *William McDougall*, p. 81. The essay was originally published in *The Case For and Against Psychical Belief*, ed. Carl Murchison (Worcester, Mass.: Clark University, 1927), pp. 149-62.

60. Gardner Murphy, "Notes for a Parapsychological Autobiography," *JP* 21 (1957): 166. Troland's initial reaction to Murphy has survived: "Of course my experience with Mr. Murphy has been brief, but I have found him to be a clear thinker and a rigorous worker. He is quick in making decisions and has good executive ability in the laboratory. I cannot yet judge as to his originality; on the whole his philosophy seems to follow conventional lines" (Troland to Ernest Hocking, 10 November 1916 [HUA]).

61. Murphy, "Notes," pp. 168-69.

62. *Ibid.*, p. 167; and interview, 5 December 1973. See also Murphy's autobiographical statement in *A History of Psychology in Autobiography*, vol. 5, ed. Edwin G. Boring and Gardner Lindzey (New York: Appleton-Century-Crofts, 1967), pp. 256-59.

63. Interview, 5 December 1973; Murphy, "Notes," pp. 167–68.
64. *JASPR* 16 (1922): 654.
65. G. Murphy, "L'Expérimentation télépathique," *RM*, 1924, pp. 334–36 (letter of 12 June 1924). Murphy described other tests (and psychic experiences) in his article "Telepathy as an Experimental Problem," in Murchison, *The Case For and Against Psychical Belief*, pp. 271–78.
66. Murphy, "L'Expérimentation télépathique," p. 335.
67. René Warcollier, *Experimental Telepathy*, ed. Gardner Murphy (Boston: Boston Society for Psychic Research, 1938), pp. 56–73.
68. McDonald played an active role in sponsoring another series of nationally broadcast telepathy tests in the fall of 1937 (see below, chap. 6).
69. J. Malcolm Bird, "Telepathy and Radio," *Scientific American* 130 (1924): 382, 433–35; and *idem*, "Experimental Telepathy—I," *Psychic Research* 23 (1929): 528–33.
70. Murphy, "Telepathy As an Experimental Problem," pp. 265–78.
71. Gardner Murphy to Franz Boas, 5 March 1923, Boas Papers, American Philosophical Society, Philadelphia, Pa. We owe this reference to the acuity of Dr. James E. McClellan.
72. J. Malcolm Bird, "Our Psychic Investigation," *Scientific American* 128 (1923): 7.
73. "Harry Helson," *A History of Psychology in Autobiography*, vol. 5, ed. Edwin G. Boring and Gardner Lindzey (New York: Appleton-Century-Crofts, 1967), p. 201.
74. Interview, 5 December 1973. Murphy would often bring third persons to these sittings to record Mrs. Piper's trance responses to their questions. One such sitting was published by Jane J. Sagendorph, in *A Vision and Its Sequel*, Boston Society for Psychic Research Bulletin no. 4 (Boston, 1926).
75. "Gardner Murphy," *A History of Psychology in Autobiography*, vol. 5, pp. 257–59, 261–63.
76. Though Cyril Burt recalled that he, McDougall, and J. C. Flugel used to carry out experiments in telepathy at Oxford shortly after McDougall's arrival there in 1904 (quoted by J. Wainwright Evans in Van Over and Oteri, *William McDougall*, p. 25).
77. McDougall to J. B. Rhine, 6 June 1936 (JBR).
78. G. H. Estabrooks, *A Contribution to Experimental Telepathy* (Boston: Boston Society for Psychic Research, 1927).
79. *Ibid.*, p. 26.
80. *Ibid.*, p. 16.
81. *Ibid.*, pp. 18–20.
82. Boring to Estabrooks, 28 June 1926 (EGB).
83. [Edwin G.] Boring, "The Paradox of Psychic Research," *Atlantic Monthly* 137 (1926): 81–87.
84. Walter Franklin Prince, "A Review of the Margery Case," *Amer. J. Psychol.* 37 (1926): 431–41; reprinted in Murchison, *The Case For and Against Psychical Belief*, pp. 199–213.
85. Boring to Estabrooks, 14 January 1927 (EGB).
86. On McDougall's move to North Carolina, see below, chap. 6.
87. McDougall to Boring, 11 January 1928 (EGB).
88. McDougall to Boring, 10 July 1928 (EGB). Kuklick, *Rise of American Philosophy*, chap. 24, discusses philosophy and psychology at Harvard in the 1920s and explains some of the forces that tended to prevent McDougall from functioning successfully within that department.
89. The genesis of the idea of the symposium is described by Murchison in his preface to *The Case For and Against Psychical Belief* (p. vii). He proceeded immediately to discuss it with President Wallace Atwood of Clark, justifying the series on the grounds that it would help make it possible legitimately to transfer the Smith-Battles funds to other uses: "It is difficult to estimate what this series of lectures will cost. The cost will lie somewhere between two and three thousand dollars. However, there will be a great demand for the book containing all these lectures. We can publish that book here at Clark. I feel certain that we can make several thousand dollars profit on the entire series. In this way money can be transferred from the Smith-Battles Fund, where it is of little value, to the Stanley Hall Fund or other funds of equal value to the University" (Murchison to Atwood, 18 December 1925 [CUA]). The eventual cost to the fund was just under three thousand dollars, and *The Case For and Against Psychical Belief*—which appeared in March 1927—was the result.
90. Murchison, *The Case For and Against Psychical Belief*, p. 159.

91. *Ibid.*, p. 161. This credo may be compared with his defense of the ASPR as the natural locus of organizational activity for psychical research in his address of May 1922, "The Need for Psychical Research" (see above, n. 56), which does not suggest any possible role for the universities.

Chapter Four

1. Where specific references are not given, this chapter has been based upon the Rhine Papers (JBR) and upon conversations with J. B. Rhine and L. E. Rhine, tape recordings of which are in the Southern Historical Collection, Wilson Library, University of North Carolina.
2. Louisa E. Rhine, "Divergence of Catalase and Respiration in Germination," which appeared in *Bot. Gaz.* 78 (1924): 46-67, is based upon Louisa's dissertation, which bore the same title.
3. Strausbaugh to Rhine, 22 June 1925 (JBR). Unless otherwise indicated, all correspondence cited in this chapter is from this collection.
4. His dissertation, "Translocation of Fats As Such in Germinating Fatty Seeds," was published in *Bot. Gaz.* 82 (1926): 154-69. He had already published "Clogging of Stomata of Conifers in Relation to Smoke Injury and Distribution," in *ibid.*, 78 (1924): 226-32, and was at work on another problem in plant metabolism, research published in 1926 as "Studies on the Oxidation of Certain Fatty Acids," in *Plant Physiol.* 1 (1926): 349-62.
5. For Mathews's own views at this time, see Shailer Mathews, *The Contributions of Science to Religion* (New York and London: D. Appleton & Co., 1924), a volume that apparently embodies the lectures in question. Mathews was a prominent defender of a "modernist" theology that would profit from an openness to liberal and scientific thought; on his views, in the context of an evolving Protestant theology, see William R. Hutchison, *The Modernist Impulse in American Protestantism* (Cambridge, Mass., and London: Harvard University Press, 1976), pp. 275-82.
6. Rhine to Edwards, 17 June 1923. The other letters were written the same day.
7. Jastrow to Rhine, 11 July 1923.
8. McDougall to Rhine, 19 September [1923].
9. Edwards to Rhine, 9 July 1923.
10. Edwards explained to Rhine (25 January 1924) that there had been difficulties between Prince and Miss Tubby and warned him: "You will not get into it; will leave politics alone; keep your own counsel; and start for fresh air and a fresh start. All you know of the past is Psychical Research; you are simply loyal to the board and the society."
11. Rhine to Bird, 26 March 1925.
12. Bird to Rhine, 31 March 1925; Rhine to Bird, 9 April 1925.
13. Rhine to Bird, 13 January 1926.
14. In this and what follows, it must not be forgotten that the Rhines were committed *as a couple* to the study of man's nature and to psychical research. Their actual decision to leave West Virginia was a joint one, made at two o'clock one morning after a long night's discussion of their future. Circumstances perhaps made it inevitable that the formal career opportunities, first in botany and then in parapsychology, should have fallen to J. B., but Louisa continued to be scarcely less actively involved in psychical research than he. They both read widely in the field and involved themselves in investigations of reported psychical phenomena, and J. B.'s laboratory research was bound to be examined in close detail in discussions at home. Our subsequent analysis of J. B. Rhine's work should not be allowed to disguise his wife's continuing contribution to the enterprise.
15. The paper on telepathy was delivered on 14 January, that on mediums on 26 March; copies of both are in JBR.
16. Alan Gauld, *The Founders of Psychical Research* (London: Routledge & Kegan Paul, 1968), p. 103.
17. Rhine to Joseph De Wyckoff and all members of the ASPR board of trustees, 15 July 1926 (copy in HPL).
18. J. B. Rhine and Louisa E. Rhine, "One Evening's Observations on the Margery Mediumship," *J. Abnorm. Soc. Psychol.* 21 (1927): 401-21. For the general setting of these events, see Thomas R. Tietze, *Margery* (New York: Harper & Row, 1973), esp. chap. 13.
19. *The Banner of Life*, 12 March 1927.
20. Rhine to Jacob Saposnekow, 9 August 1926.
21. Walter Franklin Prince, won over by the Rhines' repudiation of Margery, had endorsed J. B.

to E. G. Boring as a candidate for Hodgson fellow, and Boring passed on the recommendation to the departmental chairman, J. H. Woods, with his endorsement. Woods replied: "I do not know of any Hodgson Fellowship. My impression is that we should reserve the Fund for research and for inventions of methods of the highest order. When once Estabrooks is off our hands, I think we should follow this policy and not merely use the Fund to help students study for a degree" (Woods to Boring, 16 July 1926 [HUA]). It was Woods's attitude that determined the matter. The graduate students would have been more sympathetic to the Rhines' concerns. Bruce Kuklick, *The Rise of American Philosophy, Cambridge, Massachusetts, 1860-1930* (New Haven and London: Yale University Press, 1977), app. 3 (pp. 581-89), shows clearly that many of the graduate students in the Harvard philosophy department in the 1920s had entered the program in the hope of resolving the conflict between science and religion.

22. Rhine to "Youngster," 7 November 1926.
23. Rhine to "Folks at Home," 10 November 1926.
24. Typed note, "Summary of the Situation, Past, Present, and Perspective Future, Jan. 23, 1927."
25. Rhine to Strausbaugh, 30 March 1927.
26. Thomas to Rhine, 30 March 1927.
27. Rhine to W. F. Prince, 15 August 1927.
28. Thomas first wrote to McDougall in July, but the idea to do so had been put into his mind that spring by an English medium who in trance had told him to "go to McDougall."
29. Rhine to Thomas, 23 August 1927.
30. Thomas to Rhine, 29 August 1927 and 2 September 1927.
31. Rhine to McDougall, 11 September 1927; McDougall to Rhine, 17 September 1927.
32. Indeed, Rhine had reviewed in the November 1925 *Journal* of the ASPR an account from the *Psychische Studien* of Bechterew's experiments on telepathy with dogs, and he returned to the subject in his February 1926 reviews, calling Bechterew's work "one of the neatest and most completely convincing series of experiments in the recent literature on psychical research" (*JASPR* 20 [1926]: 126).
33. Rhine to Thomas, 31 March 1928: "I got tired of watchless waiting yesterday and told McD. I thought the Lady paper ought to go out, seeing that our case would be 'crimped,' and our priority of the project 'scooped,' in case another university study of Lady or Black Bear came out ahead of us. That is the effect in other sciences. . . ."
34. Rhine to Fonda, 8 March 1928.
35. The release was dated May 23. Frank Thone of Science Service, a friend of the Rhines' from graduate school who had been trained in plant ecology, told him: "I was told about this work on the horse by our managing editor, Mr. Watson Davis, who learned about it from Prof. Knight Dunlap, of the Johns Hopkins University. I think Dunlap picked it up at some psychological meeting as a bit of corridor gossip" (Thone to Rhine, 21 May 1928).
36. J. B. Rhine and Louisa E. Rhine, "An Investigation of a 'Mind-Reading' Horse," *J. Abnorm. Soc. Psychol.* 23 (1928-29); 449-66. Morton Prince showed no hesitation about the paper; indeed, he wrote, "I find your MS very interesting & I should be very glad to publish it in the Journal. It will have to be passed upon also by [Henry] Moore, but I think he will be of the same opinion" (Morton Prince to Rhine, 11 August 1928).
37. Rhine to Coover, 10 February 1928; Murphy to Rhine, 15 February 1928.
38. Rhine to Thomas, 29 February 1928.
39. *Ibid.*
40. Memorandum, McDougall to Rhine, 19 September 1928, defines the terms under which Rhine was to assist with the Lamarckian work.
41. For a full report of the Lamarckian experiments as they progressed, see William McDougall, "An Experiment for the Testing of the Hypothesis of Lamarck," *Brit. J. Psychol.* 17 (1926-27): 267-304; *idem*, "Second Report on a Lamarckian Experiment," *ibid.*, 20 (1929-30): 201-18; J. B. Rhine and William McDougall, "Third Report on a Lamarckian Experiment," *ibid.*, 24 (1933-34): 213-35; and William McDougall, "Fourth Report on a Lamarckian Experiment," *ibid.*, 28 (1937-38): 321-45.
42. McDougall, "An Experiment," p. 268. McDougall had certainly come under criticism by this time for daring to defend the Lamarckian hypothesis as not yet disproven, and he was to receive more. He gave a condensed version of his "Second Report" at the Ninth International Congress of

Psychology at Yale in 1929 and was bitterly attacked by James McKeen Cattell, president of the congress. Most psychologists were not so outspoken, but their views of McDougall could not help but be affected by their disbelief in Lamarckianism. In later years, this in turn made McDougall's support of Rhine somewhat the less impressive to many psychologists (see below, chap. 10).

43. Rhine to McDougall, 13 August 1932.
44. McDougall, "Second Report," p. 213.
45. Rhine to McDougall, [December 1928]. Rhine pointed out to W. F. Prince that this was still not out of keeping with experiments on man: "The horse has lost its telepathic sensitivity.... It now has a very evident habit of following signals, and I can direct it myself, even with the owner absent, by whip and movements; but I cannot direct it telepathically, as I did before. Very embarrassing indeed; it is like the case at the Univ. of Groningen, by Brugmanns. Their subject too lost his power" (Rhine to Prince, 8 December 1928).
46. J. B. Rhine and Louisa E. Rhine, "Second Report on Lady, the Mind-Reading Horse," *J. Abnorm. Soc. Psychol.* 24 (1929-30): 287-92.
47. Rhine simply asserted that in tests in which he alone knew the target, he had exerted voluntary control over possible involuntary gestures.
48. Rhine's attention was originally drawn to *The Bridge* and the question it raised by John Thomas (letter of 13 December 1927), to whose own studies it offered a number of parallels, and had finished reading it by the end of January 1928. He associated his new focus upon telepathy with Mrs. Sidgwick's review in a letter to W. F. Prince (15 May 1928) and was still emphasizing the importance of *The Bridge* in the seminar paper he gave the next year (discussed immediately below).
49. "It may be asked: In seeking evidence of survival, how does elimination of telepathy from the sitter help us if after all the source of information is perhaps telepathy from a distant living person? There are two answers to this. The first is, that it adds to our knowledge of telepathy, which is, I think, the most promising line of advance in psychical research. We at present know extremely little about it, and if we could discover its modus operandi, its conditions and its limits and possibilities, that would in itself greatly contribute to the understanding of questions of survival and communication with the dead" (*PSPR* 38 [1928-29]: 12-13).
50. Interview, Louisa E. Rhine, 15 March 1974.
51. "We have a very real appreciation for Mr. T. himself, and have been convinced that his material is in part at least, genuinely evidential, that of the hypotheses available to our knowledge certainly the spirit interpretation is the most acceptable." To be sure, Rhine went on to say they found themselves unable to accept "Mr. T's maximum interpretation" in evaluating much of the material, and during the next few years they gradually dissociated themselves from the project (Rhine to Prince, 21 March 1929).
52. The paper, in JBR, is entitled merely "Psychic Research—Seminar '29."
53. At least three drafts still exist among Rhine's papers: one dated "summer 1929," one marked simply "1928," and a third still earlier but undated.
54. For example, in Rhine to Mrs. Ellen Wood or in Rhine to Mrs. W. H. Crunden, both 16 July 1935. It is interesting to compare Rhine's confidence that positive results would tend to support religious belief with the contemporary judgment of the psychologist Knight Dunlap, throughout his life an unremitting opponent of psychical research: "The efforts of psychic researchers, who are attempting a 'scientific' proof of the 'other' world, would, if successful, destroy religion, thinks Dr. Knight Dunlap, professor of psychology at Johns Hopkins University. Dr. Dunlap, however, said emphatically that he does not think there is the remotest chance of their success" ("Religion Beyond Science," *Science News Letter* 13 [24 March 1928]: 185).
55. "Courses Suggested for 1929-30" (JBR). The proposal is accompanied by syllabi and lecture notes for the courses in question, on which the following paragraphs have been based.
56. Rhine to William K. Boyd, [undated], replying to Boyd's letter of 17 March 1930.
57. Rhine to President W. P. Few, 13 June 1930; cf. letters to Harrison Bowie Smith, 17 June 1930, and to McDougall, 14 July 1930.
58. Form letter, Ina Jephson to Rhine, April 1930. She reported that Rhine had scored high but not extraordinarily or improbably so.
59. As it turned out, Rhine was disappointed by Estabrooks, for the latter seemed to have lost all interest in experimental psychical research. Rhine wrote to McDougall in amazement that "he's just a good-natured bluff; who *wants* to administrate, where bluff is more at par. Likeable, though" (Rhine

to McDougall, 14 July 1930). Since his previous meeting with Rhine in 1926 Estabrooks had tried and failed to repeat his previous experimental success.

60. Upton Sinclair, *Mental Radio* (Monrovia, Calif.: privately published, 1930), p. vii. McDougall spent the summer of 1929 in California, conducting several experiments with Mrs. Sinclair, and he wrote back to Rhine about the "interesting stuff" but "only partial success" he was getting (McDougall to Rhine, 18 July and 29 July 1929).

61. Rhine to N. Murray, 19 October [1929]: "You may be interested in a book on telepathic experiments that is about to be published by Upton Sinclair. It is very good."

62. Stuart to Rhine, 11 June and 8 August 1930; Frick to Rhine, 12 July 1930.

63. J. B. Rhine, *Extra-Sensory Perception* (Boston: Boston Society for Psychic Research, 1934), pp. 35-36 (hereinafter to be cited as *ESP*).

64. While this is what the Rhine correspondence of 1930 reveals, Widgery recounted a somewhat different story thirty years later: "When I first went to Duke, he [Rhine] was assistant professor not only in psychology but in philosophy. Agreeing with me that he was not adequately qualified for a position in philosophy, the senior University administrators transferred him from philosophy" (Alban G. Widgery, *A Philosopher's Pilgrimage* [London: George Allen and Unwin, 1961], p. 163).

65. Rhine to Prince, 1 July 1931.

66. Harvey Lee Frick, "Extra-Sensory Cognition" (M. A. thesis, Duke University, 1931), p. 148.

67. McDougall to Rhine, 2 July 1931.

68. *ESP*, pp. 59-63.

69. Rhine to Prince, 1 July 1931.

70. *ESP*, pp. 38-39.

71. Rhine to Paul Gross, 31 March 1932.

72. McDougall to Rhine, 25 July 1932.

73. Rhine to Prince, 9 September 1932. The suggestion was apparently made by Helge Lundholm.

74. Prince to Rhine, 9 September 1932.

75. "It would seem desirable," he wrote to Prince, "to have the briefer one come out in advance of... the longer" (19 September 1932).

76. Rhine to Henry Moore, 22 June 1933.

77. J. B. Rhine, "Extra-Sensory Perception of the Clairvoyant Type," *J. Abnorm. Soc. Psychol.* 29 (1934-35): 152.

78. *Ibid.*, p. 169.

79. Rhine to Stuart, 11 July 1932.

80. Rhine to Sinclair, 16 February 1933.

81. Rhine to Prince, 14 June 1933.

82. Rhine to Moore, 22 June 1933. He added, "Some of my colleagues have urged that it belongs rather in '[the Journal of] General Psychology,' but I feel a strong historic kinship for your Journal."

83. Rhine, "Extra-Sensory Perception," p. 171.

84. Explaining the derivation of this term, Rhine has since explained: "It had been called 'perception' before, 'super-sensory perception,' and 'super-sensory perception' sounded at least two syllables towards 'super-natural'; and to call it 'cognition' was to use a word that's not very often used and not very well understood and [gives] lots of latitude" (interview, 15 March 1974).

85. Rhine used the term "extra-sensory cognition" in correspondence as late as 31 March 1932; his letter to Sinclair of 15 July 1932 marks the first use of the new terminology.

86. Rhine to Prince, 9 September 1932.

87. Two letters from John Thomas of 12 January 1933—one to McDougall and one to Rhine—make it plain that Adams was the one who had raised this point.

88. "If the thesis is, itself, well enough put together and if it makes a good enough case to the members of the committee, if they read it beforehand, it is not very likely that the examination will offer any serious difficulties" (Rhine to Thomas, [spring 1932]).

89. Rhine to Prince, 14 June 1932; Prince to Rhine, 30 June 1932.

90. H. F. Saltmarsh, "Report on the Investigation of Some Sittings with Mrs. Warren Elliott," *PSPR* 39 (1929): 47-184.

91. Rhine to Prince, 14 June 1933.
92. Rhine to McDougall, 13 July 1931; see also Rhine to McDougall, 21 April 1931, and McDougall to Rhine, 2 July 1931.
93. H. F. Saltmarsh, "Is Proof of Survival Possible?" *PSPR* 40 (1931-32): 105-22.
94. Rhine was at work on the article in March (see Rhine to Thomas, 10 March 1932); it was received by the SPR in June and published the next March as "The Question of the Possibility of Proving Survival," *JSPR* 28 (March 1933): 35-45.
95. *Ibid.*, p. 38.
96. Rhine to Thomas, 10 July 1933.
97. J. B. Rhine, *JSPR* 28 (October 1933): 127.
98. W. G. Roll, "Pagenstecher's Contribution to Parapsychology," *JASPR* 61 (1967): 219-40.
99. Thomas R. Tietze, "Ursa Major: An Impressionistic Appreciation of Walter Franklin Prince," *ibid.*, 70 (1976): 1-34.
100. See the Research Officer [Walter Franklin Prince], *The Sinclair Experiments Demonstrating Telepathy,* Boston Society for Psychic Research Bulletin no. 16 (Boston, April 1932). Another expression of Prince's enthusiasm for this work is the article he prepared for *Scientific American* on the same subject, "Mrs. Sinclair's 'Mental Radio,'" *Scientific American* 146 (March 1932): 135-38.
101. Prince to Rhine, 16 February 1932.
102. Rhine to Prince, 11 March and 30 April 1932. Others soon raised the same questions with Prince (see Prince to Saltmarsh, 30 June 1932 [SPR]).
103. Prince to Rhine, 5 May 1932.
104. *ESP*, pp. 81-84.
105. *Ibid.*, pp. 88-89, 162. The figures on pp. 88-89 do not include Zirkle's work on clairvoyance, but this can be calculated from the data on p. 162.
106. *Ibid.*, pp. 87-103.
107. F. L. Usher and F. P. Burt, "Quelques expériences de transmission de la pensée à grande distance," *Annales des sciences psychiques* 20 (1910): 14-21, 40-54.
108. See above, chap. 2.
109. *ESP*, p. 104.
110. *Ibid.*, p. 105. For a critical examination of this series of experiments, based on Rhine's account, see C.E.M. Hansel, *ESP: A Scientific Evaluation* (New York: Charles Scribner's Sons, 1966), pp. 54-56.
111. *ESP*, pp. 85-86. Because of the importance of the Pearce-Pratt experiments, they have been described many times in the parapsychological literature after 1934, most fully perhaps in J. B. Rhine and J. G. Pratt, "A Review of the Pearce-Pratt Distance Series of ESP Tests," *JP* 18 (1954): 165-77. C.E.M. Hansel (*ESP* . . . , pp. 71-85) has called attention to discrepancies in these various accounts. For our purposes, it has seemed best to quote the contemporary description provided in the original monograph.

Chapter Five

1. This chapter is a revised version of an earlier article: Michael McVaugh and Seymour H. Mauskopf, "J. B. Rhine's *Extra-Sensory Perception* and Its Background in Psychical Research," *Isis* 67 (1976): 161-89.
2. Whately Carington, *Telepathy* (London: Methuen & Co., 1945), p. 20.
3. For example: "The modern period in parapsychological experimentation started in 1927 when Dr. J. B. Rhine joined McDougall's psychological department at Duke University in North Carolina, and started a new attack on the experimental problems" (Robert H. Thouless, "Thought-Transference and Related Phenomena," *Proceedings of the Royal Institution* 34 [1947-50]: 677. See also Gardner Murphy, *Challenge of Psychical Research* [New York: Harper, 1970], pp. 65-68, or Robert H. Thouless, *From Anecdote to Experiment in Psychical Research* [London: Routledge & Kegan Paul, 1972], pp. 33-36).
4. Our use of the term is thus derived from the historical transformation of psychical research in the two decades after 1920 and applies specifically to a time in the mid-1930s. Ingemar Nilsson, "The Paradigm of the Rhinean School," *Eur. J. Parapsychol.* 1 (1975-77): pt. 1, pp. 45-59; pt. 2,

pp. 45–56, uses the term to characterize the set of guiding assumptions shared by many parapsychologists during the forty-year period following the publication of *Extra-Sensory Perception*. Our complementary approaches are in general accord as to the content of the paradigm.

5. Price to Joad, 12 January 1933 (HPL).

6. The most important items besides Warcollier that Rhine had not had the opportunity to study at the time he wrote his monograph are Coover's report (see Prince to Rhine, 5 September 1933 [JBR])—although of course he knew of its general content from other sources—and, evidently, S. G. Soal's negative papers of 1931 and 1932, since the monograph did not allude to them.

7. J. B. Rhine, *Extra-Sensory Perception* (Boston: Boston Society for Psychic Research, 1934), pp. 85, 168. The book (hereinafter referred to as *ESP*) was published a second time in 1935 by Bruce Humphries of Boston; the pagination of these two editions is the same. In the recent reprint (Boston: Bruce Humphries, 1964), the original text has been altered in a number of significant respects.

8. Charles Richet, "La Suggestion mentale et le calcul des probabilités," *Revue philosophique* 18 (1884): 609–74.

9. E. G. Gurney, "M. Richet's Recent Researches in Thought-Transference," *PSPR* 2 (1883–84): 239–64. Lodge's approach is given on pp. 257–62 (in a letter from Lodge and his brother, Alfred, to Gurney dated 8 January 1885); and Gurney's on pp. 262–64.

10. F. Y. Edgeworth, "The Calculus of Probabilities Applied to Psychical Research," *PSPR* 3 (1885): 190–99; and "The Calculus of Probabilities Applied to Psychical Research. II," *PSPR* 4 (1886–87): 189–208. On Edgeworth's achievements in statistics, see *International Encyclopedia of the Social Sciences*, s. v. "Edgeworth, Francis Ysidro."

11. Gurney, "M. Richet's Recent Researches," p. 242.

12. E. Gurney, F.W.H. Myers, and Frank Podmore, *Phantasms of the Living*, 2 vols. (London: Trübner, 1886), vol. 1, pp. 34–35, presents a criticism of card experiments along these lines. There are occasional later defenses of the statistical evaluation of card-guessing and the like as offering more satisfactory proof of psychical abilities than spontaneous occurrences (see Alice Johnson, "Coincidences," *PSPR* 14 [1898–99]: 183).

13. Malcolm Guthrie, "Further Report on Experiments in Thought-Transference at Liverpool," *PSPR* 3 (1885): 427; and Charles Richet, "Rélation de diverses expériences sur la transmission mentale, la lucidité, et autres phénomènes non explicables par les données scientifiques actuelles," *PSPR* 5 (1888–89): 148–52, where Richet reported *no* significant results in his statistical experimental work with the guessing of playing cards.

14. See, for example, Oliver J. Lodge, "Some Recent Thought-Transference Experiments," *PSPR* 7 (1891–92): 377, where Lodge reports simply that "out of 39 trials 16 were correct and 23 wrong," without trying to deal with the measurable improbability of such a series of successes; or Edmund Gurney *et al.*, "Second Report on Thought-Transference," *PSPR* 1 (1882–83): 76–77: "*excluding the second trial the successful results were rather more than 1 in 3; pure chance would have given 1 in 16.*" Mrs. A. W. Verrall, who carried on several thousand trials under varying circumstances, was the last person to pursue the ideal at all seriously. In 1894 she had C. P. Sanger apply Edgeworth's techniques to her results, which seemed to indicate a certain degree of success, especially in guessing suits (Mrs. A. W. Verrall, "Some Experiments on the Supernormal Acquisition of Knowledge," *PSPR* 11 [1895]: 174–97).

15. Charles Richet, *Traité de métapsychique* (Paris: Félix Alcan, 1922), pp. 115–16 (hereinafter *TM*); in English *Thirty Years of Psychical Research*, trans. Stanley De Brath (New York: Macmillan, 1923), pp. 96–98 (hereinafter *TY*).

16. Mrs. A. W. Verrall, "Report on a Series of Experiments in 'Guessing,'" *PSPR* 29 (1918): 97. Hyperacuity of hearing, instead of telepathy, has been put forward as a possible explanation of Murray's success in these experiments. For recent examinations of these alternatives, see E. R. Dodds, "Gilbert Murray's Last Experiments," *PSPR* 55 (1972): 371–402; and Eric J. Dingwall, "Gilbert Murray's Experiments: Telepathy or Hyperaesthesia?" *PSPR* 56 (1973): 21–39.

17. René Sudre, "Clairvoyance and the Theory of Probabilities," *Psychic Research* 22 (1928): 63–69.

18. René Sudre, "An Experiment in Card Guessing," *Psychic Research* 23 (1929): 81–86.

19. Ina Jephson, "A Reply to M. Sudre's Article, 'An Experiment in Card Guessing,'" *PSPR* 39 (1930–31): 188.

20. R. A. Fisher, "The Statistical Method in Psychical Research," *ibid.*, pp. 189–92.

21. *ESP*, pp. 35–36.

22. V. J. Woolley, "The Broadcasting Experiment in Mass Telepathy," *PSPR* 38 (1928-29): 1-9; and Ina Jephson, "Evidence for Clairvoyance in Card-Guessing," *ibid.*, pp. 269-71. Interest in the study of innate preference for one test symbol over another had been shown by the ASPR as far back as the 1880s (see "The Number-Habit," *Proceedings ASPR* 1 [1885-89]: 86-95; and "Second Report on Experimental Psychology—Upon the Diagram Tests," *ibid.*, pp. 302-17).

23. Interview, J. B. Rhine, 15 March 1974. On the rationale behind the selection of the Zener cards, see also *ESP*, pp. 50, 167; J. B. Rhine, *New Frontiers of the Mind* (New York and Toronto: Farrar and Rinehart, 1937), p. 48; and J. G. Pratt *et al.*, *Extra-Sensory Perception after Sixty Years* (New York: Henry Holt, 1940), p. 270.

24. G. H. Estabrooks, *A Contribution to Experimental Telepathy* (Boston: Boston Society for Psychic Research, February, 1927), p. 15.

25. See *ESP*, pp. 31-34, for Rhine's discussion of the mathematics used.

26. Gardner Murphy had previously written: "An attempt to control all sources of error *at the beginning* is not only futile because of the impossibility of foreseeing all sources of error, but prejudicial to obtaining the kinds of occurrences one is out to observe. Tenseness, distrust, and apathy are but three of many ways of becoming negatively conditioned to a long series of laboratory experiments" ("Telepathy as an Experimental Problem," in *The Case For and Against Psychical Belief*, ed. Carl Murchison [Worcester, Mass.: Clark University, 1927], p. 275).

27. *ESP*, p. viii. In the original version of this foreword (in the William McDougall Papers, DUA), McDougall had written that "the results of the experimentation show a lawfulness and consistency which even the most careful conspiracy of deception could not hope to achieve," thus arguing for the soundness of the work on rather different grounds, but he changed this statement at Rhine's suggestion.

28. Mrs. Henry Sidgwick, "On the Evidence for Clairvoyance," *PSPR* 7 (1891-92): 30.

29. *Ibid.*, p. 31.

30. *ESP*, p. 27. Rhine commented, "I feel particularly indebted to Miss Jephson's work in that it helped to stimulate my own interest in clairvoyance."

31. Ina Jephson, "Evidence," pp. 227-29, 235-37. Previously Rudolf Tischner had noted (in *Telepathy and Clairvoyance*, trans. W. D. Hutchinson [New York: Harcourt, Brace & Co.; London: Kegan Paul, Trench, Trubner & Co., 1925]) that "it is generally required, and justly so, of decisive experiments on clairvoyance (from which the possibility of telepathy is excluded) that nobody should know what the object under consideration is" (p. 147). Tischner described his own experiments, mostly concentrated in the period 1918-21, and argued that they provided convincing proof of the existence of pure clairvoyance. G.N.M. Tyrrell is another who tried to distinguish experimentally between the two ("The Case of Miss Nancy Sinclair," *JSPR* 20 [March 1922]: 294-327).

32. Oliver Lodge, "A Record of Observations of Certain Phenomena of Trance, (2) Part 1," *PSPR* 6 (1889-90): 453.

33. The assumption goes back to the origins of the society (see, for example, W. F. Barrett *et al.*, "Report of the Literary Committee," *PSPR* 1 [1882-83]: 118-99). It quite rapidly became an automatic, unconscious model for interpreting data—as, for example, in Balfour Stewart's presidential address of 1887 (*PSPR* 4 [1886-87]: 265).

34. *TM*, pp. 764-69; *TY*, pp. 602-4; see also above, chap. 1.

35. Tischner is one such, as is to be expected from his interest in demonstrating clairvoyance (see above, n. 31). The review of Tischner's *Telepathy and Clairvoyance* by René Sudre (in *RM*, 1925, p. 43) calls attention to Tischner's opinion on this subject but does not try to decide its merits. Another who maintained the same view was Eugène Osty. Rhine cited Osty's conviction, expressed in his *Supernormal Faculties in Man*, trans. Stanley De Brath (New York, Dutton: 1923), as "a supporting judgment of great weight" that had helped him to his own conclusion (*ESP*, p. 145; see also McVaugh and Mauskopf, "J. B. Rhine's *Extra-Sensory Perception*," n. 44).

36. See Richard Hodgson, "A Further Record of Observations of Certain Phenomena of Trance," *PSPR* 13 (1898): 393. Support for the likelihood and even frequency of clairvoyance was also adduced at about this time by W. F. Barrett in his study of dowsing, "On the So-called Divining Rod," *PSPR* 15 (1900-1901): 309-12, 359-66.

37. *ESP*, p. 133, quoting Gilbert Murray, "Presidential Address," *PSPR* 29 (1918): 58.

38. Malcolm Guthrie, "An Account of Some Experiments in Thought-Transference," *PSPR* 2 (1883-84): 27. Cf. Oliver Lodge, as reported in *Phantasms of the Living*, vol. 1, pp. 50-51; or, for similar judgments, A. Schmoll and J. E. Mabire, "Experiments in Thought-Transference," *PSPR* 5

(1888-89): 205; W. F. Barrett, "On Some Phenomena Associated with Abnormal Conditions of Mind," *PSPR* 1 (1882-83): 238-44; and Mrs. H. Sidgwick and Miss Alice Johnson, "Experiments in Thought-Transference," *PSPR* 8 (1892): 594-95.
   39. Verrall, "Report," p. 84.
   40. The relaxed atmosphere of an extremely successful run carried out with Hubert Pearce is described in Rhine's *New Frontiers of the Mind,* pp. 94-96.
   41. See, for example, Guthrie, "Further Report," p. 425; Sidgwick and Johnson, "Experiments," p. 537; and Edmund Gurney *et al.,* "Third Report on Thought-Transference," *PSPR* 1 (1882-83): 171. The last reference is based upon the society's experiences with the Creery sisters, but the fact that the Creerys were later detected in using a code of signals did not prevent the generalization, reinforced by cases where such collusion could be ruled out, from being accepted.
   42. Miss Jephson called attention (in "Evidence," pp. 253-55) to the fact that G. H. Estabrooks's data revealed the same sort of curve, although Estabrooks had not attempted to offer any interpretation of it. Another suggestion of such a pattern had been made by Richet in "La Suggestion mentale," pp. 626-27.
   43. *ESP,* pp. 54-55; see above, chap. 4.
   44. See *ESP,* pp. 135, 159-60.
   45. *Ibid.,* pp. 141-42.
   46. Oliver J. Lodge, "An Account of Some Experiments in Thought-Transference," *PSPR* 2 (1883-84): 200.
   47. *ESP,* p. 130. On the other hand, A. J. Linzmayer, Rhine's first really successful subject, remembers today that in those initial experiments a special sense of inner confidence preceded his correct guesses (interview, 21 March 1974).
   48. *ESP,* p. 142. When Pearce once scored twenty-five consecutive hits (in an obviously exceptional performance), Rhine reported that "the very strain he showed at the end was evidence of strong effort to concentrate attention. He said, 'You'll never get me to do that again!' He could not describe his feeling further, however. (And, to avoid developing self-consciousness, I do not push introspective exploration.)" (*ibid.,* p. 134).
   49. On early distance experiments at Duke, see *ibid.,* pp. 83-86, 104-7. The earliest such was perhaps the series attempted by Harvey Frick in the summer of 1930 as background for his Duke M.A. thesis in psychology (above, chap. 4); these experiments, while surely present to Rhine's mind, were not cited in *ESP.*
   50. *ESP,* p. 122.
   51. William Crookes, "Address by the President," *PSPR* 12 (1896-97), 338-55.
   52. A. J. Balfour, "Address by the President," *PSPR* 10 (1894): 10-11.
   53. Crookes, "Address," p. 352. He went on to say: "Far be it from me to say anything disrespectful of the law of inverse squares, but I have already endeavoured to show we are dealing with conditions removed from our material and limited conceptions of space, matter, form."
   54. Frederic W. H. Myers, *Human Personality and Its Survival of Bodily Death,* 2 vols. (London: Longmans, Green and Co., 1915), vol. 1, pp. 245-46.
   55. W. F. Barrett, "Note on Telepathy and Telergy," *PSPR* 30 (1920): 257-60. Barrett had begun as John Tyndall's laboratory assistant and had gone on to become professor of physics at the Royal College of Science, Dublin. Barrett's general opposition to "any physical analogy or materialistic explanation" of psychical phenomena was already clear in 1882 (*idem,* "Appendix of the Report on Thought-Reading," *PSPR* 1 [1882-83]: 62). Among the many other discussions of this issue, we might point to the detailed one by Edmund Gurney ("Hypnotism and Telepathy," *PSPR* 5 [1888-89]: 224-30), in which he showed the difficulties in either assuming a vibratory energetic transmission (no falling off with distance) or not assuming a physical cause; he opted for a psychical force.
   56. For earlier echoes of this same difficulty, see Tischner, *Telepathy and Clairvoyance,* pp. 199-203.
   57. See above, chap. 1. There were some reports of the effects of yagé and peyote on psychic abilities in the 1920s (see the report of A. Rouhier, "Métagnomie et psycho-physiologie. I. Phénomènes de métagnomie expérimentale observés au cours d'une expérience faite avec le 'Peyotl' (*Echinocactus Williamsii*)," in *RM,* 1925, pp. 144-54, or the suggestive review by L. E. Rhine in *JASPR* 20 [1926]: 124-25), but they were merely anecdotal accounts incorporating at best uncontrolled experiments on qualitative material, far less ambitious or impressive.
   58. See Malcolm Jay Kottler, "Alfred Russel Wallace, the Origin of Man, and Spiritualism,"

Isis 65 (1974): 145-92; and Wilma George, *Biologist Philosopher: A Study of the Life and Writings of Alfred Russel Wallace* (London, Toronto, and New York: Abelard-Schuman, 1964), pp. 243-46.
59. Frederic W. H. Myers, "Automatic Writing.—III," *PSPR* 4 (1887): 259.
60. Frederic W. H. Myers, "The Subliminal Consciousness. Chapter III," *PSPR* 8 (1892): 359.
61. "Address," p. 8.
62. *ESP*, p. 127.
63. *Ibid.*, p. 156.
64. *Ibid.*, p. 7.
65. *Ibid.*, p. 142.
66. *Ibid.*, p. 143.
67. Alan Gauld, *The Founders of Psychical Research* (London: Routledge & Kegan Paul, 1968), pp. 275-99; and R. Laurence Moore, *In Search of White Crows* (New York: Oxford University Press, 1977), chap. 5.
68. Frederic W. H. Myers, "The Subliminal Consciousness. Chapter VI," *PSPR* 9 (1893-94): 7, 12-15. Ernest Jones implies, in *The Life and Works of Sigmund Freud*, 2 vols. (New York: Basic Books, 1953), vol. 1, p. 250, that this was the first such account.
69. Prince to Dingwall, 24 February 1933 (ASPR); Prince to W. Drayton Thomas, 13 June 1933 (ASPR); Prince to Price, 1 February 1934 (HPL).
70. Thus W. H. Salter wrote to Prince to congratulate him "and Dr. Rhine too, on a piece of work which will rank as a notable landmark in research into mental phenomena. We have not been so lucky over here in getting successes, either in experimental telepathy or in clairvoyance. Possibly that has been due to some fault in our methods. I am sure that Rhine's success will do much to stimulate further experiments over here. Several other members of the Society have expressed their sense of the great importance of this piece of work. Miss Jephson is doing us a review." Prince sent this extract to Rhine in a letter of 4 June 1934; Salter's original had been sent on 24 May. Other praise came from Theodore Besterman (letter to Rhine, 14 May 1934); Edith Lyttleton, then president of the SPR (letter to Rhine, 4 September 1934); Lord Rayleigh (letter to Prince, 22 September 1934); and Eleanor Mildred Sidgwick (letter to Rhine, 4 March 1935).
71. W. Whately Carington, "Positive Implications of Telepathy," abridged in *JSPR* 28 (April 1933): 57-64.
72. Rhine to Carington, 10 July 1934 (JBR).
73. *JSPR* 28 (October 1933): 131-33.
74. Saltmarsh to Rhine, 6 September 1934 (JBR).
75. Carington's remarks are recorded on a slip of paper dated 19 June 1934 by Prince (JBR).
76. "It is a first-class piece of work of outstanding importance—incomparably the most important ever done. Actually, in my personal opinion, it is the *only* work to date that can fairly be regarded as coercive as opposed to merely highly suggestive..." (Carington to Rhine, 2 June 1934 [JBR]).
77. Saltmarsh to Rhine, 2 June 1934 (JBR).
78. The review, by the science editor, A. S. Russell, appeared in *The Listener* for 25 July 1934, p. 150.
79. *Ibid.*, 22 August 1934, pp. 329-30.
80. Dingwall to Price, 26 July 1934 (HPL).
81. *Nature* 134 (1 September 1934): 308. The unsigned review was attributed to Dingwall by Harry Price in a letter to S. G. Soal of 10 September: "I did not see the Nature review, but it is undoubtedly Dingwall. He reviews most psychic works for Nature—and invariably slates them" (HPL).
82. S. G. Soal, *Preliminary Studies of a Vaudeville Telepathist*, University of London Council for Psychical Investigation Bulletin no. 3 (London, 1937).
83. Soal to Prince, 14 July 1934 (JBR).
84. Soal to Price, 26 July 1934 (HPL); the letter began a long correspondence between the two over the details of manufacturing the cards.
85. Rhine to Soal, 6 September 1934; Soal to Rhine, 20 September 1934 (JBR).
86. Price to C. A. Pannett, 24 December 1934 (HPL).
87. Robert H. Thouless, "Dr Rhine's Recent Experiments on Telepathy and Clairvoyance and a Reconsideration of J. E. Coover's Conclusions on Telepathy," *PSPR* 43 (1935): 24-37.

88. F.H.G. Van Loon and R. H. Thouless, "Report of a Demonstration of Experiments by Mr. Gustav Wallenius... ," *PSPR* 36 (1928): 437–54.
89. *JSPR* 29 (April 1935): 52.
90. In the annual report of the council for 1934, read at the meeting of the society of 9 January 1935, it was announced that "several of our members have been conducting experiments in paragnosis of various kinds. The work of Dr J. B. Rhine... on extra-sensory perception is of great interest. In it he has emphasised the importance of conducting experiments in such a way that results attributable to any one particular form of paragnosis shall be distinguishable from results attributable to other forms, and this is being borne in mind by the experimenters of our Society" (*JSPR* 29 [February 1935]: 23–24).
91. Dodds to Rhine, 28 August 1934 (JBR). In his autobiography, *Missing Persons* (Oxford: Clarendon Press, 1977), chap. 11, Dodds discusses the history of his association with psychical research.
92. Thouless, "Dr Rhine's Recent Experiments," p. 31. Thouless expressed this feeling still more strongly in a letter to Rhine: "I should think it very likely that the reason for your much higher percentage of successes than those reported by other people is that the Zener cards are more satisfactory material for transmission than the ordinary playing cards" (Thouless to Rhine, 24 January 1935 [JBR]).
93. Carington to Rhine, 1 January 1935 (JBR).
94. Thouless, "Dr Rhine's Recent Experiments," p. 34.
95. *Ibid.*, p. 25.
96. *The Listener*, 1 August 1934, p. 207.
97. *ESP*, pp. 21–22, 25.
98. Thouless, "Dr Rhine's Recent Experiments," p. 25. In fact, F.C.S. Schiller's review of Coover's book also had argued that there was "a source of rightness beyond chance" in the experiments that Coover recorded (*PSPR* 30 [1920]: 261–73). The point was repeated by Richet in *TY*, pp. 93–94.
99. Thouless, "Dr Rhine's Recent Experiments," p. 27, quoting Coover, *Experiments in Psychical Research at Leland Stanford Junior University*, Psychical Research Monograph no. 1 (Stanford: Leland Stanford University Publications, 1917), p. 123.
100. Thouless, "Dr Rhine's Recent Experiments," p. 30.
101. For another example (dating from March 1935) of the stimulus provided by this reanalysis of Coover's work, see Sir Cyril Burt, *ESP and Psychology*, ed. Anita Gregory (London: Weidenfeld and Nicolson, 1975), pp. 22–24. Burt had associated himself with Harry Price's endeavor to establish psychical research in the academic community in the previous year (see below, chap. 8).
102. Thouless, "Dr Rhine's Recent Experiments," p. 37.
103. J. B. Rhine, "Note on Professor Thouless's Review of *Extra-Sensory Perception*," *PSPR* 43 (1935): 542–43.
104. Rhine to Lydia Allison, 2 May 1935 (JBR).
105. W. H. S[alter], "An Appeal for Co-operation in Further Experiments in Extra-Sensory Perception," *PSPR* 43 (1935): 38–39.
106. T[yrrell], "The Case of Miss Nancy Sinclair."
107. G.N.M. Tyrrell, "Some Experiments in Undifferentiated Extra-Sensory Perception," *JSPR* 29 (April 1935): 52–71.
108. Sidgwick to Rhine, 4 March 1935 (JBR).
109. Our treatment of Driesch has profited considerably from the comments of Dr. Elmar R. Gruber of the Institut für Grenzgebiete der Psychologie und Psychohygiene, Freiburg im Breisgau.
110. Hans Driesch, *Parapsychologie* (Munich: F. Bruckmann, 1932), p. 62. The translation is that of Theodore Besterman: *Psychical Research* (London: G. Bell and Sons, 1933), p. 69; and see also pp. 70–71 and 102–3.
111. The same theme is stressed in Driesch's presidential address (*PSPR* 36 [1928]: 178).
112. Driesch to Rhine, 17 July 1934 (JBR).
113. J. B. Rhine, *Neuland der Seele* [New Frontiers of the Mind], translated and introduced by Hans Driesch (Stuttgart: Deutschen Verlags-Anstalt, 1938), pp. 6–7.
114. See below, chap. 8.
115. *Zeitschrift für Psychologie* 135 (1935): 218–19.

116. René Sudre, "La Divination et ses probabilités," *Journal des débats* 147 (11 April 1935): 3.
117. Above, chap. 2, n. 70.
118. *RM*, 1934, pp. 262-65.
119. Murphy to Rhine, 8 June 1934 (JBR): "I have been *deeply* impressed by your epoch-making studies in E.S.P. and am most eager to cooperate in every way I possibly can." Murphy proceeded to invite Pratt or Stuart to do doctoral work at Columbia, volunteered his own money to help defray their expenses there, volunteered to direct their work, and finally offered to perform replications of the Duke work or to undertake collaborative long-distance experiments.
120. Murphy had originally intended to submit the review to the *Journal of Social Psychology* but eventually decided on the *Journal of General Psychology* in order to assure the review a wider circulation. It appeared in vol. 11 (1934), pp. 454-57.

Chapter Six

1. On Trinity College, President Few, and the founding of Duke University, see Robert F. Durden, *The Dukes of Durham, 1865-1929* (Durham: Duke University Press, 1975), esp. pp. 199-260; and Earl W. Porter, *Trinity and Duke, 1892-1924* (Durham: Duke University Press, 1964). On Few, see also Robert H. Woody, "William Preston Few, 1867-1940: A Biographical Appreciation," in *Papers and Addresses of William Preston Few*, ed. R. H. Woody (Durham: Duke University Press, 1951), pp. 3-141. Unless otherwise indicated, all manuscript material cited in this chapter is in JBR.
2. W. P. Few, "An Old College and a New University," p. 7, in "The Beginnings of an American University" (Few Papers, DUA).
3. Seashore to W. I. Cranford, 1 March 1924 (Few Papers).
4. Cranford to Mount, 7 May 1924 (Few Papers).
5. Few to George W. Cram, 11 March 1926; Few to McDougall, 13 April 1926 (Few Papers).
6. McDougall to Few, early April 1926; Few to McDougall, 13 April 1926; McDougall to Few, 16 April 1926 (Few Papers). McDougall actually spoke of "G. K." Lashley and "R. S." Hunter.
7. McDougall to Few, 23 April 1926 (Few Papers).
8. Weigle to Soper, 18 June 1926 (Few Papers).
9. See above, chap. 3.
10. McDougall to Few, 3 May 1926 (Few Papers).
11. *Chronicle*, 2 February 1927, p. 1.
12. William McDougall, *Modern Materialism and Emergent Evolution* (London: Methuen and Co., 1929), p. 99. The book consists of lectures first given at the Louisville Presbyterian Theological Seminary in 1928.
13. On 17 April 1936 Rhine wrote to H. F. Saltmarsh with some satisfaction that "Dr. L. believes he can account for all the better established parapsychical phenomena by his theory and since he is one of the few psychologists willing to speculate on these phenomena, it is a great advantage to have him do so."
14. "Dr. Zener, although he gave me express permission to use his name in the connections where I used it and although it is not used in any way that was not fully warranted by the facts, has somehow come to dislike seeing his name appear in this connection" (Rhine to Murphy, 13 July 1934). Nevertheless, use of the term "Zener cards" continued to be widespread during the rest of the 1930s, especially in England but also to some extent in the United States, and we maintain that original terminology below insofar as the individuals we discuss employed it.
15. Zener, Adams, and Lundholm to McDougall, 9 April 1934.
16. McDougall to Rhine, 22 May [1934].
17. Report of the president, "The Launching of a University," *Bulletin of Duke University*, April 1932, pp. 11-12.
18. McDougall File, Few Papers. McDougall's memorandum was entitled "Memo embodying reflections and suggestions upon a plan for promoting research in the Sciences of Man and of Society, agreed upon by some members of the faculty of Duke University during conferences held on Oct. 21 and 31, 1930." The comment about subdepartmental status was penned in in this version, and the term was used by McDougall in reference to Rhine in a later letter to Few (4 November 1930 [Few Papers]).

19. On Mrs. Garrett's career in this period, see Eileen J. Garrett, *My Life as a Search for the Meaning of Mediumship* (London: Rider & Co., 1939); and Allan Angoff, *Eileen Garrett and the World Beyond the Senses* (New York: William Morrow, 1974). For a succinct biography of Mrs. Bolton, see *Biographical Directory of the American Congress, 1774-1961*, s. v. "Bolton, Frances."
20. For an expression of Mrs. Bolton's concerns, see Bolton to Rhine, 5 September 1934; for Rhine's, Rhine to Bolton, 7 July 1934.
21. Rhine to Bolton, 25 January 1935; Bolton to Rhine, 15 February 1935.
22. McDougall to Bolton, 4 February 1935.
23. Rhine to Worcester, 6 October 1934. Rhine had just been chosen to replace Prince as research officer of the Boston SPR.
24. Worcester to Rhine, 15 October 1934; Rhine to Wood, 27 October 1934, 31 January 1935; Wood to Rhine, 12 February 1935. President Few resisted calling anything that was not permanently endowed a "fellowship", and in the end the appointments were for "research associates."
25. See the remonstrance made by Paul Gross, chairman of the research committee, in the covering letter and "Proposed Budget for Research 1934-1935 in the Graduate School of Arts and Sciences and the Undergraduate Colleges," 20 March 1934, Research Council: Correspondence, 1931-1934 (DUA).
26. From Duke funds, $20,583.05 was spent on research in the fiscal year ending 30 June 1936 ("Report on Audit, Duke University, for Fiscal Year Ended June 30 1936" [DUA]). Departmental funds and external gifts brought total research expenditures up to $46,685 (Paul Gross to R. M. Hughes, National Resource Committee, 12 April 1938; Research Council: Correspondence, 1938-39 [DUA]).
27. "R. C. Grants for Research, 1934-35" and "R. C. Grants for Research, 1935-36" (DUA).
28. "The Graduate School of Arts and Sciences, 1934-35 (with announcements for 1935-36)," *Bulletin of Duke University*, May 1935, p. 28.
29. Few to Rhine, 12 August 1935.
30. Rhine to Mrs. Few, 11 June 1936.
31. Few to Rhine, 19 July 1935; Rhine to Few, 13 December 1935; C. B. Markham to Rhine, 4 September 1936.
32. For what follows we have drawn heavily on an interview with Professor Pratt, 20 May 1974.
33. His Master's thesis was entitled "The Role of Conflict in Learning"; his doctoral dissertation was "Experimental Investigations of Discriminative Learning in Rats."
34. On Stuart's first involvement with the Duke research, see above, chap. 4. His Ph.D. dissertation, the first since John Thomas's to be granted by Duke on a psychical-research topic, was entitled "An Analysis to Determine a Test Prediction of Extra-Chance Scoring in Card-Calling Tests."
35. All three wrote Master's theses on parapsychological subjects: Smith on "The Effect of Benzedrine Sulfate on Extra-Sensory Perception" (1937); Pegram on "Some Psychological Relations of Extra-Sensory Perception" (1937); Woodruff on "Size of Stimulus Symbols in Extra-Sensory Perception" (1939). Only Smith and Woodruff passed on to the Ph.D., and both chose to write dissertations, not in parapsychology, like Charles Stuart, but on normal sensory perception. Smith's dissertation, completed for Karl Zener in 1947, was on "Effects of Sodium Amytal on Visual Perception of Forms at the Limen"; Woodruff's, for Donald Adams in 1941, was on "The Effects of Certain Factors on Visual Form Discrimination at Near-Liminal Level."
36. It is extremely difficult now to piece together Coover's reactions to these experimental claims that were so at odds with his own researches published in 1917. Apparently when he learned of Rhine's work he made his own ESP cards and tried successfully to replicate the Duke results. He did not report this to Rhine, nor did he acknowledge an inquiry from Rhine (12 February 1935) asking for his opinion of *Extra-Sensory Perception* and for the source of the cards Coover had made. Seven months later Coover at last replied, saying nothing either about his own work or about Rhine's (12 September 1935), and he did not write again, though Rhine pressed him politely. In July 1936 Rhine visited California and had a chance to meet and talk with Coover, but the results were disappointing: he wrote William McDougall (3 August 1936) to say that Coover had been polite, friendly, and utterly evasive. It seems likely that Coover had concluded that the Duke work was faulty but had decided not to enter into open controversy (see below, pp. 253-54).
37. "It is a source of continual regret to me," he wrote from Colgate at the end of 1934, "that my work renders it impossible for me to devote more time to that field." Estabrooks's general

discouragement about the prospects of bringing telepathy under experimental control is well expressed in "The Enigma of Telepathy," *North American Review* 227 (1929): 201-11, which reveals his discovery that his Harvard experiments had *not* excluded the possibility of fraud by his student subjects. He seems also to have criticized himself for not having excluded the possibility of someone's tampering with his results (Estabrooks to Rhine, 1 August 1938).

38. Rhine to Murphy, 31 May 1934; Murphy to Rhine, 9 June 1934.
39. See below, chap. 9.
40. Carpenter to Rhine, 29 September 1934.
41. Carpenter to Rhine, 27 August 1935.
42. Interview, J. B. Rhine and L. E. Rhine, 5 July 1974; *Parapsychological Notes* 1, no. 1 (1 May 1936): 4. On these techniques, see below, chap. 7.
43. This was an ambivalence that he was slow to lose. As Rhine told E. P. Gibson: "You are achieving more, I believe, than any of the academic people who are working with us, and in the history of the subject people are not going to ask, as college presidents are compelled to do, whether a man has a Ph.D. or not. They are going to ask, 'What has he done?'" (Rhine to Gibson, 30 July 1936).
44. For Rhine's reflections on the disadvantages of publishing through the societies, see Rhine to McDougall, 23 June 1936.
45. Rhine to Hoagland, 25 April 1935; Hoagland to Rhine, 2 May 1935; Murchison to Rhine, 13 May 1935.
46. Rhine to Murchison, 24 May 1935; Hoagland to Rhine, 5 June 1935.
47. Zener to Rhine, 26 June and 1 August 1935; Spearman's reply to the article is quoted in the latter.
48. Boring to Rhine, 8 August 1935.
49. Rhine to Murphy, 19 June 1936. Some tentative attempts at communication among American researchers had already begun. The Schenectady group had started to circulate mimeographed reports to its members in March 1936, and in May the Duke laboratory had initiated its own mimeographed bulletin, *Parapsychological Notes* (with Stuart as editor), to encourage the exchange of information among ESP investigators outside the Durham area.
50. Rhine to McDougall, 23 June 1936; McDougall to Rhine, 1 August 1936.
51. Alice Crunden to Rhine, 5 February 1937. In fact, Mrs. Crunden's contributions lasted for only two years. However, by that time Rhine had found someone who was to be an important future source of support for parapsychological research and publication, Charles E. Ozanne. Educated at Yale and Harvard, Ozanne had taught high school in Cleveland until his retirement in 1935 at the age of seventy. He began to support the Duke program in 1936 from his investment income and capital; by 1940 he had given well over five thousand dollars to the William McDougall Research Fund, and he continued to make generous donations to the program throughout the next decade.
52. Rhine to Ozanne, 10 June 1936.
53. Quoted by J. G. Pratt to J. B. Rhine, late September 1936.
54. Murphy to Rhine, 19 October 1936; he felt more strongly about Carpenter's than about Warner's.
55. Murphy to Rhine, 11 December [1936].
56. [William McDougall], "Editorial Introduction," *JP* 1 (1937): 1-9.
57. *Science*, 12 February 1937; *New York Times*, 14 February 1937.
58. File, Mrs. Dorothy Pope, Foundation for Research on the Nature of Man, Durham, North Carolina.
59. In one questionnaire, 73 of 108 respondents indicated some familiarity with the journal (James C. Crumbaugh, "A Questionnaire Designed to Determine the Attitudes of Psychologists toward the Field of Extra-Sensory Perception," *JP* 2 [1938]: 304); in another, 145 of 352 did the same (Lucien Warner and C. C. Clark, "A Survey of Psychological Opinion on ESP," *ibid.*, pp. 298-99). How candid these respondents were is difficult to assess.
60. The exception was a paper by Clarence Leuba, of Antioch (on which see below, p. 281). Leuba was also somewhat atypical in that he had originally displayed a certain amount of sympathy for ESP and had even done a series of experiments in which he at first obtained apparently positive results.
61. Murphy to Rhine, 17 February 1937. His two candidates were Saul Sells and L. J. Stone.
62. See below, chap. 8.

63. Rhine to Murphy, 19 March 1937. "In my opinion," he wrote, "they will choose, out there, someone who is desirable to them on other grounds and will require only that some exploration, let us say, on hypnosis or some other safer headings [be made]. You see, I am skeptical about the basic good faith of the project."
64. "We... believe that by this plan psychical research can be brought into more intimate contact with general experimental psychology" (Lewis Terman to William McDougall, 25 February 1937—a letter accompanying a form announcing the Stanford program; on the whole subject, see Frederick C. Dommeyer, "Psychical Research at Stanford University," *JP* 39 [1975]: 173-205).
65. See below, chap. 10.
66. See below, chap. 9.
67. William McDougall, *The Riddle of Life* (London: Methuen, 1974), pp. 215, 235, 245.
68. Interview, J. B. Rhine, 1 March 1974.
69. *Scientific American* 137 (1927): 210-13.
70. E. E. Free, "How YOU Can Test Telepathy," *ibid.*, 148 (1933): 140-43; "The Second Scientific American Test of Telepathy," *ibid.*, pp. 324-25.
71. "The Results of Our First Test of Telepathy," *ibid.*, 149 (1933): 10-11, 45.
72. "Is Telepathy Indicated..?" *ibid.*, p. 152.
73. "Test For Telepathy," *ibid.*, 150 (1934): 64.
74. Walter Franklin Prince, "Extra-Sensory Perception," *ibid.*, 151 (1934): 5-7.
75. For example, the editorial in *ibid.*, 156 (1937): 361, praising the appearance of the *Journal of Parapsychology*.
76. Allison to Rhine, 12 May 1934; "The Week in Science: Telepathy and Clairvoyance," *New York Times*, 20 May 1934, sec. 8; p. 6.
77. *New York Herald Tribune*, 29 May 1934, p. 16. The article is unsigned, but Free identified himself as the author in correspondence with Rhine. In a science newsletter that he published, *The Week's Science*, he had already featured Rhine's work as the lead story for the first week in May.
78. James Jeans, *The Mysterious Universe* (New York: Macmillan, 1930); Arthur Eddington, *The Nature of the Physical World* (New York: Macmillan, 1930).
79. Robert Andrew Millikan, *Evolution in Science and Religion* (New Haven: Yale University Press, 1927), p. 18.
80. Arthur H. Compton, *The Freedom of Man* (New Haven: Yale University Press, 1935).
81. Joseph Wood Krutch, *The Modern Temper* (New York: Harcourt, Brace & Co., 1929); Carl Becker, *The Heavenly City of the Eighteenth Century Philosophers* (New Haven: Yale University Press, 1932).
82. Waldemar Kaempffert, "Searching Out the Mind's Mysteries," *New York Times Magazine*, 17 October 1937, p. 24. To the psychologist Joseph Jastrow, Kaempffert wrote privately: "To reject these results because they do not happen to fit in the old-fashioned physics gets us nowhere. They do happen to fit the new-fashioned physics very well indeed, by which I mean relativity and quantum theory. In other words, the door is now wide open for work of the kind that Dr. Rhine is doing" (Kaempffert to Jastrow, 3 December 1937). In this connection, it is worth mentioning that Albert Einstein was dubious about the validity of ESP precisely because it did not fit the "old-fashioned physics" in that the phenomena did not appear to depend on distance. Out of friendship for Upton Sinclair, Einstein wrote a sympathetic introduction to a proposed German translation of his *Mental Radio* (it may be found in the 1952 edition, published in Springfield, Illinois, by Charles C Thomas), but privately he was less confident of the soundness of telepathy. When in 1940 J. B. Rhine sent him a copy of his most recent book, Einstein replied politely but noncommittally. "I could not yet read your book because printed matters are not forwarded to my summer-address. I have read your former book and that of Upton Sinclair. I must confess that I am very sceptical about the reality of the phenomena in question although I have no explanation for the positive results you have obtained together with your collaborators. In any case I don't feel able to contribute effectively to the elucidation of the problems concerned." (Einstein to Rhine, 23 July 1940; by permission of the Estate of Albert Einstein.) Einstein explained his feelings more fully to Jan Ehrenwald in 1946, acknowledging nevertheless the need for the scientist always to keep an open mind. Einstein's correspondence with Ehrenwald has been published by Martin Gardner in "Einstein and ESP," *Zetetic* 2 (1977): 53-56, and "A Second Einstein ESP Letter," *Skeptical Inquirer* 2 (1978): 82-83; and, in different detail, by Jan Ehrenwald, "Einstein Skeptical of ESP? Postscript to a Correspondence," *JP* 42 (1978): 137-42.

83. John J. O'Neill, *Prodigal Genius: The Life of Nikola Tesla* (New York: Ives Washburn, 1944), p. 4.

84. Two well-known books of the 1920s reflect this widely diffused vision of the scientist as pioneer: Paul de Kruif's *Microbe Hunters* (New York: Harcourt, Brace and Co., 1926), and Sinclair Lewis's *Arrowsmith* (New York: Harcourt, Brace and Co., 1925). On the same theme, and these two authors in particular, see Charles E. Rosenberg, "Martin Arrowsmith: The Scientist As Hero," *American Quarterly* 15 (1963): 447-58, reprinted among his essays in *No Other Gods* (Baltimore and London: The Johns Hopkins University Press, 1976), pp. 123-31.

85. O'Neill to Rhine, 4 August 1934.

86. "Blind Sight," *Time,* 10 December 1934, pp. 39-42; J. B. Rhine, "Are We Psychic Beings?" *Forum,* December 1934, pp. 369-72; idem, "The Practical Side of Psychism," *ibid.,* January 1935, pp. 51-54; idem, "After Death—What?" *ibid.,* February 1935, pp. 114-18; idem, "The Gift of Prophecy," *ibid.,* July 1935, pp. 50-53; idem, "The Evidence for Prophecy," *ibid.,* August 1935, pp. 120-23; idem, "Don't Fool Yourself: Pitfalls in Psychic Research," *ibid.,* September 1935, pp. 187-89; *The American Weekly,* 13 January 1935; Will Irwin, "Can We Mortals See Without Eyes?" *Liberty,* 11 May 1935, condensed in *Reader's Digest* 26 (June 1935): 104-6; and J. B. Rhine, "Telepathy and Clairvoyance in a Trance Medium," *Scientific American* 153 (1935): 12-14.

87. Alexis Carrel, *Man, The Unknown* (New York: Harper Brothers, 1935), p. 125 (*metaphysical* should probably have read *metapsychical*). On Carrel, see *Dictionary of Scientific Biography.* The *New York Times* hailed Carrel's return from Europe to America with the following headline: "Everybody Has Telepathic Power, Dr. Carrel Says After Research" (19 September 1935, p. 25). The *New York Herald Tribune* took the same line: "Telepathy, Orphan Adopted by Science, Defies Research Efforts to Identify It" (22 September 1935, sec. 2; pp. 1, 5).

88. E. H. Wright, "The Case for Telepathy" and "The Nature of Telepathy," in *Harper's Magazine* 173 (1936): 575-86; and 174 (1936): 13-21, respectively; summary in *Reader's Digest* 30 (January 1937): 53-57.

89. Gardner Murphy, "Things I Can't Explain," *American Magazine* 72 (1936): 40-41, 130-32.

90. Rhine to Few, 25 January 1936.

91. Lester Hutter, "America's New Pastime," *New Current Digest,* January 1937, p. 38.

92. A. A. Wilkinson to Rhine, 8 February 1937.

93. Rhine to Free, 6 April 1937.

94. "Parapsychology," *New York Times,* 12 April 1937, p. 16. See also J. J. O'Neill in the *New York Herald Tribune* of 18 April; the Associated Press story, "16 Proofs of ESP," in the *Boston Herald* of 17 April; and *Time,* 26 April.

95. The Zenith broadcasts of 1937 also associated H. B. English, of Ohio State University, with the earlier "ground-breaking pioneer effort," but English disavowed the connection and angrily declared, "In 1924 Dr. Gardner Murphy and R. H. Gault planned an experiment in telepathy in which they used radio as a means of getting subjects. As I was spending the afternoon with Murphy he invited me to join in the experiment. As I remember it, my function was to hold up to the studio audience the stimulus cards. I had only a negligible part in planning the project and none at all in considering the data" ("Report of Committee," p. 3, enclosure in Goodfellow to Wolfle, 11 October 1938 [Wolfle Papers. We are grateful to Dr. Wolfle for permitting us to consult his correspondence on extra-sensory perception from the 1930s]).

96. McDonald to Rhine, 27 July 1937.

97. Zenith had done its best to see to it that the broadcasts commanded a large audience, with tantalizing hints to its radio dealers of what was to come, direct mailings to the public (see P. H. Erbes, Jr., "Broadcasting Silence," *Printer's Ink* 181 [28 October 1937]: 32), and magazine advertisements (see *New Yorker,* 18 September 1937, p. 65).

98. On the contents of *New Frontiers of the Mind,* see below, chap. 7.

99. C. E. Stuart and J. G. Pratt, *A Handbook for Testing Extra-Sensory Perception* (New York: Farrar and Rinehart, 1937). This handbook was not a commercial success.

100. Louis D. Goodfellow, "A Psychological Interpretation of the Results of the Zenith Radio Experiments in Telepathy," *J. Exper. Psychol.* 23 (1938): 601-32.

101. Few to Rhine, 20 July 1937.

102. Rhine to Paul Bryant [McDonald's secretary], 18 August 1937.

103. For example, E. E. Free to Rhine, 13 October 1937; and John Farrar to Rhine, 15 October 1937.

104. Baker to Rhine, 5 November 1937.
105. "Report of Committee," pp. 1-2, enclosure in Goodfellow to Wolfle, 11 October 1938 (Wolfle Papers).
106. "Rhine Question," *Time,* 4 October 1937, p. 44.
107. On Kellogg's critique and psychologists' reaction, see below, chap. 9.
108. On the genesis of Science Service, see Ronald C. Tobey, *The American Ideology of National Science, 1919-1930* (Pittsburgh: University of Pittsburgh Press, 1971), chap. 3; see also David J. Rhees, "A New Voice of Science: Science Service under Edwin E. Slosson, 1921-29" (M.A. thesis, University of North Carolina at Chapel Hill, 1979). The data given here are reproduced from "Report to the Annual Meeting of Board of Trustees, Science Service, May 1, 1930, by Watson Davis, Managing Editor," in *Corporation Records of Science Service, Inc.* (Washington, D.C.), pp. 636-39, 642, and were supplied to us through the kindness of David J. Rhees.
109. Thone to Rhine, 19 November 1936.
110. By 1931 Watson Davis had directed his writers *not* to use stories on "'supernatural' stuff," including "telepathy and mind reading, spirit manifestations of any sort, . . . animals that 'think,' 'read minds,' etc." A printed copy of his 1931 instructions in the Archives of the Smithsonian Institution is reproduced very closely in "Hints for Writing Science," *Science News Letter* 58 (1 July 1950): 12-13. In this respect Davis was faithfully maintaining the hostility towards psychical research that Science Service's founder and first director, Edwin Slosson, had always displayed (see Rhees, "A New Voice," pp. 75-76), which no doubt accounts for its silence on the subject during the 1920s.
111. Rhine to Thone, 28 April 1937; Van de Water to Rhine, 11 May 1937.
112. "Epidemic of Telepathy is Now Raging in the U.S.," *Boston Evening Transcript,* 13 October 1937, p. 1 (including a photograph of Watson Davis); "Telepathy and Clairvoyance Being Boomed as a 'Science,'" *Science News Letter,* 6 November 1937, pp. 298-99.
113. One exception to this general tendency is the review by Clifton Fadiman in the *New Yorker,* 9 October 1937, pp. 72-73.
114. Lewis Gannett, "Books and Things," *New York Herald Tribune,* 4 October 1937, p. 13.
115. For example, W. J. Cash, "Buck Duke's University," *American Mercury* 30 (1933): 102-10. A particularly malicious specimen—by a bitter ex-employee of Duke—is Ernest Seeman, "Duke: But Not Doris," *New Republic* 88 (1936): 220-22.
116. Compare the approach of the following two stories, separated in time by the controversy over ESP: *Social Frontier* 3 (1937): 102; and *Time,* 26 June 1939, p. 55.
117. *New Republic* 93 (1937): 49. Guterman was reacting to the following passage: "I am driven to believe that the most urgent problem of our disillusioned and floundering society is to find out more about what we are, in order to discover what we can do about the situation in which we exist today. In the conduct of our personal and our group affairs, our various outward and inward lives, we recognize more and more the need for a profounder kind of self-knowledge than any former age has had" (*New Frontiers of the Mind,* p. 4).
118. The sociopolitical orientation of early parapsychology is far more complex than R. Laurence Moore's remark, "Its founders had mostly been people of conservative tendencies," would suggest (*In Search of White Crows* [New York: Oxford University Press, 1977], p. 219). Moore's sketch of Rhine's attitudes in 1957 (as those of "a fierce cold warrior") is certainly no sound index to his convictions twenty years before.
119. "Books of the Times," *New York Times,* 5 October 1937, p. 21.
120. Kaempffert, "Searching Out the Mind's Mysteries."
121. Waldemar Kaempffert, *Science Today and Tomorrow* (New York: Viking Press, 1939), p. 272.
122. *Ibid.,* p. 110.
123. Kaempffert to Rhine, 29 September 1938.

Chapter Seven

1. Our brief outline of the development of the first Duke work on precognition and psychokinesis (1934-36) is founded upon a systematic examination of the correspondence and materials of the laboratory during that period. We have not always felt it necessary here to give

detailed references to the specific materials upon which we have based our account. Unless otherwise indicated, all manuscript material cited in this chapter is in JBR.

2. Before either Tyrrell or Rhine, L. T. Troland had attempted to incorporate sensitivity to possible motor responses into his experimental design. For a fuller account of Tyrrell's work, see below, chap. 8.

3. Rhine to Bolton, 7 July 1934.

4. Rhine to Allison, 15 March 1934.

5. J. B. Rhine, "Telepathy and Clairvoyance in the Normal and Trance States of a 'Medium,'" *Character & Pers.* 3 (1934): 91-111.

6. J. B. Rhine, "After Death—What?" *Forum*, February 1936, 117.

7. "I do not wish to put the ultimate objective, the survival hypothesis, prominently forward to the psychological world at this time. That is an attitude which is taken for the best interest of the work itself. We must win many important battles before we can lead our following into the still larger ones. To make our ambitions too large now or to make them *too conspicuous* would simply be to frighten off the group that we are building up here and abroad. I have many things up my sleeve or filed away in my files that the world just has to get along without for awhile" (Rhine to Eileen Garrett, 15 October 1934).

8. Rhine to Tyrrell, 11 March 1936.

9. On the publication of the precognition work, see below, pp. 224-26.

10. Unaware of the difficulty, Lydia Allison wrote: "Personally I think that your position in psychic research would increase in proportion to the reports you made of your experiments as you added to the subject, to which you have made such important contributions. I think it weakens your position at this time to generalize too much and write from the historical or philosophical angle. Everyone is doing that and it is much too easy" (Allison to Rhine, 24 May 1935).

11. William McDougall wrote Rhine from England on 2 July that he had seen Tyrrell ("a very attractive person, and, I judge, trustworthy") and that he had received a copy of Bender's report, "very similar in many respects to your book." Rhine himself had seen the report by 22 July.

12. Interview, Hans Bender, 26 August 1976. The dissertation was published as *Psychische Automatismen* (Leipzig: Johann Ambrosius Barth, 1935).

13. Bender's report, "Zum Problem der aussersinnlichen Wahrnehmung," first appeared in *Zeitschrift für Psychologie* 135 (1935): 20-130; it was then published in booklet form under the same title by Barth in 1936, and our references are to that edition. A summary of certain aspects may be found in Donald K. Adams, "Bender on Extra-Sensory and Sensory Form Perception," *JP* 1 (1937): 52-62.

14. Adams, "Bender," pp. 59-60, pointed to this loose method of evaluation: "In trial 13 (with the letter X) the subject drew K, A, V, W, X, C and G in that order as well as two additional figures. This is scored as a hit. If all repetitions, both erroneous and correct, are counted in this series the percentage of correct hits is 12.7." In retrospect, Bender wrote, "I do not disagree with Dr. Adams' opinion that the quantitative evaluation of my results to which I attributed little importance, is in some places rather too optimistic" (Bender to Rhine, 5 August 1937).

15. Bender, *Zum Problem*, pp. 26-28.

16. Interview, Hans Bender, 26 August 1976.

17. On the review, see above, p. 129. Bender chose to label the subject of his own researches "aussersinnlichen Wahrnehmung," which might seem to be a translation of *Extra-Sensory Perception*. In fact, however, Bender derived the phrase from Gustav Pagenstecher's *Aussersinnliche Wahrnehmung: experimentelle Studie über den sogennanten Trancezustand* (Halle: C. Marhold, 1924) (personal communication, 4 February 1977).

18. Bender, *Zum Problem*, p. 11. Our translation is that prepared by Donald Adams in 1936 under the title "Toward the Problem of Extra-Sensory Perception," a copy of which is in the library of the Foundation for Research on the Nature of Man, Durham, North Carolina.

19. Rhine to Pratt, Rhine to McDougall, and Rhine to Few—all 22 July 1935. Others, too, saw Bender's work as merely confirmatory of Rhine's (cf. *RM*, 1936, p. 58), which is ironic, since of course they were entirely independent.

20. Rhine to C. Hilton Rice, 22 July 1935.

21. Rhine to McDougall, 22-24 July 1935.

22. J. G. Pratt and Margaret M. Price, "The Experimenter-Subject Relationship in Tests for ESP," *JP* 2 (1938): 84-94; see also the editorial in the same volume, pp. 79-81.

23. Margaret H. Pegram, "Some Psychological Relations of Extra-Sensory Perception," *JP* 1 (1937): 194-95.
24. J. G. Pratt, "A Report Upon a Case of Ability in Extra-Sensory Perception," p. 19.
25. J. G. Pratt, "Clairvoyant Blind Matching," *JP* 1 (1937): 15.
26. *Ibid.*, pp. 12-13. The effect of this qualification was seriously compromised in the report by the way in which the data were tabulated, since no distinction was made between those data obtained under the "loose" STM, in which the subject could see the target cards, and those data from the STM in which the target cards were screened.
27. J. B. Rhine, "The Hypothesis of Deception," *JP* 2 (1938): 151-52.
28. Thomas to McDougall, 30 August 1938; Rhine to McDougall, 5 October 1938; McDougall to Thomas, 10 October 1938.
29. Carpenter to Rhine, 20 April 1935.
30. C. R. Carpenter, "Report on a Series of 'E.S.P.' Experiments," p. 4 (attached to Rhine to Carpenter, 19 December 1935).
31. C. R. Carpenter, "An Experiment in Card Guessing," p. 15 (attached to Carpenter to Rhine, 18 September 1936). The text was eventually published as C. R. Carpenter and H. R. Phalen, "An Experiment in Card Guessing," *JP* 1 (1937): 31-43.
32. "The fact that your results declined with advancing precautions has to be disentangled so far as the results justify, from obscuring the fact that at the same time the conditions were increased in difficulty according to the subject's probable belief as to what was difficult. It is important that the reader has before him those alternatives. We know that if the subject believes he cannot do a thing under a certain condition, he cannot. You can safely point out, as I think perhaps you do, that some of your subjects fail to score under conditions in which others succeeded, with no apparent difference except mental attitude or capacity on the part of the subjects" (Rhine to Carpenter, 10 November 1936).
33. Carpenter to Rhine, 3 November 1936; cf. Pratt to Rhine, 9 November 1936.
34. Rhine to Carpenter, 18 October 1935; Rhine to Pratt, 16 October 1935.
35. Rhine to Warner, 25 May 1934. Neither Warner nor Rhine was ever able to track this quotation to its source (Rhine believed he had seen it quoted by Coover), and we have been no more successful. The original report is attached to a letter from Warner to Rhine dated 4 March 1936; a final revision was published as Lucien Warner and Mildred Raible, "Telepathy in the Psychophysical Laboratory," *JP* 1 (1937): 44-51.
36. Murphy to Rhine, 17 June 1936.
37. Murphy to Rhine, 26 September 1936.
38. Woodworth to Warner, 20 January 1937.
39. Boring to Rhine, 8 February 1937.
40. Louisa E. Rhine, "Some Stimulus Variations in Extra-Sensory Perception with Child Subjects," *JP* 1 (1937): 102-13.
41. Rhine to George, 19 July 1934, in which Rhine encouraged George to set interested students to work on experiments in the field.
42. E.g., Rhine to Woodruff, 24 April and 3 September 1935.
43. Rhine to Woodruff, 8 October 1935.
44. J. L. Woodruff and R. W. George, "Experiments in Extra-Sensory Perception," *JP* 1 (1937): 18-30.
45. Bernard F. Riess, "A Case of High Scores in Card Guessing at a Distance," *ibid.*, pp. 260-63. Some of our detail as to events is based on the account in "The ESP Symposium at the A.P.A.," *JP* 2 (1938): 270-71.
46. Betty M. Humphrey and John A. Clark, "A Comparison of Clairvoyant and Chance Matching," *ibid.*, pp. 31-37.
47. On the general attitude of the Colorado department towards the subject, see below, chap. 9.
48. Martin to Rhine, 6 February 1938. See Dorothy R. Martin, "Chance and Extra-Chance Results in Card Matching," *JP* 1 (1937): 185-90.
49. This work was summarized in D. R. Martin and F. P. Stribic, "Studies in Extra-Sensory Perception, I," *JP* 2 (1938): 23-30; and *idem*, "Studies in Extra-Sensory Perception, II," *ibid.*, pp. 287-95.
50. Rightly or wrongly, the report circulated among psychologists that Murphy was kept back from promotion because of his interest in psychical research (P. R. Farnsworth [Farnsworth-Hilgard

memoirs, AHAP], who gives Carney Landis, of the Columbia psychology department, as his source).

51. "A Report to the Members of The Boston Society for Psychic Research," 10 May 1935.

52. "First Draft of Research Plan for Study of Three Functions Possibly Related to E.S.P. Ability," enclosure in Rhine to Murphy, 8 January 1937.

53. "My first impression is that it is too comprehensive for the purpose of initiating successful E.S.P. work with a group of subjects. I am a little inclined to think that you are inclined to over-reflect before action. . . . Such a plan as you have outlined would be splendid to have, as it were, in the back of your mind to be used when success is promised by a simpler approach. I feel that both the subject and the actual experimenter must have a very simple, clearly grasped working scheme in mind as he begins" (Rhine to Murphy, 8 January 1937).

54. See below, chaps. 9, 10.

55. C. E. Stuart and J. G. Pratt, *A Handbook for Testing Extra-Sensory Perception* (New York: Farrar and Rinehart, 1937).

56. Cf. "Editorial Introduction," *JP* 1 (1937): 4-7.

57. J. B. Rhine, *Extra-Sensory Perception* (Boston: Boston Society for Psychic Research, 1934), p. 74, table 18 (hereinafter abbreviated as *ESP*).

58. *Ibid.*, p. 88.

59. Robert H. Thouless, "Dr Rhine's Recent Experiments on Telepathy and Clairvoyance and a Reconsideration of J. E. Coover's Conclusions on Telepathy," *PSPR* 43 (1935): 36-37.

60. "It is perhaps to be regretted that a chronological and literary organization was imposed upon the report instead of the clearer and more customary logical one. It is very difficult to determine from the presentation exactly what was done or to be certain one has not overlooked important conditions; . . . we are still in some trepidation lest some basic control whose absence we have criticized should turn up presently in some unexpected place" (R. R. Willoughby, "A Critique of Rhine's 'Extra-Sensory Perception,'" *J. Abnorm. Soc. Psychol.* 30 [1935]: 202, n. 4).

61. See J. B. Rhine, "Note on Professor Thouless's Review of *Extra-Sensory Perception*," *PSPR* 43 (1935): 542.

62. See above, pp. 183-84.

63. *Parapsychological Notes* 1, no. 5 (1 October 1936): 17.

64. On these criticisms by academic psychologists, see below, chap. 9.

65. Rhine to Clark, 15 November 1937; cf. Rhine to Clark, 22 June 1937.

66. J. B. Rhine, "The Question of Sensory Cues and the Evidence," *JP* 1 (1937): 276-91; Riess, "A Case of High Scores," pp. 260-63.

67. Lucien Warner, "A Test Case," *JP* 1 (1937): 234-38.

68. "Notes," *ibid.*, p. 305.

69. "Research Notes," *JP* 2 (1938): 73.

70. *ESP*, p. 32. Rhine failed to give formula (.6745 $\sqrt{npq}$, where n = number of trials, p = probability of success, and $q = [1 - p]$) for the PE in the first edition of *ESP*, as a number of readers complained; the formula was supplied in the English edition (1935).

71. *ESP*, p. 33. The term "critical ratio" does not appear in *ESP*; it is used by R. R. Willoughby in "A Critique," p. 199.

72. *ESP*, p. 110.

73. Carington to Rhine, 2 June 1934.

74. Willoughby to Rhine, 1 August 1934. Willoughby incorporated this point in his ensuing paper ("A Critique," p. 199). See R. A. Fisher's interesting discussion of the logic of this kind of reaction in *Statistical Methods and Scientific Inference* (Edinburgh: Olivers and Boyd, 1956), pp. 40-41.

75. Willoughby to Rhine, 2 August 1934.

76. Willoughby, "A Critique," p. 201; C.. E. Stuart, "In Reply to Willoughby's Critique," *J. Abnorm. Soc. Psychol.* 30 (1935): 384-85. See also R. R. Willoughby,"The Use of Probable Error in Evaluating Clairvoyance," *Character & Pers.* 4 (1935): 79-80; and C. E. Stuart, "A Reply to Dr. Willoughby," *ibid.*, p. 80.

77. Chester E. Kellogg, "Dr. J. B. Rhine and Extra-Sensory Perception," *J. Abnorm. Soc. Psychol.* 31 (1936): 190-93. The problem was given more attention in Kellogg's "The Problems of Matching and Sampling in the Study of Extra-Sensory Perception," *ibid.*, 32 (1937): 462-79.

78. J. A. Greenwood and C. E. Stuart, "Mathematical Techniques Used in ESP Research," *JP* 1 (1937): 206-25.
79. Willoughby, "A Critique," p. 201.
80. Kellogg, "Dr. J. B. Rhine," pp. 191-92.
81. Greenwood and Stuart, "Mathematical Techniques," pp. 211-12.
82. The shift to the use of the standard deviation is first to be found in Pratt, "Clairvoyant Blind Matching," pp. 10-17. In their review of the Duke mathematics, Greenwood and Stuart acknowledged the change in practice, explaining that "this usage ... follows a general trend in statistical method" ("Mathematical Techniques," p. 208).
83. "Mathematical Techniques," pp. 214-18. An attempt at a definitive survey of the Pearce-Pratt experiment is J. B. Rhine and J. G. Pratt, "A Review of the Pearce-Pratt Distance Series of ESP Tests," *JP* 18 (1954): 165-77.
84. Chester E. Kellogg, "New Evidence (?) for 'Extra-Sensory Perception,'" *Scientific Monthly* 45 (1937): 336.
85. Edward V. Huntington, "Exact Probabilities in Certain Card-Matching Problems," *Science* 6 (26 November 1937): 499-500; T. E. Sterne, "The Solution of a Problem in Probability," *ibid.*, pp. 500-501.
86. Greville and Brown are credited with this work on a sheet of paper attached to a reprint of Huntington's article (JBR). Brown communicated the information to Huntington in a letter of 27 December 1937, while Greville is there said to have computed it "at a slightly earlier date." See T.N.E. Greville, "Exact Probabilities for the Matching Hypotheses," *JP* 2 (1938): 55-59.
87. Edward V. Huntington, "A Rating Table for Card-Matching Experiments," *JP* 1 (1937): 292-94.
88. C. E. Stuart and J. A. Greenwood, "A Review of Criticisms of the Mathematical Evaluation of ESP Data," *ibid.*, pp. 295-304.
89. J. A. Greenwood, "Variance of the ESP Call Series," *JP* 2 (1938): 60-64.
90. See, for example, Chester E. Kellogg, "A Note in Reply to Mr. Charles E. Stuart," *J. Abnorm. Soc. Psychol.* 33 (1938): 521-26; and *idem*, "The Statistical Techniques of ESP," *J. Gen. Psychol.* 19 (1938): 383-90. Kellogg's concession appears in J. G. Pratt *et al.*, *Extra-Sensory Perception after Sixty Years* (New York: Henry Holt, 1940), p. 229. The remark is, perhaps, mildly equivocal.
91. These views are summarized from lecture notes prepared for a course entitled "Survey of Science," which Rhine taught in 1930, outlining a generally antimaterialistic world view that reflected the new physics of the 1920s and the impact of philosophers such as Bergson and Whitehead. "Flux" and "process" rather than stasis were at the root of things; "waves of energy" were fundamental, rather than "particles of matter."
92. *ESP*, p. 119. The same objection was made much of by C. D. Broad in his attempt to sketch out a theoretical structure for mental phenomena (see below, pp. 228-29).
93. Rhine to E. E. Free, 12 December 1934. He was understandably less naturalistic in tone when writing to his correspondents who hoped to find proof of post-mortem survival.
94. *New Frontiers of the Mind* (New York and Toronto: Farrar and Rinehart, 1937), p. 190.
95. *Ibid.*, p. 213.
96. See Gardner Murphy, "Psychical Phenomena and Human Needs," *JASPR* 37 (1943): 163-91.
97. *ESP*, p. 2; *New Frontiers*, pp. 174-88. Rhine's conviction was shared by many English psychical researchers, such as H. F. Saltmarsh and G.N.M. Tyrrell (see, for example, the latter's "Normal and Supernormal Perception," *JSPR* 29 [January 1935]: 3-20).
98. Boring to Rhine, 30 November 1936 (EGB). Boring admitted that psychologists had not yet adequately pursued these problems as they related to difficult SP and suggested that they might be studied at Harvard "with the intent of finding out whether it can approximate the conditions and results of ESP."
99. For the Harvard work, see below, pp. 269-72.
100. Burke M. Smith, "Effects of Sodium Amytal on Visual Perception of Forms at the Limen" (Ph.D. diss., Duke University, 1947), pp. 3-4; his research had begun by 1939-40, however.
101. Even the work of Hans Bender, which seemed to reveal parallels between clairvoyant guessing and image formation in SP under a wide variety of conditions, did not convince Rhine. As he wrote to H. F. Saltmarsh: "I feel that his [Bender's] tendency to a too facile comparison of the

mere imagery in ESP with sensory experiences in faint visual perception will ultimately be found to be very misleading. Our own work, certainly, and much other work, to a considerably [sic] extent supports the view that there is a rather sharp departure between the ESP phenomenon [and sensory perception] and that the image formation is concerned merely with the subject's translation. The similarities, that is, between SP and ESP are superficial. The differences are much more fundamental" (Rhine to Saltmarsh, 4 August 1936).

Chapter Eight

1. The adjectives are those of Rudolf Lambert, "My Debt of Gratitude to Dr. W. F. Prince," in *Friends and Colleagues in Psychical Research, Walter Franklin Prince: A Tribute to His Memory* (Boston: Boston Society for Psychic Research, 1935), pp. 40–41; the article gives a good picture of Prince's European reputation (which Lambert came eventually to view as too severe). For a bibliography of Prince's work, see *Walter Franklin Prince*, pp. 90–96.
2. The reorganization was arranged in a council meeting on 13 October 1934 and was communicated to the membership in an undated announcement later that month (JBR).
3. Rhine to Allison, 23 November 1934 and 7 January 1935, for example (JBR).
4. J. B. Rhine, "The Pratt-Garrett Study of Mediumistic Trance Utterances," postscript to J. G. Pratt, *Towards a Method of Evaluating Mediumistic Material*, Boston Society for Psychic Research bulletin 23 (Boston, 1936), pp. 54–59.
5. E.g., Rhine to Ellen Wood, 28 July 1936 (JBR).
6. Rhine to Thomas, 31 July 1936 (JBR).
7. Thomas to Rhine, 4 June 1936; Rhine to Thomas, 11 July 1936 (JBR).
8. Strickland to Rhine, 24 January 1937 (JBR).
9. Rhine to Thomas, 8 June 1937 (JBR). The book was published as *Beyond Normal Cognition* (Boston: Bruce Humphries, 1937).
10. René Warcollier, *Experimental Telepathy*, ed. Gardner Murphy (Boston: Boston Society for Psychic Research, 1938). The book was largely subsidized by a gift of E. F. McDonald of the Zenith Radio Corporation, who remained very much interested in studies of telepathy.
11. Since becoming research officer in 1925 Bird had been a vocal supporter of Margery, but paradoxically he had been as well a very shrewd and sympathetic exponent of the experimental study of mental phenomena (see, for example, his articles "Probabilities and Metaphysics," *Psychic Research* 22 [1928]: 69–82; "Experimental Telepathy—I," *ibid.*, 23 [1929]: 517–33; and "Experimental Telepathy. II," *ibid.*, pp. 672–81). The series of discussions of previous work on experimental telepathy was cut short by his resignation from the ASPR. Harry Price, in London, continued to serve as foreign research officer until in 1931 he exposed Mrs. Duncan, a medium to whom a wealthy member of the society was devoted; the member protested, and Price lost his title (Price to Rhine, 28 July 1932 [JBR]).
12. Quoted by Thomas R. Tietze in *Margery* (New York: Harper & Row, 1973), p. 119, from a February 1927 letter from Bond to Elwood Worcester. On Bond's life, see William W. Kenawell, *The Quest at Glastonbury: A Biographical Study of Frederick Bligh Bond* (New York: Helix Press, 1965).
13. Bond to A. E. Schaaf, 15 May 1935, quoted in "Mr. Bond and the 'Margery' Mediumship," *JASPR* 29 (1935): 161.
14. Events can be followed in "Editorial Notes," *ibid.*, pp. 130–31; "Statement to Members," *ibid.*, pp. 153–58; and "Mr. Bond and the 'Margery' Mediumship," *ibid.*, pp. 159–62. The trustees announced in the January 1936 issue that they were assuming editorial control themselves, but they were at least temporarily assisted in this by Seward Collins, editor of the *North American Review*.
15. Fodor's appointment was announced in *JASPR* 30 (1936): 44, and his first letter appeared in February. On the founding of the IIPR, see *ibid.*, 28 (1934): 244; see also below. On Fodor, see the biographical introduction written by Leslie Shepard for the reprint of Fodor's *Encyclopedia of Psychic Science* (1933; reprint ed., [Secaucus, N.J.]: University Books, 1966).
16. E.g., John J. O'Neill, "Peculiar Properties of the Human Mind," *JASPR* 30 (1936): 205–19, 237–50 (where Rhine is discussed on pp. 246–47).
17. Jocelyn Pierson, "Psychical Research in Recent Periodicals," *ibid.*, pp. 365–74.
18. Jocelyn Pierson, "A Study of Clairvoyance," *ibid.*, 31 (1937): 129.

19. "'Parapsychology,'" *ibid.*, p. 154.
20. William H. Button, "The Margery Mediumship," *ibid.*, 32 (1938): 1-4; see also "Editorial Notes," *ibid.*, pp. 33-38.
21. The complete review, by Herbert Nichols, is in *ibid.*, pp. 31-32.
22. The historical genesis of this work and its format are described in Gardner Murphy and Ernest Taves, "Tests of Extra-Sensory Perception Among A.S.P.R. Members," *Proceedings ASPR* 23 (1939-40): 1-10. Murphy's original proposal, headed "Notes regarding a plan of research for the American Society for Psychical Research" and dated 21 September 1938, is in ASPR.
23. William H. Button and John J. O'Neill, "Editorial Comment," *Proceedings ASPR* 23 (1939-40): iii. O'Neill was of course the science editor for the *New York Herald Tribune* who had done so much to publicize Rhine's work.
24. H. A. Robinson to Rhine, 15 October 1935; Bender to Rhine, 10 January 1936 (JBR). Robinson's letter passed on the rumor that the Germans were kept home "because it was discovered that a prominent Communist was to be present"; Bender, however, attributed the decision to "personal intrigues of occultist cliques which had not been invited."
25. Eugène Osty, "Les Pouvoirs inconnus de l'esprit sur la matière," *RM*, 1932, pp. 121-22. The control of the legacy from Jean Meyer, which ensured the institute's continued work, had been tied up in litigation. The appeal was repeated at year's end ("Reprise prochaine d'expériences sur les pouvoirs inconnus de l'esprit sur la matière," *ibid.*, p. 418).
26. Charles Richet made the first appeal in "Pour le progrès de la métapsychique," *RM*, 1933, pp. 345-46; the first meeting was announced in *RM*, 1934, p. 120. The society's *statuts* appear in *RM*, 1935, pp. 454-58.
27. See the report in *RM*, 1934, pp. 70-71, where "15 Fellows of the Royal Society and 4 Nobel Prize laureates" are spoken of as supporting the IIPR.
28. Price to W. F. Prince, 12 February 1934 (HPL).
29. However, in collaboration with Hereward Carrington's recently refounded American Psychical Institute, it did issue a series of bulletins containing some experimental material (see the summaries in *JSPR* 29 [March 1936]: 203-4, and [November 1936] 296-97; 30 [March 1938]: 192).
30. University of London, Minutes of the Senate, 24 January 1934, pp. 29-34. Price described the equipment of the laboratory in *Fifty Years of Psychical Research* (London, New York, and Toronto: Longmans, Green, 1939), pp. 318-23.
31. Minutes of the Senate, 21 February 1934, pp. 64-65.
32. Burt to Price, 21 February 1934 (HPL). Burt had come to University College, London, in 1931 to head its psychology department, having established himself as a student of child development and statistical methodology during the previous decade. His involvement with Price's group apparently initiated his increasingly sympathetic association with psychical research, but the influence of McDougall (with whom he had actually done telepathic experiments while a student at Oxford early in the century) had already made this association an easy one. For his writings on psychical research, see the essays edited by Anita Gregory, *ESP and Psychology* (London: Weidenfeld and Nicolson, 1975). The biography by L. S. Hearnshaw, *Cyril Burt: Psychologist* (Ithaca: Cornell University Press, 1979), pp. 221-26, discusses his later involvement in the field in some detail.
33. Gregory to Price, 3 April 1934 (HPL).
34. Price, *Fifty Years*, p. 64; see also Price to A. M. Low, 15 May 1934 (HPL). Macnamara died shortly thereafter and was eventually replaced on the council by Charles Pannett.
35. Cf. *JSPR* 28 (October 1934): 277.
36. Price to C. A. Pannett, 24 December 1934 (HPL).
37. Joad to Price, 23 August 1934 (HPL).
38. What may be the final draft of this paper by Soal is "Experimental Telepathy and Clairvoyance in England, 1881-1933" (SPR).
39. Brown to Price, 5 February 1935 (HPL).
40. A popular account of the fire walk, held in September 1935, is in Price, *Fifty Years*, pp. 250-62; a longer report is contained in Harry Price, *A Report on Two Experimental Fire-Walks* (London: University of London Council for Psychical Investigation, 1936).
41. Brown to Price, 19 August 1935. J.C. Flugel had already warned Price (19 June 1934) that "a good many members of the new Council are clearly anxious to avoid all unnecessary publicity at present" and to be cautious in his dealings with the press (HPL).
42. Cyril Burt to Price, 8 May 1934 (HPL).

43. For a description of his doctoral research, see J. Hettinger, *The Ultra-Perceptive Faculty* (London: Rider and Co., [1940]). This work is discussed and criticized in Christopher Scott, "Experimental Object-Reading: A Critical Review of the Work of Dr. J. Hettinger," *PSPR* 49 (1949): 16-50.

44. Burt had proposed putting a research student to work immediately when the council was formed in May 1934 (Price to Burt, 9 May 1934), and three years later he reported to Price that one of his senior students, D. C. Russell, was "thinking of choosing some problem in psychical research for his Ph. D. thesis; and McDougall has already offered to receive him at Duke University so as to study Rhine's methods at first hand. He is, however, not interested solely in 'extra-sensory perception,' but would like to extend his knowledge of the whole field" (Burt to Price, 4 April 1937 [both in HPL]).

45. Burt to Price, 6 March 1937 (HPL).
46. Aveling to Price, 14 October 1936 (HPL).
47. Brown to Price, 9 September 1938; Joad to Price, 27 September; Brown to Price, 11 October (HPL).
48. C.E.M. Joad, "Adventures in Psychical Research," *Harper's Magazine* 177 (1938): 35-41, 202-10; reviewed by W. H. Salter in *PSPR* 45 (1939): 217-22.
49. Dingwall to Price, 1 February 1939 (HPL).
50. Bender to SPR, 11 October and 28 October 1932 (SPR).
51. Bender to SPR, 26 November 1933 (SPR).
52. Price to Bender, 21 April 1936 (HPL).
53. Bender to Price, 2 November 1936 (HPL).
54. "The matter was very carefully examined by the German Government, i.e., by the 'Reichs- und Preussische Ministerium für Wissenschaft, Erziehung und Volksbildung' (Board of Education), by the 'Innenministerium' (Home Office), by the 'Auswärtiges Amt' (Foreign Office), and by the officials of Bonn University. The importance of your offer and of the scientific principles involved in a decision needed long deliberations.

I am glad to repeat you now what our collaborator has already been telling you in London: that the German Government and the University have decided to accept in principle your original offer. They authorize the establishment of a Department for Abnormal Psychology and Parapsychology (Forschungstelle für psychologische Grenzwissenschaften) and think of special interest to concern this Department, besides the Research work, with questions of social hygiene in occult matters. As acknowledgement of your possible gift and considering its importance for public health the German Government would confer to you the 'Rote-Kreuz-Medaille I-Klasse' (Medal of the Red Cross, Irst Class)" (Bender to Price, 20 March 1937 [HPL]).

55. Bender to Rhine, 10 January 1936 (JBR).
56. Rhine to Bender, 11 March 1937; Bender to Rhine, 3 April 1937 (JBR).
57. See below, chap. 10.
58. Cf. Rhine to Thomas, 10 November 1927 and 29 February 1928 (JBR).
59. Minutes of Council, 30 May 1934 (SPR).
60. *Ibid.*, 24 October 1934; 9 January, 30 January 1935 (SPR); his resignation was reported in "Annual Report of the Council for 1934," *JSPR* 29 (February 1935): 25, but not in full detail.
61. Minutes of Council, 1 November 1934.
62. The council considered at least three candidates for the position on 30 January and 27 February 1935: J. G. Maby of Oxford, turned down because he lived too far away; the biologist Frank Baker, turned down because of too little preparation in psychology; and a psychologist at Kings College, London, turned down because he could work only part-time.
63. Herbert was proposed in a council meeting 29 May, and he accepted the position 26 June; his apppointment was announced in *JSPR* 29 (July 1935): 105, and his promotion in *ibid.* (November 1936): 286.
64. W. Whately Smith, *The Measurement of Emotion* (New York: Harcourt, Brace & Co., 1921). The help of Mr. Fraser Nicol has been invaluable to us in our attempt to assess Carington's accomplishments.
65. Minutes of Council, 25 April 1934.
66. Mr. T[yrrell], "The Case of Miss Nancy Sinclair," *JSPR* 20 (March 1922): 294-327.
67. G.N.M. Tyrrell, "Some Experiments in Undifferentiated Extra-Sensory Perception," *JSPR* 29 (April 1935): 52-71; *idem*, "Further Research in Extra-Sensory Perception," *PSPR* 44 (1936): 99-166.

68. Minutes of Council, 1 November 1934 and 30 January 1935.
69. Letter published in *JSPR* 29 (February 1935): 35-36.
70. On the "Fisk effect," see Tyrrell "Further Research," pp. 153-57. In 1974 Miss Johnson remembered little about the tests except that she did not enjoy them (interview, 7 June 1974).
71. The request is recorded in Minutes of Council, 27 February 1935; and the results are reported in Tyrrell, "Further Research," pp. 114-19. Miss Johnson and one other participant had the best individual scores ($p = .013$).
72. See Tyrrell's letter in *JSPR* 29 (May 1935): 80-81. Thouless and Dingwall reply in *ibid*. (July 1935), pp. 107-9.
73. See the exchange between Soal and Tyrrell in *JSPR* 30 (October 1938): 274-76 and (November 1938): 288-90.
74. G.N.M. Tyrrell, "The Tyrrell Apparatus for Testing Extra-Sensory Perception," *JP* 2 (1938): 107-18.
75. Minutes of Council, 1 May 1935; 24 January 1936.
76. In "A Suggested New Test for Evidence of Survival," *JSPR* 19 (February 1920): 163-64; see Smith, *Measurement of Emotion*, p. 121.
77. Hereward Carrington, *An Instrumental Test of the Independence of a "Spirit Control,"* American Psychical Institute Bulletin 1 (New York, [1933]). Carrington had reestablished his "institute" in 1932, after nearly a decade of inactivity, because of his dissatisfaction with the ASPR's devotion to Margery.
78. *PSPR* 42 (1934): 241-49.
79. Rhine to Prince, 27 March 1934 (JBR).
80. Whately Carington, "The Quantitative Study of Trance Personalities I," *PSPR* 42 (1934): 173-240; "The Quantitative Study of Trance Personalities II," *PSPR* 43 (1935): 319-61; "The Quantitative Study of Trance Personalities III," *PSPR* 44 (1936): 189-222.
81. For the exchange between Carington and Maby, see J. Cecil Maby, "Note on Mr Carington's Investigation," *PSPR* 43 (1935): 362-66; Whately Carington, "Reply to Mr Maby's Note," *ibid.*, pp. 367-70; J. Cecil Maby, "A Further Note on Mr Whately Carington's Investigation," *ibid.*, pp. 520-32; and Whately Carington, "Some Comments on Mr Maby's 'Further Note,'" *ibid.*, pp. 533-36.
82. Letter in *JSPR* 29 (February 1936): 185. On Ridley, see the account in *Biographical Memoirs of Fellows of the Royal Society,* vol. 3 (London: The Royal Society, 1956), pp. 141-59.
83. A summary of the paper is given in *JSPR* 30 (March 1937): 34-36.
84. Minutes of Council, 20 January 1937.
85. *Ibid.*, 24 March 1937.
86. *Ibid.*, 28 April, 26 May 1937.
87. Robert H. Thouless, "Review of Mr Whately Carington's Work on Trance Personalities," *PSPR* 44 (1937): 223-75.
88. *Ibid.*, p. 272.
89. Whately Carington, "Note on Professor Thouless' Paper," *ibid.*, p. 277.
90. J. W. Dunne, *An Experiment with Time* (London: A. & C. Black, 1927); the second edition appeared in 1929.
91. C. D. Broad, "Mr. Dunne's Theory of Time in 'An Experiment with Time,'" *Philosophy* 10 (1935): 168-85; Soal review in *JSPR* 24 (October 1927): 119-23; review by Tyrrell of revised edition in *JSPR* 28 (July 1934): 270-74.
92. Theodore Besterman, "Reports of an Inquiry into Precognitive Dreams," *PSPR* 41 (1933): 186-204.
93. At Duke, J. B. Rhine was quite interested when Dunne's book appeared (Rhine to Thomas, 13 December 1927); eventually he was moved to give John Thomas a brief summary of the work, adding that William McDougall too was "much interested" but warning that Dunne's "engineering is surely unsound as to his Serialism of Time and the Observer" (Rhine to Thomas, 23 January 1928 [both JBR]).
94. H. F. Saltmarsh, "Report on Cases of Apparent Precognition," *PSPR* 42 (1934): 49-103.
95. Whately Carington, "Preliminary Experiments in Precognitive Guessing," *JSPR* 29 (June 1935): 86-104.
96. W. Whately Carington, "Note on Precognitive Guesses," *ibid.* (October 1935): 117-18.

97. Whately Carington, "Precognitive Guessing, II: Revised and Extended Analysis," *ibid.* (January 1936): 158–67.
98. Tyrrell, "Further Research," pp. 149–50.
99. See above, pp. 174–75; see also Rhine to Saltmarsh, 23 April 1937 (JBR).
100. Broad's autobiography, from which these details are taken, appears in *The Philosophy of C. D. Broad,* ed. Paul Arthur Schilpp (New York: Tudor Publishing Co., 1959), pp. 3–68.
101. *Ibid.,* p. 58.
102. C. D. Broad, "The Present Relations of Science and Religion," in *Religion, Philosophy and Psychical Research* (New York: Harcourt, Brace, 1953), p. 243; the article originally appeared in *Philosophy* 14 (1939): 131–54.
103. C. D. Broad, *The Mind and Its Place in Nature* (London: Routledge & Kegan Paul, 1925), chap. 12.
104. Quoted (from *Mind*) in John Passmore, *A Hundred Years of Philosophy* (London: Duckworth, 1957), p. 350.
105. C. D. Broad, "Normal Cognition, Clairvoyance, and Telepathy," *PSPR* 43 (1935): 397–438.
106. Price had supervised the participation of a group of Oxford undergraduates in Besterman's 1932 experiment testing for precognitive dreams and had continued to take an interest in the problem.
107. Broad, "Normal Cognition," p. 416.
108. *Ibid.,* p. 437.
109. Aldous Huxley to Rhine, 2 February 1937 (JBR); cf. J. Cecil Maby, "Some Observations on Extra-Sensory Perception," *PSPR* 44 (1936): 169–82.
110. The stages of his investigation are summarized in S. G. Soal, "Fresh Light on Card-Guessing—Some New Effects," *PSPR* 46 (1940): 152.
111. Soal to Rhine, 25 April 1936 (JBR).
112. Soal to Rhine, 31 August 1936 (JBR).
113. Soal, "Fresh Light," pp. 155–56.
114. Soal to Rhine, 31 August 1936 (JBR).
115. Soal to Rhine, 25 April 1936 (JBR); S. G. Soal, "A Repetition of Dr J. B. Rhine's Work in Extra-Sensory Perception," *JSPR* 30 (April 1937): 55–58.
116. K. M. Goldney and S. G. Soal, "Report on a Series of Experiments with Mrs Eileen Garrett. Part II. A Repetition of Dr J. B. Rhine's Work with Mrs Eileen Garrett," *PSPR* 45 (1938): 69–87.
117. Mrs. Garrett did not interpret these disappointing results as had Soal (see her *My Life as a Search for the Meaning of Mediumship* [London: Rider & Co., 1939], pp. 172–87).
118. Rhine to Carington, 26 February 1935 (JBR), where his initial attitude is made very clear. "Our point of view could best be expressed with the remark, 'Here it is and here are the conditions. This is just a beginning. We are not asking you to believe this; we are asking you to be stimulated by it into repeating it or doing something else about it by way of active research.' "
119. C. D. Broad, "Discussion: G.N.M. Tyrrell, Science and Psychical Phenomena," *Philosophy* 13 (1938): 467–68.
120. Allison to Rhine, 6 July 1937 (JBR).
121. E. J. Dingwall, " 'Extra-Sensory Perception' in the United States," *JSPR* 30 (December 1937): 141.
122. Minutes of Council, 26 January, 23 February, 23 March, 27 April 1938.
123. In a review of *JP* articles in *JSPR* 30 (February 1938): 171. His remark elicited a sharp exchange with Rhine: Rhine to editor, *JSPR,* 4 March 1938; Herbert to Rhine, 15 March; Rhine to Herbert, 29 March (JBR).
124. S. G. Soal, " 'Snags' in Extra-Sensory Perception" (SPR). Joseph Rinn to Rhine, 1 June 1938 (JBR), discusses the Ghost Club reaction.
125. *PSPR* 45 (1938): 88–96.
126. *Ibid.,* p. 95.
127. Soal to Rhine, 12 June 1938 (JBR).
128. Soal to Rhine, 8 July 1938 (JBR).
129. Saltmarsh to Rhine, 21 June 1938 (JBR).
130. Minutes of Council, 25 May 1938.
131. Soal to Price, 21 June 1938 (HPL).

132. Minutes of Council, 29 June 1938; for Soal's private defense, blustering in tone, see Soal to Price, 4 July 1938 (HPL).
133. See Saltmarsh's characterization of Soal in his letter to Rhine dated 21 June 1938 (JBR).
134. Soal to Salter, 22 September 1938 (SPR).
135. Soal to Salter, 7 October 1938 (SPR).
136. Salter to Soal, 2 November 1938 (SPR).
137. Salter to Lord Rayleigh, 11 October 1938 (SPR).
138. A summary of Carington's talk was printed as "Some Early Experiments Providing Apparently Positive Evidence for Extra-Sensory Perception," *JSPR* 30 (December 1938): 295-305.
139. Whately Carington, "Experiments on the Paranormal Cognition of Drawings," *PSPR* 46 (1940): 34-151.
140. Interview, R. H. Thouless, 26 June 1974.
141. Soal, "Fresh Light," pp. 152-98.
142. W. Whately Carington and S. G. Soal, "Experiments in Non-Sensory Cognition," *Nature* 145 (9 March 1940): 389-90.
143. C. D. Broad, "Introduction to Mr Whately Carington's and Mr Soal's Papers," *PSPR* 46 (1940): 25-33.
144. An announcement of the studentship was published in *ibid.*: 23-24.

Chapter Nine

1. Interview, J. B. Rhine, 18 December 1975.
2. In this work, published in 1930 by the Boston Society for Psychic Research, Prince analyzed what he saw as the consistently unscientific approach adopted by scientists venturing out of their own fields to consider psychical research. The second part of the volume consists of selections from Prince's own debates in correspondence with a number of these.
3. Prince to Rhine, 30 January 1934 (JBR).
4. Prince to Rhine, 14 May 1934 (JBR).
5. Rhine to Prince, 18 May 1934 (JBR).
6. *Character & Pers.* 3 (1934-35): 85-89; *J. Abnorm. Soc. Psychol.* 29 (1934-35): 350-52. On Murphy's review, see above, chap. 5.
7. The information comes from correspondence to Rhine from Ralph McGrath, a graduate student in psychology at Chicago who had done some work with George Estabrooks at Colgate as an undergraduate (McGrath to Rhine, 8 June, 12 July, 16 August, 24 August 1934 [JBR]).
8. McGrath to Rhine, 29 June 1935 (JBR).
9. Jastrow to Rhine, 4 May 1934 (JBR).
10. Rhine to Jastrow, 10 May 1934 (JBR).
11. Murchison to President Wallace W. Atwood, 19 January 1934 (CUA).
12. Hunter to Rhine, 14 June 1934 (JBR). On Hunter, see the autobiographical statement in *A History of Psychology in Autobiography*, vol. 4, ed. E. G. Boring *et. al.* (Worcester, Mass.: Clark University Press, 1952), pp. 163-87.
13. Rhine to Hunter, 3 July 1934 (JBR). Two days later Rhine wrote ruefully to his Harvard friend Hudson Hoagland, who was teaching physiology at Clark, "I would have written this book differently if I had known it was to be sent out free to scientific men to the extent of 400 extra copies." Hoagland had written to Rhine on 30 June to praise the book and to urge him henceforth to publish his results in a psychological journal, since work published through psychical-research societies was bound to be "overlooked and discounted" by psychologists.
14. Hunter to Rhine, 17 July 1934 (JBR).
15. Murchison to Atwood, 24 March 1928 (CUA).
16. Willoughby to Rhine, 17 August 1934 (JBR).
17. R. R. Willoughby, "A Critique of Rhine's 'Extra-Sensory Perception,'" *J. Abnorm. Soc. Psychol.* 30 (1935): 207.
18. Raymond Royce Willoughby, "Prerequisites for a Clairvoyance Hypothesis," *J. Appl. Psychol.* 19 (1935): 543-50.
19. "A Critique," p. 207.
20. Willoughby to Rhine, 17 August 1934 (JBR).

21. Willoughby to Rhine, 24 November 1934; Rhine to Willoughby, 4 December 1934 (JBR).
22. Willoughby to Rhine, 7 December 1934 (JBR).
23. Rhine to McDougall, 20 April 1935 (JBR); interview, J. B. Rhine, 24 January 1975.
24. Willoughby to Rhine, 22 June 1935 (JBR).
25. Raymond Royce Willoughby, "Further Card-Guessing Experiments," *J. Gen. Psychol.* 17 (1936): 3–13.
26. Willoughby to Rhine, 8 September 1936 (JBR).
27. Munzinger to Rhine, 14 April 1935 (JBR).
28. Schoolland to Rhine, 19 November 1935 (JBR).
29. Munzinger to Rhine, 19 December 1935 (JBR).
30. Karwoski to Rhine, 24 January 1935; Rhine to Karwoski, 11 March, 5 April 1935 (JBR).
31. Estabrooks to Rhine, 12 January 1935; Adams to Rhine, 25 January, 28 February 1935 (JBR).
32. Leuba to Rhine, 8 February 1935; Rhine to Leuba, 11 March 1935; Frances E. Lemcke to Rhine, 1 November 1935 (JBR).
33. Kahoe to Rhine, 13 February 1936 (JBR).
34. Brown to Rhine, 2 March 1936 (JBR).
35. Brown to Rhine, 7 October 1936, includes a report of the data (JBR).
36. Reichenberg to Rhine, 23 October 1936 (JBR).
37. Rhine to Brown, 5 November, 6 December 1936; Brown to Rhine, 9 December 1936. Gardner Murphy discovered independently how cold Brown's feelings were. "'I saw J. F. Brown at the Lewin meetings at Harvard. He is quite antagonistic to you, feels very doubtful what his own results mean, is very hostile to the idea of their being quoted as support of ESP, etc. . . . I think he is continuing experiments, & may in time swing with us if no pressure is put on him" (Murphy to Rhine, 5 January [1937] [all in JBR]).
38. Kenneth H. Baker, "Report of a Minor Investigation of Extra-Sensory Perception," *J. Exper. Psychol.* 21 (1937): 120–25.
39. Baker to Rhine, 26 February 1937 (JBR).
40. Adams to Yerkes, 11 December 1930 (Adams Papers, AHAP).
41. "The Natural History of a Prejudice" (Adams Papers, AHAP).
42. *Ibid.*, p. 21. To Rhine, Adams wrote: "It was quite a thing—B's article. Impresses me more than any amount of statistical accumulation because of a) wealth of detail as to conditions b) wealth of detail as to introspections and drawings c) analogies with those processes about which something is known. These explain nothing but do give a comfortable feeling" (Adams to Rhine, 19 August 1936 [JBR]. Adams reviewed Bender's work in "Bender on Extra-Sensory and Sensory Form Perception," *JP* 1 [1937]: 52–62).
43. Coover's views of Rhine's work are summed up in a letter from Lucien Warner, written after a visit to Stanford. "He is rather condescending regarding your work. I tried to pin him down to specific criticisms but without much luck. Apparently you made a few minor statistical mistakes. He objects to your delegation of responsibility to others. . . . Your work was too informal; you defend this by saying it is necessary especially at the start, but he wants to know just how long it was before strict control superceded informality. It seems that you haven't written fully and exactly enough to permit repetition of your work. He wound up by saying that your book showed evidence of having been written hastily; that it was all right for the public but not for the scientist. I didn't bother to defend your work because it stands on its own merits" (Warner to Rhine, [June 1935?] [JBR]. Coover's correspondence with Wilbur is in SUA; on his concern for unfavorable publicity, see, e.g., Coover to Wilbur, 3 June 1937 [Coover Correspondence (4512) SUA]).
44. Coover to Wilbur, 6 May 1935 (Coover Correspondence [4512], SUA).
45. Rhine to Wilbur, 24 May 1935; Coover to Wilbur, 11 June 1935 (Coover Correspondence [4512], SUA).
46. See above, chap. 6, n. 36.
47. Coover to Wilbur, 19 January 1937 (Coover Correspondence [4512], SUA).
48. Terman to McDougall, 25 February 1937 (JBR).
49. Jastrow to Rhine, 31 April 1937; Rhine to Jastrow, 5 May 1937 (JBR). We have italicized *younger* in Rhine's letter to make his emphasis clear.
50. William S. Cox, "An Experiment on Extra-Sensory Perception," *J. Exper. Psychol.* 19 (1936): 429–37.
51. Rhine to Baker, 29 March 1937 (JBR).

52. Vernon W. Lemmon, "Extra-Sensory Perception," *J. of Psychol.* 4 (1937): 227.
53. Chester E. Kellogg, "New Evidence (?) for 'Extra-Sensory Perception,'" *Scientific Monthly* 45 (1937): 331-41.
54. Chester E. Kellogg, "Dr. J. B. Rhine and Extra-Sensory Perception," *J. Abnorm. Soc. Psychol.* 31 (1936-37): 190-93.
55. Kellogg to Rhine, 26 October 1937 (JBR).
56. Abbot to Rhine, 16 October 1937 (JBR). Cattell had been a disbeliever in psychical phenomena for over forty years, as his article, "Esoteric Psychology," *Independent* 30 (9 March 1893): 316-17, makes plain; he was also a bitter opponent of McDougall and had delivered a personal attack upon him at the Yale International Congress (1929). It is curious and perhaps significant that Jastrow had approached Cattell just that summer about an exposé of the Duke work. "When I get back," he wrote, "I want to talk with you about the awful mess McDougall, Rhine etc are making of things they call psychology. I am sure that you agree that it deserves some plainspoken exposure" (Jastrow to Cattell, 19 August 1937 [Cattell Papers, Library of Congress, Washington, D.C.]).
57. *New York Times,* 28 November 1937, sec. 4; p. 8; Kellogg responded sharply the next day (published 7 December).
58. Cattell to Rhine, 5 November and 20 December 1937; 25 January, 18 February 1938 (JBR), grow successively more negative. The final rejection is printed in *JP* 2 (1938): 71.
59. Robert Berg to Rhine, 13 November 1937 (JBR).
60. Kellogg, "New Evidence," p. 332.
61. E. V. Huntington, "Is It Chance or ESP?" *American Scholar* 7 (1938): 201-10. On Huntington's career, see *Dictionary of Scientific Biography,* S. V. "Huntington, Edward Vermilye."
62. Fry to Rhine, 20 December 1937 (JBR).
63. Reproduced in *JP* 1 (1937): 305.
64. *Washington Evening Star,* 2 January 1938, p. A2. On the genesis of this quotation, see Huntington to Rhine, 12 January 1938 (JBR).
65. We have discussed these professional reactions in somewhat greater detail in our essay "The Controversy over Statistics in Parapsychology 1934-1938," in *The Reception of Unconventional Science by the Scientific Community,* ed. Seymour H. Mauskopf, AAAS Selected Symposium 25 (Boulder: Westview, 1979), pp. 105-23.
66. B. F. Skinner, "Is Sense Necessary?" *Saturday Review of Literature* 16 (9 October 1937): 5-6.
67. Skinner to Rhine, 15 November 1937; Rhine to Skinner, 18 November 1937; Skinner to Rhine, 23 November 1937 (JBR).
68. Skinner's report to the council, dated 6 October 1937, is in the APA archives (Library of Congress); see also the report of the Committee on Press Relations (which Skinner chaired) of 4 September 1939, which reveals more of his views on scientific publicity.
69. Blakeslee to Edmund Conklin (University of Indiana), 10 January 1938 (APA archives [Library of Congress]).
70. Skinner to Dael Wolfle, [20 November 1937] (Wolfle Papers).
71. Willoughby to Wolfle, 16 November 1937 (Wolfle Papers).
72. Skinner to Wolfle, 16 November 1937 (Wolfle Papers); Dael L. Wolfle, "A Review of the Work on Extra-Sensory Perception," *Am. J. Psychiatry* 94 (1938): 943-55.
73. Wolfle, "A Review," p. 948.
74. Interview, Dael Wolfle, 10 September 1975.
75. Harold O. Gulliksen, "Extra-Sensory Perception: What Is It?" *Am. J. Sociol.* 43 (1938): 623-31.
76. News release of 12 October 1937, in Science Service to Rhine, 13 October (JBR).
77. This is perhaps an allusion to the undergraduate thesis of Raymond Brooks, "Some Experiments in Clairvoyance and Telepathy," presented at Reed College in June 1937, which may have been directed by Griffith, although Brooks made no acknowledgment of a faculty advisor.
78. "Extra-Sensory Perception Discussed by Mathematicians at A.A.A.S. Meetings," News release, 31 December 1937 (Smithsonian Institution Archives).
79. Huntington to Davis, 10 January 1938 (copy in JBR).
80. Abbot to Rhine, 15 January 1938 (JBR).
81. Rhine to Davis, 28 January 1938; Davis to Rhine (telegram), 31 January (JBR).
82. Howell to Huntington, 1 February 1938 (copy in JBR).
83. Abbot to Rhine, 7 February 1938 (JBR).

84. Huntington to Rhine, 11 March 1938 (JBR).
85. James C. Crumbaugh, "A Questionnaire Designed to Determine the Attitudes of Psychologists toward the Field of Extra-Sensory Perception," *JP* 2 (1938): 302-7, provides the questions and summarizes the responses. In a personal communication of 16 September 1975, Dr. Crumbaugh reported that the original responses to the questionnaire had probably been destroyed.
86. Edwin G. Boring, *Psychologist At Large* (New York: Basic Books, 1961), pp. 39-60, describes his career at Harvard from 1922-1942; Bruce Kuklick, *The Rise of American Philosophy, Cambridge, Massachusetts, 1860-1930* (New Haven and London: Yale University Press, 1977), pp. 459-61, discusses Boring's role in the creation of a separate psychology department.
87. See Edwin G. Boring, "The Paradox of Psychic Research," *Atlantic Monthly* 137 (1926): 81-87.
88. Boring to Rhine, 21 June 1935 (EGB).
89. See Boring to Murphy, 3 May 1937 (EGB).
90. Boring to Rhine, 15 October 1936 (EGB).
91. Boring to Rhine, 30 October 1936 (EGB).
92. Boring to Rhine, 30 November 1936 (EGB).
93. Boring to Rhine, 27 November 1936 (EGB).
94. Boring to Rhine, 18 December 1936 (EGB).
95. See, for example, his 1920 paper "The Logic of the Normal Law of Error in Mental Measurement", reprinted with a brief introductory statement in his *Psychologist At Large,* pp. 143-79.
96. Boring to Rhine, 8 February 1937 (EGB).
97. Boring to Rhine, 30 April 1937 (EGB).
98. Rhine to Boring, 14 January 1937; Boring to Rhine, 8 February 1937 (EGB).
99. "Our Department had Rhine up to visit us last fall, and we had a very interesting time with him. We thought in advance that we might try to reduplicate the Rhine effect here and see if we could not get at its conditions.... When we found how many people in places other than Duke were beginning to check the results, the Department shied off" (Boring to James, 27 May 1937 [EGB]). It will be remembered that Boring had had occasion to read Lucien Warner's paper—sent to him from New York—a few months before (see above, pp. 185-86).
100. Interview, James Grier Miller, 6 May 1976.
101. Boring to James, 27 May 1937; James to Boring, 14 June 1937; Boring to James, 16 June 1937 (EGB).
102. Murphy to Boring, 28 April 1937 (EGB).
103. Boring to James, 3 June 1938 (EGB).
104. Boring to James, 8 February 1939 (EGB).
105. Boring to Murphy, 3 November 1938 (EGB).
106. Donald R. Griffin, personal communication, 28 May 1976.
107. Boring to James, 3 June 1938 (EGB).
108. Donald Redfield Griffin, "The Homing Ability of Birds: A Problem in Sensory Physiology" (Ph.D. diss., Harvard University, 1942). The summer's work is described in *Auk* 57 (1940): 61-74.
109. J. G. Pratt, "The Homing Problem in Pigeons," *JP* 17 (1953): 34-60, offers a parapsychological perspective on the problem.
110. Interview, James Grier Miller, 6 May 1976.
111. James to Boring, 13 June 1938; Boring to James, 20 June 1938 (EGB). Boring wrote James: "I have myself been making this winter judgments which seemed to me very much like ESP; but the experimenter is able to prove to me that they are a fairly exact function of a subliminal illumination! It is the most astonishing thing to have happen to one!" The reference is to Miller's research.
112. James Grier Miller, "Discrimination Without Awareness," *Amer. J. Psychol.* 52 (1939): 562-78. For an extension of this research see *idem,* "The Role of Motivation in Learning Without Awareness," *ibid.,* 53 (1940): 229-39.

Chapter Ten

1. Much of what follows is based upon a tape-recorded account kindly prepared for us by Dr. Kennedy in April 1976.

2. John L. Kennedy, "The Visual Cues from the Backs of the ESP Cards," *J. of Psychol.* 6 (1938): 149-53.
3. John L. Kennedy, "Suggestions Concerning the Nature and Production of 'Extra-Sensory' Perception," mimeographed, mailed out 25 February 1938 (John L. Kennedy Faculty File, SUA).
4. Charles E. Stuart, "A Review of Recent Criticisms of ESP Research," *JP* 2 (1938): 310.
5. Knight Dunlap, "Extra-Sensory Perception" (Paper delivered at meetings of the Western Psychological Association, Eugene, Oregon, 17-18 June 1938), pp. 17-18 (Dunlap Papers, AHAP).
6. Rogosin's paper was published as H. Rogosin, "An Evaluation of Extra-Sensory Perception," *J. Gen. Psychol.* 21 (1939): 200-217; a highly colored account of the meeting appeared in *Time*, 11 April 1938, p. 54. The critical view of Britt was expressed by John Gray Peatman (on Peatman, see below, n. 14).
7. At Rhine's suggestion, Jastrow sent a copy of this questionnaire to Duke in a letter of 17 May 1938 (JBR).
8. Jastrow to C. C. Clark, 18 July 1938 (JBR); Joseph Jastrow, "ESP, House of Cards," *American Scholar* 8 (1939): 13-22.
9. J. B. Rhine, *New Frontiers of the Mind* (New York and Toronto: Farrar and Rinehart, 1937), p. 176.
10. Thus Kaempffert alluded to psychology as "a sorry pseudo-science . . . in no position to give itself airs" (*New York Times*, 30 January 1938).
11. Adams to Rhine, 28 February and 10 March 1938 (JBR).
12. Rhine to Donald Paterson (secretary of the APA), 12 February 1938 (JBR).
13. The results of this questionnaire are published in Lucien Warner and C. C. Clark, "A Survey of Psychological Opinion on ESP," *JP* 2 (1938): 296-301. The discussion below has been based on the original questionnaires that were returned (JBR).
14. Burks was a Stanford Ph.D. (1929) working for the Carnegie Institute at Cold Stream Harbor; Peatman was a Columbia Ph. D. (1931), presently an assistant professor at CCNY. Bliss had taken his Ph.D. from Yale in 1893 but had gone into the Congregational ministry, from which he had retired in 1927.
15. Cf. the remarks of J. J. Gibson of Smith: "I would rather say that E.S.P. at present stands for a set of experimental facts which are unexplained. The hypothesis proposed by Rhine to explain them (the hypothesis they are supposed to verify) is irrational, and no other hypothesis proposed by any one else is at present capable of explaining them."
16. The remark is from a reply to the Clark-Warner questionnaire, unsigned but sent out from Ann Arbor, which displays advance knowledge of the forthcoming APA meetings that only Olson is likely to have had (see below).
17. According to Kennedy in his invitation to Rhine, 21 June 1938; in a letter to Rhine of 8 June, Kennedy did not mention the symposium, so that Olson's appeal postdates the eighth.
18. The correspondence between Kennedy and Rhine (JBR) works out the details; initially Kellogg was to have taken part, but he decided against it.
19. The paper was eventually published as "An Experiment to Test the Role of Chance in ESP Research," *JP* 2 (1938): 217-21.
20. "Notes on ESP at the Columbus Meeting of the American Psychological Association," mimeographed, p. 1 (JBR).
21. This at least is the memory of Dael Wolfle (interview, 10 September 1975).
22. Two other sessions (all were officially called "roundtables") met at the same time, one on "Experimentation in Hypnosis" and one on "Testing Personality of Children" ("Proceedings of the Forty-Sixth Annual Meeting... ," *Psychol. Bull.* 35 [1938]: 707). Some of the papers are published in whole or in part in "The ESP Symposium at the A.P.A.," *JP* 2 [1938]: 247-72.
23. "The ESP Symposium," pp. 266-72, provides a transcript of the discussion, printed from a manuscript offered to all the participants for corrections.
24. Sanborn to Charles E. Stuart, 13 October 1938 (JBR).
25. "The ESP Symposium," p. 269.
26. Murphy to Rhine, 15 September 1938 (JBR).
27. "Memorandum for Information of President Few on ESP Symposium at Columbus," 23 September 1938 (JBR).
28. Murphy to Rhine, 20 October 1938; Rhine to Murphy, 24 October; Rhine to Murphy, 3 November (JBR).

29. Rhine to Murphy, 8 November; Murphy to Rhine, 10 November; Rhine to Murphy, 15 November (JBR).
30. Murphy to Rhine, 2 December; Rhine to Murphy, 8 December (JBR).
31. "The ESP Symposium," p. 271. Pratt had written earlier to Rhine of Murphy's endorsement of Sells. Murphy was going to propose Sells to succeed the retiring Coover as psychical-research fellow at Stanford, on the ground that he "seems to have a genuine interest in the field, appreciates the importance of ESP (if true), and has said that he wants the place" (Pratt to Rhine, 12 February 1937 [JBR]).
32. Murphy to Rhine, 15 September 1938 (JBR).
33. "Editorial," *JP* 3 (1939): 1-2.
34. John E. Coover, "Reply to Critics of the Stanford Experiments on Thought-Transference," *ibid.*, pp. 17-28.
35. Rhine to Dunlap, 22 August 1938; Dunlap to Rhine, 1 September (JBR).
36. Rhine had circulated a letter to psychologists whom he knew to have strong opinions on his research the previous month, asking them what they would like to see done "in the interests of . . . a crucial testing of the hypothesis of extra-sensory perception." He received nearly two dozen replies (with almost as many different recommendations), but Dunlap's was by far the most systematic response (JBR).
37. Dunlap Papers, AHAP.
38. K. Dunlap *et al.*, "Adequate Experimental Testing of the Hypothesis of 'Extra-Sensory Perception' Based on Card-Sorting," *JP* 3 (1939): 29-37.
39. Dallenbach to Dunlap, 24 August 1938 (Dunlap Papers, AHAP).
40. Dallenbach to Dunlap, 21 November 1938, *ibid.*
41. "It is somewhat unusual, I should judge, in the history of science for this degree and extent of attention to be paid to so foreign a hypothetical problem so soon after its publication. While this is partly attributable, no doubt, to the fact that the hypothesis is essentially an old one, it is, I think, evidence of a markedly more tolerant scientific spirit" (Rhine to Dunlap, 25 January 1939 [JBR]).
42. Gardner Murphy and Ernest Taves, "Covariance Methods in the Comparison of Extra-Sensory Tasks," *JP* 3 (1939): 38-78.
43. Murphy to Rhine, 14 April 1939 (JBR).
44. Murphy and Taves, "Covariance Methods," p. 69.
45. J. G. Pratt and J. L. Woodruff, "Size of Stimulus Symbols in Extra-Sensory Perception," *JP* 3 (1939): 121-58. The experiments also formed the basis for Woodruff's 1939 A.M. thesis at Duke.
46. Pratt and Woodruff, "Size of Stimulus Symbols," p. 156.
47. C.E.M. Hansel, *ESP: A Scientific Evaluation* (New York: Charles Scribner's Sons, 1966), pp. 86-103. See also *idem*, "A Critical Analysis of the Pratt-Woodruff Experiment," *JP* 25 (1961): 99-113, with following reply by Pratt and Woodruff; and R. G. Medhurst and Christopher Scott, "A Reexamination of C.E.M. Hansel's Criticism of the Pratt-Woodruff Experiment," *JP* 38 (1974): 163-84, immediately followed by an exchange of comments between Pratt and Scott.
48. "Letters and Notes," *JP* 3 (1939): 246, 248.
49. Murphy to Rhine, 8 September 1939 (JBR).
50. J. G. Pratt *et al.* (J. B. Rhine, Burke M. Smith, Charles E. Stuart, and Joseph A. Greenwood), *Extra-Sensory Perception after Sixty Years* (New York: Henry Holt, 1940). We will hereinafter refer to this work as *ESP-60*. A reprint of the original was published by Bruce Humphries (Boston) in 1966. On the publication arrangements, see Sloane to Rhine, 8 November 1939; Rhine to Sloane, 14 November; Sloane to Rhine, 16 November, 21 December (JBR).
51. Murphy to Rhine, 8 and 22 September 1939 (JBR).
52. Rhine to Murphy, 26 September 1939 (JBR).
53. Form letter, Rhine to [A. Sophie Rogers], 3 October 1940 (JBR). Cf. Rhine to Sloane, 2 January 1940, for the details of the arrangement with Holt.
54. Murphy to Rhine, 15 June [1940] (JBR).
55. Paul Terry at Alabama, for example (11 October 1940 [JBR]).
56. Among others, John Seward of Connecticut College (2 July 1940); P. E. Vernon of Glasgow (9 August); and Jack Dunlap of Rochester (18 October [all JBR]).
57. Conklin to Rhine, 3 October 1940 (JBR).
58. *Character & Pers.* 9 (1940): 86-89; *J. Appl. Psychol.* 24 (1940): 373-74; *Psychoanalytic*

Review 28 (1941): 292-93. The unfavorable review in *Psychol. Bull.* 37 (1940): 823-25 was written by Douglas Ellson, John Kennedy's successor as Fellow in Psychical Research at Stanford.
   59. *Amer. J. Psychol.* 54 (1941): 449-53.
   60. Jastrow (who died in 1944) had gone so far as to imply that apparently unexceptionable experiments like the Pearce-Pratt series had to be explained in terms of fraud by student participants. "Students are not angels; other experimenters report cases of collusion among their subjects . . . it is precisely in this strongest evidential series that the absence of witnesses makes collusion possible" (Jastrow, "ESP, House of Cards," pp. 17-18; cf. Jastrow to Rhine, 9 July 1938 [JBR]).
   61. Allport to Rhine, 30 August 1940 (JBR). It is worth noting that Allport discussed the effect of William McDougall's teaching upon his own psychology in his autobiographical statement in *A History of Psychology in Autobiography*, vol. 5, ed. Edwin G. Boring and Gardner Lindzey (New York: Appleton-Century-Crofts, 1967), pp. 12-13, 15.
   62. Boring to Rhine, 19 July 1940 (EGB). A copy of the syllabus for Psychology A indicates that students were to read chapters 1-6 and 10-18 of *ESP-60*—that is, all but the polemical exchanges with psychologists.
   63. The work was also used by Goodwin Watson of Teachers College, Columbia, that fall in his course on advanced educational psychology (Watson to Rhine, 6 June 1940 [JBR]).
   64. *JASPR* 34 (1940): 266-68.
   65. "This is unquestionably the most important book yet published in the field of Psychical Research and Parapsychology" (*PSPR* 46 [1940]: 265-70).
   66. W. H. Salter (17 July 1940), Edith Lyttleton (30 July), C. Drayton Thomas (31 August), and Saltmarsh (4 October), among others.
   67. Soal to Rhine, 30 July 1940 (JBR).
   68. Soal to Saltmarsh, 23 July [1940] (SPR). Soal's remark affords one more illustration of the difficulty one feels in trying to understand his thought, for it is in utter contradiction to his detailed critique of Warcollier's experiments ten years before (see above, chap. 2).
   69. Lodge to Rhine, 20 June 1940 (JBR).

Epilogue

   1. A general survey of this work and its context may be found in S. G. Soal and F. Bateman, *Modern Experiments in Telepathy* (London: Faber and Faber, 1954). Soal's honesty in reporting the results of his work with Basil Shackleton has recently been brought very much into doubt (see Betty Markwick, "The Soal-Goldney Experiments with Basil Shackleton: New Evidence of Data Manipulation," *PSPR* 56 [1978]: 250-81, for the most recent indictments and discussions of the evidence of dishonesty). Whether or not Soal falsified his data, however, would not of course affect the way in which his published reports influenced their readers in the 1940s.
   2. Robert H. Thouless, "The Present Position of Experimental Research into Telepathy and Related Phenomena," *PSPR* 47 (1942): 19.
   3. J. B. Rhine and Betty M. Humphrey, "The PK Effect: Special Evidence from Hit Patterns. 1. Quarter Distributions of the Page," *JP* 8 (1944): 18-60; *idem,* "2. Quarter Distributions of the Set," *ibid.* (1944): 254-71; and, with J. G. Pratt, "III. Quarter Distributions of the Half-Set," *JP* 9 (1945): 150-68.
   4. Rhine's reports are published in *JP* 7 (1943) and thereafter. On the English response, see Donald J. West, "A Critical Survey of the American PK Research," *PSPR* 47 (1945): 281-90; and Robert H. Thouless, "Some Experiments on PK Effects in Coin Spinning," *ibid.*, pp. 277-81.
   5. For a general survey of this work, see Gertrude Raffel Schmeidler and R. A. McConnell, *ESP and Personality Pattern* (New Haven: Yale University Press, 1958).
   6. Cf. the contrasting moods of the editorials in *JP*: "Progress of Parapsychology as a University Study," 6 (1942): 237-42; and "Is Parapsychology a Profession?" 8 (1944): 174-75.
   7. A group of New-York-area physicians and psychiatrists had also been attracted to the field, in sufficient numbers to constitute a "medical section" of the ASPR.
   8. J. B. Rhine, "Psi and Psychology: Conflict and Solution," in *Progress in Parapsychology,* ed. J.B. Rhine (Durham: Parapsychology Press, 1971), pp. 239-50.
   9. Interview, D. J. West, 17 July 1978. He had assumed the post of research officer in May 1946 hoping to make a full-time career in psychical research, but he experienced much the same sort of

reaction to his experimentalist orientation and "scientific" skepticism that Dingwall and Besterman had met in the 1920s.

10. Robert H. Thoules[s], *Psychical Research Past and Present: Eleventh Frederic W. H. Myers Memorial Lecture* (London: Society for Psychical Research, 1952), p. 20.

11. For a recent, sophisticated treatment of the complexity (and relativity) of "replication" as applied to parapsychology, see H. M. Collins, "Upon the Replication of Scientific Findings: A Discussion Illuminated by the Experiences of Researchers into Parapsychology," Proceedings of the 4S/ISA First International Conference on Social Studies of Science, Ithaca, New York, November 1976.

12. DeWolf to Rhine, 23 November 1938 (JBR). Another example from this period is Ernest Taves, Murphy's student and research associate in parapsychology. Although Taves never felt any professional commitment to parapsychology (personal communication, 10 November 1979), he did do an M.A. thesis on a parapsychological topic under Murphy in 1938, and in 1939 he assumed the position of managing editor of the *Journal of Parapsychology* when Murphy took over the editorship. He went on to do a Ph.D. in experimental psychology and subsequently moved to medicine and psychiatry. Given his initial involvement in parapsychology, it is conceivable that his career might have developed differently had his research yielded what he considered to be unequivocally positive results.

13. For a good analysis of the problems, experimental and social, besetting parapsychology in the mid-1940s, see J. B. Rhine, "The Source of the Difficulties in Parapsychology," *JP* 10 (1946): 162-68.

14. John L. Kennedy, "Evidence for the Recording Error Criticism of ESP Data," *Psychol. Bull.* 36 (1939): 501; see also *idem*, "A Methodological Revue of Extra-Sensory Perception," ibid., pp. 59-103.

15. Rhine to Mrs. Randall Whaley, 17 December 1940.

16. "ESP, PK, and the Survival Hypothesis," *JP* 7 (1943): 223-37.

17. G. Murphy, "The Freeing of the Intelligence" (presidential address to the APA, Cleveland, Ohio, 11 September 1944), *Psychol. Bull.* 42 (1945): 11-12.

18. Memorandum, "To the Duke University Trustees," 26 November 1947; file, "Parapsychology Laboratory" (Flowers Papers, DUA).

19. Lucien Warner, "A Second Survey of Psychological Opinion on ESP," *JP* 16 (1952): 284-95.

Index

Abbot, C. G., 263, 264
Ackerson, Luton, 281
Adams, Donald, 95, 140, 177, 203, 249; learning theory, 134; and Rhine, 135, 251-53, 277-78
Adams, Eugene, 249
Alcohol, 115-16, 172, 252
Allison, Lydia, 153, 205, 206, 233, 243
Allport, Gordon, 265, 283, 292, 295
American Journal of Psychiatry, 261
American Journal of Psychology, 68, 69, 146, 295
American Journal of Sociology, 262
American Magazine, 157
American Mercury, 165
American Psychical Institute, 322 n. 17, 347 n. 29, 349 n. 27
American Psychological Association, 142, 167, 188, 240, 260, 276, 278, 281-83, 292, 303-4, 305
American Society for Psychical Research, 76, 103, 142, 156, 157, 296, 300; Advisory Scientific Council, 19, 243; experiments, 30, 209-10; founding of, 13, 46; and IMI, 18; interests of, 206-10, 304; *Journal,* 20, 21, 61, 74, 207, 208, 296, 303; leadership, 16-21, 59-60, 65, 73, 74, 206-7, 298; revival of, 16-17; and SPR, 17
American Weekly, 157
Ames, Adelbert, 248-49
Angell, Frank, 45, 46
Angell, James Rowland, 48-49
Animals, telepathy in, 81-82, 83, 84-85, 252
Animism, 58, 59, 78
Annales des sciences psychiques, 1
Antioch College, 249
Aristotelian Society, 228
ASPR. *See* American Society for Psychical Research
Automatic writing, 16, 118, 176, 207; cross-correspondences, 3, 5, 8, 27, 28, 59, 86
Aveling, F.A.P, 212, 213, 214

Bailey, June, 100
Baker, Kenneth H., 162, 249, 250-51, 255
Balfour, A. J., 25, 115, 116
Balfour, Gerald, 14

Bard College, 143
Baril, Henri V., 322 n. 17
Barrett, William F., 25, 60, 115; and ASPR, 16
Behaviorism, 223, 240, 249, 259, 271, 300; opposition to, 132-33
Bender, Hans, 149, 215-17; clairvoyance experiments, 175-77, 177 (fig.); on Rhine, 129; *Zum Problem der aussersinnlichen Wahrnehmung,* 253
Benzedrine, 178
Besterman, Theodore, 43, 219, 225, 233; experiments, 39-40; and SPR, 27-28
Bevan, John, 305
Bird, J. Malcolm, 26; and ASPR, 22-23, 207; and Margery, 21-23, 207; and Rhine, 74, 76; and *Scientific American,* 152; telepathy experiments, 63
Birds, homing of, 271
Birkhoff, G. D., 269
Blakeslee, Howard, 260
Bliss, C. B., 278
BM (blind matching), 171, 187, 311
Boas, Franz, 64
Bolton, Frances, 137-38, 173, 218
Bond, Frederick Bligh, 207
Bonn, University of, 176, 215-16
Book-of-the-Month Club, 161, 256
Boring, Edwin G., 133, 185-86, 203; at Harvard, 67-69, 265-72, 299, 303, 327 n. 21; and Rhine, 78, 79, 146, 265-69, 295-96
Boston Society for Psychic Research, 65, 103, 138, 298; experiments, 189; founding of, 23; leadership, 205-6; mediumistic studies, 78; publications of, 69, 92, 101, 145, 189, 204, 205, 206, 241
Boston University, 302
Bradley, H. Dennis, 28
Brain waves, 254
British Broadcasting Corporation, telepathy experiments, 36-37, 38, 235
British College of Psychic Science, 21, 22, 211
British Columbia, University of, 249
Britt, Steuart H., 276, 281, 282, 283
Broad, C. D., 220, 224, 226-29, 232, 236, 238-39, 298
Bromide, 116
Brown, Andrew W., 281
Brown, Bancroft H., 199

359

Brown, Guy B., 212, 213, 214
Brown, J. F., 249-50
Brown, William (psychologist), 41
Brown University, 44
Bruce, H. Addington, 322 n. 17
Brugmans, H.I.F.W., 18 (fig.), 62, 85; experiments, 8, 27, 43, 115-16, 236
BSPR. *See* Boston Society for Psychic Research
BT (before touching), 98, 100, 109, 125, 187, 261, 311
Buck, G. E., 145
Bull, Titus, 74
Burks, Barbara, 278
Burt, Cyril, 58, 62, 100, 212, 214, 236, 325 n. 76, 335 n. 101
Button, William H., 206, 207, 209-10
Bux, Kuda, 213

Caffeine, 100, 116, 172
Cambridge University, 237-39
Camp, Burton, 200, 258, 263
Cannon, Alexander, *Powers That Be*, 27
Card-guessing, 77, 90-92, 184, 270; attention curves, 113; below-chance results, 179, 231; clairvoyant, 33, 34-36, 40, 90, 91-92, 98-101, 125-26, 179-81, 193-94, 209, 274; decline effect, 34, 35, 36, 40, 67, 93, 100, 113, 290; displacement effect, 237-38, 298; emotional responses, 36; games, 30, 32; introspection, 179; matching hypothesis, 198, 199-200; patterns in, 47; performance curves, 189; precognitive, 169-72, 298; "psychic shuffle," 172; sensory contact, possibility of, 109, 175, 180-81, 193-94, 274, 282; statistics of, 34, 40, 50-53, 67, 104, 106-8, 126, 150, 189, 195-201, 230, 249-50, 256-59, 261, 263. *See also* ESP cards; Experimental techniques
Carington, W. Whately (Smith), 37, 103, 108, 110, 123, 239 (fig.); *The Death of Materialism*, 120; experiments, 222-26, 236-39; on Jephson, 36; and Rhine, 102, 119-21, 124, 196, 197, 198, 296; and SPR, 219-20, 235, 236
Carmichael, Leonard, 273, 278, 283
Carpenter, C. R., 143, 146, 147, 175, 203, 247; experiments, 183-85, 192, 261
Carrel, Alexis, *Man, the Unknown*, 157
Carrington, Hereward, 18 (fig.), 22, 49, 222. *See also* American Psychical Institute
Cason, Hulsey, 323 n. 33
Catholic University, 349
Cattell, J. McKeen, 242, 257, 263, 322 n. 12, 328 n. 42
Character and Personality, 145-46, 242, 295
Chicago, University of, 44, 45, 243, 261-63
Chicago Psychological Club, 163

Chi-square method, 198-99, 246
City College of the University of New York, 300-301
Clairvoyance: definition of, 5, 111; drugs, effects of, 93, 94; experimental study, 39-40, 85, 274 (*see also* Card-guessing); factors affecting, 93; introspection, 176; long-distance tests, 101, 234; parallels with visual perception, 176; vs. telepathy, 5-6, 33, 35, 52, 86, 98-101, 110-11, 124; theories of, 228; as type-phenomenon for ESP, 112. *See also* Card-guessing; Cryptesthesia; Extra-sensory perception; PC (pure clairvoyance); Telepathy
Clark, C. C., 278
Clark, John A., 188, 193, 249
Clark University, 44, 244-47; Smith-Battles Fund, 49, 69, 244, 245, 246
Cobb, Percy, 286
Colgate University, 249
Colorado, University of, 144, 188-89, 248
Columbia University, 44, 60, 141-42, 149, 189, 190, 209
Compton, Arthur Holley, 155
Comstock, Daniel Frost, 22
Conant, James B., 265, 268, 270, 271
Conklin, Edmund, 294
Cooper, Blanche (medium), 38
Cooper, T. Coleman, 100
Coover, J. E., 70, 75, 108, 117, 245, 264, 285; and ASPR, 19; experiments, 125-26, 236; and Rhine, 82, 141, 143, 253-54; at Stanford, 45, 46, 49-56, 149, 251, 253-55, 273, 274
Cornell University, 44, 45
Cosmos Club, 158
Covert (student), 249
Cox, William S., 255
Crandon, LeRoi, 70
Crandon, Mina. *See* Margery
Cranford, W. I., 90, 132
Creery sisters, 121, 182, 333 n. 41
Critical ratio (CR): combining of, 197; definition of, 195, 311
Crook, Dorothea, 280
Crookes, William, 2, 6, 12, 115
Cross, Charles, 49
Cross-correspondences. *See* Automatic writing
Crumbaugh, James C., 264
Crunden, Alice B., 147
Cryptesthesia: definition of, 2; evidence of, criteria for, 11-13. *See also* Clairvoyance
Cutten, G. B., 249

D., Miss (ESP subject), 176
Dallenbach, Karl, 279, 286
Dana, Charles L., 19

Index

Dartmouth College, 248-49
Dashiell, John, 279
Data: evaluation of, 198-200; recording of, 193, 194, 261, 274, 282, 289, 357 n. 1; selection errors, 39, 40, 275, 276, 281. *See also* Card-guessing, statistics of
Davis, Watson, 81, 164-65, 262, 263-64
Dawson, Miles Menander, 16
Dearborn, George, 242-43
Deland, Margaret, 20, 21
Desmond, Shaw, 211
Dessoir, Max, 99
DeWolf, Harold, 302
Dice, experiments with, 170, 225, 299
Dick, Lillian, 285
Dietz, P. A., 215
Dingwall, E. J., 14, 36, 37, 43, 119, 122, 123, 215, 221, 298; and Margery, 23 (fig.); on Rhine, 120, 233; and SPR, 26, 27, 28
Dodds, E. R., 124
Doyle, Sir Arthur Conan, 15, 20, 21, 22, 25, 70, 77; and SPR, 26, 28; spiritualism, 73
Dreams, 225
Driesch, Hans, 70, 115, 215; *Parapsychologie*, 127-28; and Rhine, 128-29
DT (down through), 98, 100, 104, 125, 171, 188-89, 202, 245, 311
Duke, Ben, 131
Duke, James B., 131
Duke University, 203, 217, 300, 304-5; critics of, 165-66; founding of, 131-32; psychical research, first thesis on, 90-91; psychical research lectures, 94; psychology department, dissension within, 135-36, 251; psychology department, establishment of, 132-34; research budget, 138-39; statistics at, 197; Walter Franklin Prince Fellowships, 138, 139; William McDougall Research Fund, 138, 147, 160
Dumas, Georges, 10
Dunlap, Knight, 275-76, 282, 285-86, 288, 291, 328 n. 54
Dunne, J. W., *An Experiment with Time*, 225

Earlham College, 188, 249
Ectoplasm, 2, 5, 10, 11, 21, 23 (fig.)
Edgeworth, F. Y., 105
Edwards, Frederick, 65; and ASPR, 19, 21, 22-23; and Rhine, 73, 74
Ehrenwald, Jan, 339 n. 82
Einstein, Albert, 339 n. 82
Eisenhart, Churchill, 197
Elliot Smith, Grafton, 211
Ellson, D. G., 150, 357 n. 58
England, psychical research, 25-29, 32-43, 110, 111, 210-15, 218-39, 296-97, 298-99, 301. *See also* Society for Psychical Research

English, H. B., 340 n. 95
ESP. *See* Extra-sensory perception
ESP cards, 135; commercial manufacture, 160-61; introduction of, 90-92, 108-9; readable from the back, 165, 193, 232, 233, 260-61, 263, 274, 276. *See also* Card-guessing
Estabrooks, George H., 77, 89, 103, 249, 261, 265, 351 n. 7; experiments, 35, 85, 108, 109, 236, 252; at Harvard, 66-69, 68 (fig.); and Rhine, 141, 242
Eva C. (Marthe Béraud) (medium), 5, 10
Experimental techniques, 145, 311; BM (blind matching), 171, 187, 311; BT (before touching), 98, 100, 187, 311; BT-5, 109, 125, 261; BT-25, 109; DT (down through), 98, 100, 104, 125, 171, 188-89, 202, 245, 311; for GESP (general extra-sensory perception), 104, 127, 187-88, 230, 311; mechanical shuffler, 145, 170, 172; OM (open matching), 171, 179, 187, 274, 311; for PC (pure clairvoyance), 104; PDT (precognitive down through), 170, 172; for PK, 170; for PT (pure telepathy), 99, 191, 311; QT (quadruple task), 287-88; STM (screened touch matching), 179, 180-81, 230, 288-90, 289 (fig.), 311; UT (up through), 189
Extra-sensory perception: definition of, 280; distance, effects of, 100-101, 104, 114; drugs, effects of, 100, 104, 115-16, 178, 203; and evolution, 54, 116-17; introspection, 114; long-distance testing, 187-88; mechanized tests, 178, 220-21; motor element, 171, 187; naming of, 94; percipient as active entity, 111-12; and personality traits, 117-18, 300; pictorial material, 189; and psychological context, 7, 100, 112-13, 117-18, 178-81, 190, 192, 221, 290-91, 293, 299-300; radio tests, 160-63; and sensory perception, comparison with, 180, 185-86, 202-3, 228, 253, 266; spontaneous occurrences, 116; subject-experimenter effect, 178, 290, 293; and target size, 180, 290; theories of, 114-17, 128, 201-3; as widespread ability, 37, 41, 43, 117, 125, 302. *See also* Card-guessing; Clairvoyance; Precognition; Psychical research; Psychokinesis; Telepathy
Extra-Sensory Perception (book), 208; distribution of, to scientists, 241-43; influence of, 102, 104, 130, 148, 169, 183, 230, 302; reception of, by American press, 153-57; reception of, by American psychologists, 242-47; reception of, on Continent, 127-30, 177; reception of, in England, 119-27, 191, 231-35; structure and aims of, 103-4

Faculté des Sciences (Paris), 10
Farrar and Rinehart (publishers), 161

Fernberger, Samuel, 286
Few, William Preston: and Duke University, 131–34, 137; and Rhine, 139, 158, 162, 217, 283, 304
Fisher, Doris, 16
Fisher, R. A., 34, 39, 40, 107, 197, 225, 344 n. 74
Fisk, G. W., 221, 235
Fletcher, John, 279
Flugel, J. C., 58, 212, 325 n. 76
Fodor, Nandor, 208, 211
Fonda, Claudia, 81, 84
Forthuny, Pascal (Georges Cochet), 43
Forum, 157
France, psychical research, 1–13 passim, 15, 29–32, 43, 129, 210
Fraser-Harris, D. F., 211
Fraud: in ESP testing, 175, 181–83, 191, 338 n. 37, 357 n. 60; in mental phenomena, 42; in physical phenomena, 5, 11, 23, 76, 207
Free, E. E., 144, 152, 154, 155, 156, 157, 159
Freeman, Frank N., 292
Frick, Harvey, 135; experiments, 89, 93; thesis, 90–91, 94
Fry, Thornton, 258

Gale, Harlow, 44
Gannett, Lewis, 165
Gardiner, H. Norman, 44, 316 n. 74
Garrett, Eileen (medium), 137, 140, 146, 157, 173–74, 222, 231, 305–6
Garrett, Henry J., 295
Gatty, Oliver, 236
Gault, Robert H., 160
Gavett, Irving G., 197
Geley, Gustave, 5, 15, 18 (fig.), 25, 31
George, R. Wilfred, 186, 187
George Washington University, 276
Germany, psychical research, 5, 15, 127–29, 175–77, 210, 215–17
GESP (general extra-sensory perception), 104, 127, 187–88, 230, 311
Ghost Club, 234
Gibson, E. P., 144, 170
Gibson, J. J., 285, 355 n. 15
Goldney, K. M., 298
Goldstein, Martin, 183–84
Goodfellow, Louis D., 161, 163, 281, 282
Greenwood, Joseph, 197–200, 281
Gregory, Sir Richard, 212
Greville, T.N.E., 199, 281
Griffin, Donald R., 271
Griffith, William, 263
Guessing, psychology of, 64, 282. *See also* Card-guessing
Gulliksen, Harold O., 262–63, 281, 282

Gundlach, Ralph, 263
Gurney, Edmund, 14, 105–6
Guterman, Norbert, 166, 167

Hall, G. Stanley: at Clark University, 49; on psychical research, 44, 46–47
Hallucinations, 2, 3, 150
Hansel, C.E.M., 291
Harper's Magazine, 157, 214, 256, 258
Harvard University, 44, 203, 252, 257, 265–72, 295–96, 299–300; Hodgson fund, 45, 54–57, 60–67, 69, 75, 77, 142, 150–51, 209, 265–66, 268–71, 296, 300, 303, 322 n. 11
Hauntings, 2, 3, 86
Helson, Harry, 65, 278
Hemenway, Mrs. Augustus, 205
Herbert, C.V.C., 219, 233, 235
Herrick, C. Judson, 243
Hettinger, John, 213–14
Higginson, G. D., 261
Hilgard, Ernest R., 285
Hinton, C. H., *The Fourth Dimension*, 225
Hisaw, F. L., 269
Hoagland, Hudson, 75, 76, 77, 146, 252, 351 n. 13
Hodgson, Richard, 16, 45, 48
Hodgson fund. *See* Harvard University, Hodgson fund
Holism, 134
Holt, Henry (publishers), 292
Hotelling, Harold, 259
Houdini, Harry, 22, 69, 75
Howell, W. H., 264
Howells, T. S., 248
Humphrey, Betty M., 188, 300, 305
Hunter, W. S., 133, 196, 244–45, 246, 278, 279
Hunter College, 144, 282
Huntington, E. V., 199, 258, 263, 264
Huxley, Julian, 211
Hyperaesthesia, 42, 47, 261
Hypnosis, 118; as psychological phenomenon, 13; and telepathy, 1, 10, 89
Hyslop, James Hervey, 44; and ASPR, 16–17, 20
Hysteria, 118

IIPR. *See* International Institute for Psychical Research
Institut Métapsychique International (IMI), 15, 18, 31, 157, 210, 211, 318 n. 14
Institute of Mathematical Statistics, 258
International Congresses for Psychical Research: First (Copenhagen, 1921), 17, 18 (fig.), 322

Index

n. 17; Second (Warsaw, 1923), 32, 62; Third (Paris, 1927), 18; Fourth (Athens, 1930), 27; Fifth (Oslo, 1935), 210
International Institute for Psychical Research (IIPR), 208, 210-11
Irwin, J. O., 236

James, Henry, 268-69, 270, 271
James, William, 44, 45, 48, 57, 117, 268; and ASPR, 16; on psychical ability, 52
Janet, Pierre, 1, 118, 176, 192; on Richet's *Traité*, 10-13
Jastrow, Joseph, 70, 339 n. 82, 353 n. 56; and ASPR, 19; on psychical research, 45-46, 48, 295; and Rhine, 73, 243-44, 255, 276-77
Jencks (ESP subject), 188-89
Jephson, Ina, 103, 123, 124, 298, 320 n. 54; experiments, 32-33, 34-36, 39-40, 85-86, 89, 98, 111, 113; statistics, 107
Joad, C.E.M., 103, 211, 212, 213, 214-15
Johns Hopkins University, The, 45
Johnson, Alice, 14
Johnson, Gertrude, 126-27, 220-21
Johnson, Harvey, 286
Jones, Harold Ellis, 283
Jones, L. J., 28
Jordan, David Starr, 45
Journal of Abnormal and Social Psychology, 77, 145, 242-43
Journal of Educational Research, 261
Journal of General Psychology, 130, 146, 242
Journal of Parapsychology, 183, 188, 190-91, 194, 256, 260, 264, 278, 284-87, 303, 305; founding of, 147-49, 159

Kaempffert, Waldemar, 153-54, 155, 156, 159, 166-67, 257, 277; and ASPR, 21, 316 n. 78; *Science Today and Tomorrow*, 167
Kahoe, Walter, 249
Kansas, University of, 249-50
Karwoski, Theodore, 248-49
Kellogg, Chester E., 163, 165, 234, 261, 279, 292; and ESP statistics, 197, 198, 199, 200, 256-58, 263
Kennedy, John L., 149-50, 254-55, 273-75, 281-82, 285, 288, 303
Kentucky, University of, 249
Kingsbury, Forrest, 243

Lady (horse), 81-82, 84-85, 145, 164
Lamarckianism, 82-84
Landis, Carney, 280
Lapicque, Louis, 10

Lashley, K. S., 133, 243, 265, 269, 271, 272
Leiden, University of, 215
Lemmon, Vernon, 255-56
Leonard, Mrs. Osborne (medium), 28, 95, 222, 227; and Lodge, 3
Leuba, Clarence, 249, 281
Lewis, C. I., 77
Liberty, 157
Linzmayer, A. J., 91-92, 93, 98, 108, 113, 333 n. 47
Listener, 121, 125
Lodge, Oliver, 3, 13, 14, 25, 70, 73, 105, 317, n. 10; data evaluation, 105; on Richet's *Traité*, 1, 4, 6, 11; on telepathy, 31, 36, 112, 297
London, University of, 211-15, 216, 231. See also University of London Council for Psychical Investigation
London Spiritualist Alliance, 20
Long, Louis, 285
Lorge, Irving, 285
Lucidité, Lucidity. See Clairvoyance
Lund, F. H., 276
Lundholm, Helge, 90, 95, 134, 329 n. 73; experiments, 89; and Rhine, 135

M., Mrs. (ESP subject), 179-80
Maby, J. Cecil, 223, 233, 348 n. 62
McComas, Henry, 275, 282-83, 286
McDonald, Eugene F., 63, 160, 162, 346 n. 10
McDougall, William, 14, 119, 136 (fig.), 158, 165, 210, 285, 348 n. 44, 353 n. 56, 357 n. 61; and ASPR, 17-23, 59-60; *Body and Mind*, 58, 59, 78; death of, 151; education and early career, 57-59; on fraud in experimentation, 181-82; at Harvard, 57-69; Lamarckianism, 82-84; and Margery, 22, 23 (fig.); *Modern Materialism and Emergent Evolution*, 134; moves to Duke, 79, 132-34, 265; philosophy, 58-59; on psychologists and psychical research, 8; and Rhine, 73-96 passim, 109, 134-38, 217, 218, 241, 242, 251, 266, 276, 279; *The Riddle of Life*, 151; on scientific standards, 20-21; and SPR, 58-59; on telepathy, 89; on university as setting for psychical research, 70, 91, 147-48
Mace, C. A., 212
McGarvey, John, 285
McGeoch, John A., 283
McGill University, 256
Macnamara, E. D., 212
Marion (telepathist), 121, 123 (fig.)
Margery (Mina Crandon) (medium), 21-23, 23 (fig.), 65, 69, 75-77, 118, 145, 152, 206, 207, 209, 252, 265, 269, 275
Marsh, J. E., 249

Martin, Dorothy, 144, 188–89, 194, 248, 282
Materialism, 6, 40, 60, 72, 120, 166, 202, 297; opposition to, 58–59, 129, 133–34, 304; undermined by modern physics, 154–56, 167
Mathews, Shailer, 72, 87
Mechanism, 72, 226–27
Mediums: absent-sitter technique, 85; book tests, 42, 86; ESP studies, 173–74, 209, 231; mental, 38, 64–65, 78, 79–80, 85, 86, 95, 137; physical, 10, 22–23, 26–28, 215; physical vs. mental, 2, 3; spirit controls, 2, 4, 173–74, 222–23; and survival hypothesis, 3–4, 95–97, 173–74, 223, 304. *See also individual mediums by name*
Menninger Clinic, 250
Merritt, O. K., 144–45, 170
Metapsychics (*métapsychique*), 1, 2
Meyer, Jean, 15, 25
Miles, E.R.C., 197
Miller, James Grier, 271–72
Millikan, Robert, 72, 155
Mind Association, 228
Miner, J. B., 249
Minnesota, University of, 44, 249, 250–51, 255, 259, 260
Mirabelli (medium), 219
Mitchell, T. W., 315 n. 39
Monism, psychical, 50, 54–55
Moore, Henry, 94, 145
Mount, George I., 132
Müller, G. E., 57
Multiple personalities, 16, 118, 228
Munn, Orson D., 152
Münsterberg, Hugo, 45, 47–48, 57, 61, 256
Munzinger, Karl, 188, 248, 292
Murchison, Carl, 69, 146, 244, 245, 273
Murphy, Gardner, 70, 140, 179, 202, 208, 217–18, 282; at C.C.N.Y., 300–301; experiments, 39–40, 150, 189–90, 209, 287, 292, 300; experiments, long-distance, 32, 100; experiments, radio telepathy, 160; *Historical Introduction to Modern Psychology,* 87–88; and Hodgson fund, 60–66, 150, 269–70, 296, 300; and *Journal of Parapsychology,* 284–85, 303; on methodology, 64, 185; on psychical research, 157–58, 299–300, 304; and psychical-research societies, 18, 21, 65, 206, 296, 298, 300; 316 n. 74; on psychologists' attitudes, 149, 185, 186, 278, 283, 292, 293–94; and Rhine, 74, 75, 82, 130, 141–42, 147, 242, 293–95; and Warcollier, 32, 41, 180, 318 n. 23
Murray, Gilbert, 106, 112, 113, 213
Myers, F.W.H., 5, 12, 14, 29, 48, 75, 117; theories of, 115, 116, 118

National Broadcasting Company, 154
National Laboratory for Psychical Research (England), 28, 42, 211–15, 229; establishment of, 26–27
Nature, 121
Nazi government, 210, 215
Netherlands, psychical research, 215
New Republic, 165, 166
New York Herald Tribune, 154, 165
New York Times, 153–54, 156, 159, 166, 257
New York University, 144, 278
Newton, Isabel, 32, 61

Oesterreich, Traugott, 15
Olson, Willard, 281, 283
OM (open matching), 171, 179, 187, 274, 311
O'Neill, John J., 154, 155, 156–57, 159, 209–10
Operationism, 265, 280
Ossowiecki, Stefan, 42, 85, 107
Osty, Eugène, 31, 32, 129; and IMI, 210, 318 n. 14; on telepathy, 41–42, 332 n. 35
Ownbey, Sara. *See* Zirkle, Sara Ownbey
Ozanne, Charles E., 338 n. 51

Pagenstecher, Gustav, 99
Palladino, Eusapia (medium), 2, 5, 182
Pannett, Charles, 347 n. 34
Parapsychological Notes, 338, n. 49
Parapsychology, origin of term, 117. *See also* Extra-sensory perception; Psychical research
Parapsychology Foundation, 305–6
PC (pure clairvoyance), 104
PDT (precognitive down through), 170, 172
Pearce, Hubert, 92, 93, 98, 99, 101, 104, 114, 140, 171 (fig.), 191, 333 n. 48; long-distance tests, 199, 234, 282, 292; precognition, 170
Peatman, John Gray, 278
Pegram, Margaret, 141, 218; experiments, 178, 203
Pennsylvania, University of, 44
Perception, sensory: and ESP, comparison with, 180, 185–86, 202–3, 228, 253, 266; subliminal, 203, 271–72. *See also* Extra-sensory perception; Hyperaesthesia
Peterson, Frederick, 19
Peyton, W. C., 17, 19
Phalen, H. R., 184
Phantasms of the Living, 46, 75, 208–9
Philosophers, and psychical research, 44, 226–29
Philosophy, 228

Physics, modern, and antimaterialistic implications, 154-56, 167, 202
Piddington, J. G., 14, 25-26, 223, 233; on scientific method, 8
Piéron, Henri, 10
Pierson, Jocelyn, 208, 298
Pierson, Thomas H., 206
Piper, Leonora (medium), 2, 4, 5, 16, 34, 48, 65, 86, 138
PK. *See* Psychokinesis
Poltergeists, 86
Positivism, 4, 84, 92, 201, 280
Pratt, J. Gaither, 139, 146, 183, 218, 222, 278, 281, 283, 300; education, 140; experiments, 92, 170, 178, 179-81, 271, 288-92; experiments, long-distance, 101, 199, 234, 282, 292; *Handbook,* 161, 190; in New York, 142, 150, 179-81; *Towards a Method of Evaluating Mediumistic Material,* 205
Precognition, 159, 202, 224-26; experimental studies, 169-75, 225-26, 298
Premonitions, 2, 3
Price, H. H., 228
Price, Harry, 36, 233, 234, 235, 334 n. 81, 335 n. 101, 346 n. 11; and ASPR, 103; experiments, 122, 123 (fig.); National Laboratory, 26-27, 28, 42, 211-15, 229; on Rhine, 125; and University of Bonn, 216
Price, Margaret, 178, 194
Prince, Morton, 19, 49, 77, 78, 81
Prince, Walter Franklin, 18 (fig.), 70, 121-22; *Enchanted Boundary,* 241; experiments, 39-40; and Margery, 68; and psychical research societies, 16-23 passim, 69, 119, 204-5; and Rhine, 77, 78, 82, 89, 92, 94, 95, 241-42, 326 n. 21; and *Scientific American,* 152-53; on telepathy, 99
Princeton University, 255
Probable error (PE), definition of, 195
Psi phenomena, naming of, 299
Psychical research: amateur investigators, 144-45; career, problems of, 143-44; choice of subject-matter, 5-6; historical stages of, 1-2; mental vs. physical phenomena, 2, 103; scientific method, attempts at, 6-10; spiritists vs. nonspiritists, 3-5, 16, 20-23, 26-28; surveys and questionnaires, 44, 149, 264, 276-77, 278-81, 286, 305. *See also* Card-guessing; Clairvoyance; England; Extrasensory perception; France; Germany; Metapsychics; Netherlands; Parapsychology
Psychische Studien, 15
Psychoanalytic Review, 295
Psychokinesis (PK), 202, 304; drugs, effects of, 172; experimental studies, 169-75; QD (quarter distribution) effect, 299

Psychological Abstracts, 149, 256
Psychologists, and psychical research, 8-9, 17, 19, 44-70, 117-18, 124, 134-36, 142-44, 159-64, 191-99, 212, 240-97, 301-5
Psychology: abnormal, 145, 212, 217, 279; early development of field, 13-14, 45, 58; experimental, 45, 48, 50-52, 118, 275-76, 279, 291, 301-3; Freudian, 54; Gestalt, 134, 176, 177; hormic, 58, 83, 84; introspective, 51-52, 114; reflex-arc concept, 55
Psychometry, 99, 213, 262
Psychophysical parallelism, 31, 47-48
Psychophysics, and telepathy, 185-86
Psychophysiology, 55
PT (pure telepathy), 99, 191, 311
Purdy, Lawson, 206

QT (quadruple task), 287-88

Rauth, J. E., 249
Rayleigh, John William Strutt, 3rd baron, 236
Reader's Digest, 157
Reed College, 263
Reichenberg, Wally, 250
Reiser, Oliver, 144
Religion, and science, 72-73, 75, 86-87, 88, 134, 227, 327, n. 21
Revue métapsychique, 15, 31, 43, 129, 210
Rhine, Joseph Banks, 136 (fig.), 217, 220, 222, 281-82, 283, 299, 301; and amateur investigators, 144-45; and BSPR, 205-6; in Cambridge, 75-79; and colleagues at Duke, 134-36, 251; courses taught, 87-88, 90, 94; education and early career, 71-75; experiments, 81-84, 88-92, 98-101, 169-75, 171 (fig.), 226; *Extra-Sensory Perception After Sixty Years,* 292-97; financial support, 137-39, 205; and *Journal of Parapsychology,* 147-49, 284, 286; moves to Duke, 79-80; *New Frontiers of the Mind,* 156, 160-66 passim, 202, 203, 209, 256, 259, 277, 278; philosophy, 166; philosophy of science, 88, 96-97, 304; precognition studies, 169-75, 226; on psychology, 241; research institute, 136-39, 304-5; statistical methods, 195-201; theories of ESP, 201-3; and Zenith radio tests, 160-63. See also *Extra-Sensory Perception* (book)
Rhine, Louisa E., 71-72, 74, 180, 186, 326 n. 14
Rice, C. Hilton, 144
Richardson, Mark W., 206
Richet, Charles, 8, 13, 25, 29, 32, 33, 35, 84, 104, 115, 192, 335 n. 98; experiments,

Richet, Charles (*continued*)
 105–6; on fraud, 182; and IMI, 210; statistics, 43, 95, 106, 107; on telepathy, 52, 112; *Traité de métapsychique*, 1–13 passim, 20, 31, 102, 106, 297
Riddle, Mrs. John Wallace, 45
Ridley, Henry, 223
Riess, Bernard, 144, 187–88, 193, 282, 284, 292
Rogosin, Hyman, 276, 282
Rothacker, Emil, 176, 215–16
Rothera, Ralph, 247, 261
Röthy, Charles, 317 nn. 9, 10
Ruckmick, Christian, 279
Rulon, P. J., 197
Russell, Bertrand, 227
Russell, D. C., 348 n. 44

S., Miss (ESP subject), 188, 193
Salter, Helen, 14, 18 (fig.), 32, 123, 218, 220
Salter, W. H., 14, 18 (fig.), 28, 123, 126, 215, 218, 220, 334 n. 70; on Soal, 235–36; on statistics, 223–24
Saltmarsh, H. F., 37, 220; evaluation methods, 95; on Jephson, 36; on precognition, 225; and Rhine, 119–21, 233; on Soal, 235; on survival hypothesis, 96, 97
Sanborn, Herbert, 282–83
Santoliquido, Rocco, 15, 25
Saturday Review of Literature, 259
Saudek, Robert, 242
Schiller, F.C.S., 14, 57, 70, 226, 285, 318 n. 16, 335 n. 98
Schmeidler, Gertrude, 299–300
Schneider, Rudi (medium), 210, 222
Schoolland, John, 188, 248
Schrenck-Notzing, Albert von, 5, 15, 18 (fig.), 25
Science News Letter, 164, 165
Science Service, 81, 164–65, 167, 260, 262–64
Scientific American, 157; prize offered, 22, 23, 64–65; telepathy testing, 152–53
Scientific Monthly, 163, 256, 257, 263
Scripps, E. W., 164
Seashore, Carl E., 132
Seashore, R. H., 163
Sells, Saul, 269, 284–85, 338 n. 61
Shackleton, Basil, 238, 298
Shapley, Harlow, 269
Shull, Charles A., 72, 243
Sidgwick, Eleanor Mildred, 14, 25, 28, 85, 104, 218, 225; on clairvoyance, 110–11; on Rhine, 127; on telepathy, 34
Sidgwick, Henry, 14, 75, 226
Sinclair, Upton, 94; *Mental Radio*, 89, 99, 153, 339 n. 82

Skinner, B. F., 193, 259–61, 263, 274
Sloane, William, 292
Smith, Burke, 141; experiments, 178, 203
Smith, Carl, 273
Smith, W. Whately. *See* Carington, W. Whately (Smith)
Smith College, 44
Snyder, James, 295
Soal, S. G., 89, 99–100, 103, 108, 110, 211, 212; on displacement effect, 237–38; experiments, 37–43, 123 (fig.), 125, 213, 214, 229–32, 298; and Rhine, 121–22, 197, 230–31, 234–35, 296–97; and SPR, 233–36, 320 n. 54
Society for Psychical Research, 13–14, 61, 110, 144, 157, 298–99, 301; aims of, 3, 6, 105; and ASPR, 17; critics of, 211, 213; experiments, 30, 32–43, 53, 89, 218–26, 233; factions within, 25–28, 59; *Journal*, 27, 32; leadership, 8, 13–15, 16–17, 28, 58–59, 103, 128, 210, 227; and Margery, 207; phenomena studied, choice of, 5–6; *Proceedings*, 46, 122–23, 126, 296; publications of, 14, 85, 145; and Rhine, 118–27, 231–33; and Soal, 233–36; on telepathy, 112, 115
Sociometry, 284
Sodium amytal, 93, 100, 116, 203
Soper, E. D., 133
Soule, Minnie M. (medium), 78, 79
Southern Methodist University, 264
Spearman, Charles, 146
Spirit hypothesis, Spiritism. *See* Survival, post-mortem, hypothesis of
Spiritualism, 3, 16, 18, 19, 20, 21, 45, 50, 73, 112, 207, 211
SPR. *See* Society for Psychical Research
Springfield College, 141
Stanford, Thomas Welton, 44–45, 50, 53, 150
Stanford University, 253–55; Psychical Research Fellowship, 44–45, 50–54, 149–50, 254–55, 273–75, 300
Statisticians, reaction of, to criticism of Rhine's mathematics, 258–59
Statistics, debates over usefulness of, 107–8, 223–24, 267. *See also* Card-guessing, statistics of; Data
Stern, William, 134
Sterne, T. E., 199, 200
Stewart, Gloria, 230, 238, 298
STM (screened touch matching), 179, 180–81, 230, 288–90, 289 (fig.), 311
Stone, L. J., 338 n. 61
Stratton, G. M., 282–83, 319 n. 34
Strausbaugh, P. D., 72, 218
Stribic, F. P., 194
Strickland, Francis, 206

Stuart, Charles E., 113, 140, 142, 147, 150, 178, 245, 255, 257, 274, 300, 338 n. 49; experiments, 89, 92, 170, 172 (fig.); *Handbook,* 161, 190; statistics, 196–200, 246
Sudre, René, 8, 18 (fig.); on Jephson, 107; on method, 7, 107; on Rhine, 129; on Warcollier, 31
Survival, post-mortem, hypothesis of, 3–6, 14, 15, 17, 20, 28, 37, 48, 59, 61, 70, 86, 95–97, 102–3, 119, 128, 151, 155, 173–75, 208, 211, 223, 304

Tarkio College, 170, 186–87
Taves, Ernest, 150, 284, 286, 296, 300, 358 n. 12; experiments, 190, 209, 270, 287, 292
Taylor, W. S., 132–33
Telekinesis, 2, 4
Teleology, 58
Telepathy (sometimes including clairvoyance): in animals, 81–82, 83, 84–85, 252; below-chance results, 56, 67; definition of, 128; evaluation of results, difficulties of, 36–37; experimental studies, 7, 19, 30–34, 46–47, 50–53, 55–56, 61–64, 91, 99, 106; factors affecting, 30, 32, 42, 61–62, 67, 82, 84; under hypnosis, 1, 10, 89; introspection, 51–52; long-distance tests, 8, 32, 38–39, 40–41, 61, 62–63, 64 (fig.), 89, 100–101, 114, 236–37, 282 (*see also* Telepathy, radio tests); mass testing, 152–53; mechanized tests, 55–56; naming of, 5; physical analogies, 31, 34, 115; pictorial material, 30–31, 39, 61–64, 89, 99, 236–37; psychophysical tests, 185–86; radio tests, 36–37, 38, 63, 161, 235, 282; semiexperimental, 34; spontaneous, 100, 128; statistics of, 56; theories of, 5, 20, 29–31, 33–34, 54, 85–86, 120, 228–29, 254; as widespread ability, 52, 66, 85, 152. *See also* Card-guessing; Clairvoyance; Extra-sensory perception; PT (pure telepathy)
Temple University, 276
Tenhaeff, W.H.C., 215
Terman, Lewis M., 254, 273, 274, 278
Thomas, John F., 79–86 passim, 89, 94–97, 135, 181, 182, 205, 206, 214
Thompson, Ralph, 166
Thone, Frank, 164
Thought-transference. *See* Telepathy
Thouless, Robert H., 219, 221, 233, 236, 237, 285, 298–99, 301; on Carington, 224; on Rhine, 123–26, 191–92
Thurman, Thomas, 250
Thurstone, L. L., 262
Tijdschrift voor Parapsychologie, 215
Tillyard, R. J., 215

Time, dimensionality of, 225
Time, 157, 163
Tischner, Rudolf, 15, 85, 89, 332 n. 35
Titchener, E. B., 44, 50, 53–54, 265, 285
Tolman, Edward E., 133, 240
Trance, and telepathy, 42
Trinity College, Cambridge, Perrott Studentship, 238–39
Trinity College, North Carolina, 131, 132
Troland, L. T., 108, 117, 248, 269; and ASPR, 19; experiments, 109, 236, 342 n. 2; at Harvard, 45, 54–57, 60, 322 n. 11
Tryon, R. C., 285
Tubby, Gertrude Ogden, 16, 19–20, 74
Turner, May Frances, 100, 101, 114
Tyrrell, G.N.M., 126–27, 174, 187, 234; experiments, 171, 175, 220–22, 226

Uhrbrock, Esther, 280
University of London Council for Psychical Investigation (ULCPI), 212–15, 235
Usher, F. L., 62, 100, 236
UT (up through), 189
Utrecht, University of, 215
Uvani (trance control), 173–74, 222

Valentine, Willard L., 292
Van Dam (student), 8
Van de Water, Marjorie, 164, 165, 260, 263
Verrall, Helen, 113
Vesme, César de, 129, 321 n. 70
Vett, Carl, 18 (fig.), 32
Vitalism, 127
Volkmann, John, 285, 296

Walker, Nea, *The Bridge,* 85
Wallace, Alfred Russel, 115, 116
Walther, Gerda, 176
Walton, Daniel Day, 206
Warcollier, René, 85, 89; articles published by BSPR, 189; long-range telepathy, 62, 63 (fig.), 100; on percipience, 33–34; *La Télépathie,* 29–38 passim, 180; on telepathy, 8, 38, 39, 41–42
Warner, Lucien, 143, 146, 147, 183, 254, 278, 305; experiments, 185–86, 193–94, 203
Washburn, Margaret Floy, 322 n. 12
Washington, University of, 263
Waterhouse, E. S., 212
Watson, John B., 271
Weigle, Luther Allan, 133
Wellman, Adele, 209
West, D. J., 357 n. 9
Western Psychological Association, 274–76, 285

Whitehead, Alfred North, 77, 78
Widgery, A. G., 90
Wieman, Henry Nelson, 87, 243
Wilbur, Ray Lyman, 253, 254
Wilks, S. S., 258
Willoughby, R. R., 146, 255, 278, 285; on Rhine, 192, 196–97, 198, 199, 245–47, 261
Wilson, E. B., 269
Wolfle, Dael, 261–63, 264
Wood, Mrs. William M. (Ellen A.), 138, 205, 241
Woodruff, Joseph L., 141, 296; experiments, 172 (fig.), 183, 186–87, 203, 288–92, 289 (fig.)
Woods, J. H., 327 n. 21
Woodworth, R. S., 185–86, 278
Woolley, V. J., 36–37, 38
Worcester, Elwood, 19, 21, 138, 205
Wright, E. H., 157, 158, 208, 256, 258

Wright, George E., 20
Wundt, Wilhelm, 13, 16, 45, 50, 185

Yale University, 44
Yerkes, Robert, 61, 251, 252, 278

Zeitschrift für Parapsychologie, 15
Zeitschrift für Psychologie, 129, 177
Zener, Karl, 107, 108, 203; experiments, 89–90, 134; and Rhine, 135, 145–46
Zener cards. *See* ESP cards
Zenith Radio Corporation, 63, 160–63, 193, 256, 263, 281, 282, 346 n. 10
Zirkle, George, 140; experiments, 100, 101, 170, 178, 191
Zirkle, Sara Ownbey, 140, 178; experiments, 100, 101, 114, 170, 191

Library of Congress Cataloging in Publication Data

Mauskopf, Seymour H.
  The elusive science.

  Includes bibliographical references and index.
  1. Psychical research—History. I. McVaugh,
Michael R., joint author. II. Title.
BF1028.M38   133.8′01′5   80-7991
ISBN 0-8018-2331-5